AF249038

Trends in Mathematics

Trends in Mathematics is a series devoted to the publication of volumes arising from conferences and lecture series focusing on a particular topic from any area of mathematics. Its aim is to make current developments available to the community as rapidly as possible without compromise to quality and to archive these for reference.

Proposals for volumes can be submitted using the Online Book Project Submission Form at our website www.birkhauser-science.com.

Material submitted for publication must be screened and prepared as follows:

All contributions should undergo a reviewing process similar to that carried out by journals and be checked for correct use of language which, as a rule, is English. Articles without proofs, or which do not contain any significantly new results, should be rejected. High quality survey papers, however, are welcome.

We expect the organizers to deliver manuscripts in a form that is essentially ready for direct reproduction. Any version of TEX is acceptable, but the entire collection of files must be in one particular dialect of TEX and unified according to simple instructions available from Birkhäuser.

Furthermore, in order to guarantee the timely appearance of the proceedings it is essential that the final version of the entire material be submitted no later than one year after the conference.

Felix L. Schwenninger • Marcus Waurick

Editors

Systems Theory and PDEs

Open Problems, Recent Results, and New Directions

 Birkhäuser

Editors
Felix L. Schwenninger
Department of Applied Mathematics
University of Twente
Enschede, The Netherlands

Marcus Waurick (iD)
Institut für Angewandte Analysis
TU Bergakademie Freiberg
Freiberg, Germany

ISSN 2297-0215　　　　　　　ISSN 2297-024X　(electronic)
Trends in Mathematics
ISBN 978-3-031-64990-5　　　ISBN 978-3-031-64991-2　(eBook)
https://doi.org/10.1007/978-3-031-64991-2

Mathematics Subject Classification: 93-xx

This book is published under the imprint Birkhäuser, www.birkhauser-science.com by the registered company Springer Nature Switzerland AG
The registered company address is: Gewerbestrasse 11, 6330 Cham, Switzerland

If disposing of this product, please recycle the paper.

Preface of Systems Theory and PDEs—Open Problems, Recent Results, and New Directions

The theory of partial differential equations is profound and rich and has many applications in diverse areas of mathematics. A modern perspective addressing problems in the area of partial differential equations (PDEs) heavily uses functional analysis. This point of view necessitates structural insights in the equations at hand and, in turn, leads to a deeper understanding of the considered PDE other methods fail to offer. On the other hand, mathematical systems theory was initially developed for models described by ordinary differential equations and even linear problems. This finite-dimensional theory is thus strongly based on linear algebra, whereas infinite-dimensional systems, described by PDE models, require more (functional) analysis and operator theory. At this junction of infinite-dimensional linear algebra (i.e., functional analysis), infinite-dimensional systems theory, and (applications to) partial differential equations is where this book is positioned.

In July 2022, the possibility that results and insights for the one, infinite-dimensional systems theory, and for the other, modern treatment of partial differential equations, can be mutually benefitting was shown during the Workshop on Systems Theory and PDEs held at TU Bergakademie Freiberg. The research-intensive working atmosphere and the mixture of experienced researchers as well as future experts in the field led to many interesting discussions and results. This volume marks this workshop as a great success.

We briefly summarise some of the topics touched upon next. The theory of evolutionary equations, that is, a universal Hilbert space perspective towards (time-dependent) partial differential equations is applied to a model from thermo-piezo-electromagnetism allowing for (time-dependent, generalised) boundary conditions of Leontovich type for all unknowns in question is presented by Buchinger and Doherty. By Picard and Trostorff, a particular Hilbert space perspective is offered for the understanding of non-local boundary conditions in the connection of Beltrami-fields, which yields to a general mechanism for certain natural boundary conditions leading to a self-adjoint realisation of curl. In a certain one-dimensional manifold, boundary conditions are also treated by means of δ- and δ'-couplings for Schrödinger operators by Rohleder and Seifert. In a one-plus-one-dimensional setting, the theory of port-Hamiltonian systems is used and extended by characterising asymptotic stability via boundary conditions by Waurick and Zwart. In a higher-dimensional set-up, the applicability of port-Hamiltonian systems to fluid mechanics is addressed by show-casing a port-Hamiltonian formulation for Oseen flows; which is detailed by Reis and Schaller. For finite-dimensional port-Hamiltonian systems, the equivalence for geometric and descriptor representation is addressed using tools from multi-valued linear algebra by Gernandt et al. An open problem for (infinite-dimensional) differential-algebraic equations (DAEs) in Hilbert spaces is solved and deeper insights are readily formulated for DAEs with bounded spectrum on Banach spaces by Philipp. The relation between BIBO stability and funnel control for semilinear systems with unbounded control and observation operators is addressed by Hastir et al. Finally, some extensions to well-known Carleson embeddings as a tool to characterize admissible control operators is treated by Preussler and Schwenninger.

We are very grateful to having hosted all the WOSTAP participants in Freiberg in 2022 with many fantastic discussions and new insights.

<table>
<tr><td>Enschede, The Netherlands</td><td>Felix L. Schwenninger</td></tr>
<tr><td>Freiberg, Germany</td><td>Marcus Waurick</td></tr>
<tr><td>December, 2023</td><td></td></tr>
</table>

Contents

On some Impedance Boundary Conditions for a Thermo-Piezo-Electromagnetic System

Andreas Buchinger ⓘ and Michael Doherty ⓘ

1 Introduction

Ultrasonic transducers are measurement devices which enjoy frequent application across a range of different fields including medical imaging and non-destructive testing. Most mathematical models of these devices focus on their piezo-electromagnetic properties, with the impact of a high-temperature regime often being neglected. Issues in the manufacturing and testing processes can account for this. Such physical considerations motivate the use of—and need for—abstract mathematical modelling approaches. The focus of this chapter is one such abstract approach.

We will formulate a thermo-piezo-electromagnetic model (which could be used to model an ultrasonic device) and consider its well-posedness when considered together with a set of impedance boundary conditions for full thermo-piezo-electromagnetic data.

$$n \times H - n \times \widetilde{Q}^* v + n \times (E \times n) = 0 \text{ on } \partial\Omega,$$

$$T \cdot n - \widetilde{Q}\,(n \times E) + \left(1 + \widetilde{\alpha}\partial_t^{-1}\right) v = 0 \text{ on } \partial\Omega, \tag{1}$$

This chapter is dedicated to Rainer Picard, without whom there would be no evolutionary equation perspective to work with.

A. Buchinger
Institute for Applied Analysis, TU Bergakademie Freiberg, Freiberg, Germany

M. Doherty (✉)
Department of Mathematics and Statistics, University of Strathclyde, Glasgow, UK
e-mail: m.doherty@strath.ac.uk

© The Author(s), under exclusive license to Springer Nature Switzerland AG 2024
F. L. Schwenninger, M. Waurick (eds.), *Systems Theory and PDEs*,
Trends in Mathematics, https://doi.org/10.1007/978-3-031-64991-2_1

were considered together with a piezo-electromagnetic system without any thermal input. Here u, E and $H \colon \mathbb{R} \times \Omega \to \mathbb{R}^3$ are the displacement of the elastic body Ω, the electric field and the magnetic field, respectively. Moreover, $T \colon \mathbb{R} \times \Omega \to \mathbb{K}_{\mathrm{sym}}^{3\times3}$ is the stress tensor taking values in symmetric 3×3 matrices and $v := \partial_t u$. We also have $\widetilde{Q}$ and $\widetilde{\alpha}$ as the given (bounded and linear) boundary mappings with

$$\widetilde{Q} \colon V_{\gamma_t} \to H^{1/2}(\partial\Omega)^3 \quad \text{and} \quad \widetilde{\alpha} \colon H^{1/2}(\partial\Omega)^3 \to H^{1/2}(\partial\Omega)^3.$$

The boundary traces and spaces V_{γ_t} and $H^{1/2}(\partial\Omega)^3$ are later recalled in Sect. 2.2. Specific regularity assumptions are made in [1] which ensure that the boundary equations (1) are well-defined as equations on $L^2(\partial\Omega)$. These boundary conditions were later generalized to the setting of *abstract boundary data spaces* (cf. [2]), and it is this generalization which we take as the starting point for the formulation of our own impedance boundary conditions (we recall abstract boundary data spaces in Sect. 2.3). We will obtain our new boundary conditions after suitably extending the above boundary equations to allow for the influence of a high-temperature regime (Sect. 3.2). Whilst the set of newly formed boundary conditions is useful to us as an example, we highlight that they are abstract in nature. The task of finding and formulating a physically relevant set of boundary conditions—which fits within the schema of these new boundary conditions—remains an avenue of future research.

The basis for our extended model is a thermo-piezo-electromagnetic system (cf. [3]) which was originally shown to be well-posed as an *evolutionary equation* (cf. [4] or the more recent [5]) under the influence of homogeneous Dirichlet and Neumann boundary conditions. We recall the main components of this system in Sect. 3.1 as well as at the beginning of Sect. 3.3. Well-posedness in this context means Hadamard well-posedness and causal dependence on the given data, which we will review first in Sect. 2.1. We will extend this system in such a way as to be able to accommodate for the novel impedance boundary conditions formulated. This will be achieved primarily by applying the methodology used in [2]. We will focus on addressing whether our extended system is well-posed as an evolutionary equation under our new boundary conditions. A proof of the evolutionary well-posedness of the system is presented in Sect. 3.4, with our main solution result, Theorem 3.6.

2 Preliminaries

2.1 Evolutionary Equations

First, we introduce some notation and definitions based on [5]. Let H be a complex Hilbert space (linear in the second argument) and let $C \in L(H)$. We say that C is *positive-definite* iff

$$\forall x \in H : \langle x, (C + C^*)x \rangle_H \geq 2c_0 \|x\|_H^2$$

for some $c_0 \in \mathbb{R}_{>0}$. We can rephrase this requirement as $\operatorname{Re} C \geq c_0$. If ever we are not concerned with the actual value of $c_0 \in \mathbb{R}_{>0}$, we shall instead write $C \gg 0$.

For an open $U \subseteq \mathbb{C}$, we call a holomorphic $M \colon U \to L(H)$ a *material law* iff there exists a $\nu \in \mathbb{R}$ with $\mathbb{C}_{\operatorname{Re}>\nu} \subseteq U$ and

$$\sup_{z \in \mathbb{C}_{\operatorname{Re}>\nu}} \|M(z)\| < \infty.$$

In that case, $s_b(M)$ denotes the infimum of all such ν. Considering the Hilbert space

$$L_{2,\nu}(\mathbb{R}; H) := \left\{ f \colon \mathbb{R} \to H \text{ Bochner-meas.}; \int_{\mathbb{R}} \|f(t)\|_H^2 \, e^{-2\nu t} \, dt < \infty \right\},$$

we define the weak derivative $\partial_{t,\nu} \colon \operatorname{dom}(\partial_{t,\nu}) \subseteq L_{2,\nu}(\mathbb{R}; H) \to L_{2,\nu}(\mathbb{R}; H)$ in the classical way

$$(f, g) \in \partial_{t,\nu} :\Longleftrightarrow \forall \varphi \in C_c^\infty(\mathbb{R}) : -\int_{\mathbb{R}} \varphi'(t) f(t) \, dt = \int_{\mathbb{R}} \varphi(t) g(t) \, dt,$$

and the (unitary) *Fourier–Laplace transform*

$$\mathcal{L}_\nu := \mathcal{F} \exp(-\nu m) \colon L_{2,\nu}(\mathbb{R}; H) \to L_2(\mathbb{R}; H),$$

where $\mathcal{F}$ is the classical (unitary) Fourier transform on $L_2(\mathbb{R}; H)$ and

$$\exp(-\nu m) \colon \begin{cases} L_{2,\nu}(\mathbb{R}; H) & \to & L_2(\mathbb{R}; H) \\ f & \mapsto & [t \mapsto \exp(-\nu t) f(t)] \end{cases}.$$

For a material law M, a $\nu > s_b(M)$ and

$$M(\operatorname{im} + \nu) \colon \begin{cases} L_2(\mathbb{R}; H) & \to & L_2(\mathbb{R}; H) \\ f & \mapsto & [t \mapsto M(it + \nu) f(t)] \end{cases},$$

we call

$$M(\partial_{t,\nu}) := \mathcal{L}_\nu^* M(\operatorname{im} + \nu) \mathcal{L}_\nu \in L(L_{2,\nu}(\mathbb{R}; H))$$

the associated *material (law) operator*.

For a densely defined and closed operator $A \colon \operatorname{dom}(A) \subseteq H \to H$, the graph inner product makes $\operatorname{dom}(A)$ a Hilbert space, and basic calculations show that the operator

$$\begin{cases} L_{2,\nu}(\mathbb{R}; \operatorname{dom}(A)) \subseteq L_{2,\nu}(\mathbb{R}; H) & \to & L_{2,\nu}(\mathbb{R}; H) \\ f & \mapsto & [t \mapsto Af(t)] \end{cases} \tag{2}$$

is well-defined, densely defined and closed. Using a mollifying argument, we can easily see that (2) is skew-self-adjoint for a skew-self-adjoint A. Hence, we will not distinguish between A and its extension (2).

With these tools, we can define *evolutionary equations* as

$$\left(\partial_{t,\nu} M(\partial_{t,\nu}) + A\right) U = F. \tag{3}$$

The solution theory for the class of these equations is encapsulated in the following [5, Theorem 6.2.1].

Picard's Theorem *Let $\nu_0 \in \mathbb{R}$ and H be a Hilbert space, let $M\colon \mathrm{dom}(M) \subseteq \mathbb{C} \to L(H)$ be a material law with $s_b(M) \leq \nu_0$ and let $A\colon \mathrm{dom}(A) \subseteq H \to H$ be skew-self-adjoint. Assume that there exists a constant $c > 0$ such that*

$$\mathrm{Re}\, z M(z) \geq c$$

for all $z \in \mathbb{C}_{\mathrm{Re} > \nu_0}$. Then for all $\nu \geq \nu_0$, the operator $\partial_{t,\nu} M(\partial_{t,\nu}) + A$ is closable and

$$S_\nu := \left(\overline{\partial_{t,\nu} M(\partial_{t,\nu}) + A}\right)^{-1} \in L\left(L_{2,\nu}(\mathbb{R}; H)\right).$$

Moreover, S_ν is causal, i.e. for $F \in L_{2,\nu}(\mathbb{R}; H)$ and $a \in \mathbb{R}$

$$\mathrm{spt}\, F \subseteq [a, \infty) \implies \mathrm{spt}\, S_\nu F \subseteq [a, \infty),$$

and S_ν satisfies $\|S_\nu\| \leq \frac{1}{c}$. For all $F \in \mathrm{dom}(\partial_{t,\nu})$, we have

$$S_\nu F \in \mathrm{dom}(\partial_{t,\nu}) \cap \mathrm{dom}(A),$$

i.e. $U := S_\nu F$ solves the evolutionary equation in the sense of (3). Furthermore, for $\eta, \nu \geq \nu_0$ and $F \in L_{2,\nu}(\mathbb{R}; H) \cap L_{2,\eta}(\mathbb{R}; H)$, we have $S_\nu F = S_\eta F$.

Finally, we recall three useful results which we will use in the sequel. The first can be found as [5, Theorem 6.2.3 (b)]. The second can be found as [6, Lemma 3.2], whereas the third can be found as [5, Theorem 5.2.3].

Lemma 2.1 *Let $a \in L(H)$ and $c \in \mathbb{R}_{>0}$. Assume $\mathrm{Re}\, a \geq c$. Then $a^{-1} \in L(H)$ with $\|a^{-1}\| \leq \frac{1}{c}$ and $\mathrm{Re}\, a^{-1} \geq c \|a\|^{-2}$.*

Lemma 2.2 *Let H be a Hilbert space and $V \subseteq H$ be a closed subspace. Let*

$$\iota_V : \begin{cases} V & \to & H \\ x & \mapsto & x \end{cases}$$

denote the canonical embedding of V into H. Then, $\iota_V \iota_V^* \colon H \to H$ is the orthogonal projection on V and $\iota_V^* \iota_V \colon V \to V$ is the identity on V.

Lemma 2.3 *For $\nu \in \mathbb{R}$, we have $\partial_{t,\nu} = \mathcal{L}_\nu^*(\mathrm{im} + \nu)\mathcal{L}_\nu$.*

2.2 Differential Operators and Classical Trace Spaces

We now turn our attention to recalling the standard classical traces and associated spaces. In the following, let $\Omega \subseteq \mathbb{R}^d$ be a bounded Lipschitz domain for $d \in \mathbb{N}$. We will denote the usual continuous (*Dirichlet*) trace by $\gamma : H^1(\Omega) \to H^{1/2}(\partial\Omega)$, where $H^{1/2}(\partial\Omega)$ stands for $\mathrm{ran}(\gamma) \subseteq L_2(\partial\Omega)$ considered as the Hilbert space $H^1(\Omega)/\ker\gamma$.

Next, we define the weak divergence div: $\mathrm{dom}(\mathrm{div}) \subseteq L_2(\Omega)^d \to L_2(\Omega)$ as

$$(f, g) \in \mathrm{div} :\Longleftrightarrow \forall \varphi \in C_c^\infty(\Omega) : -\int_\Omega f(x) \cdot \mathrm{grad}\,\varphi(x)\,\mathrm{d}x = \int_\Omega g(x)\varphi(x)\,\mathrm{d}x,$$

and write $H(\mathrm{div}, \Omega)$ for the Hilbert space $\mathrm{dom}(\mathrm{div})$ endowed with the graph inner product. Following [7] and applying γ to each component, we define the continuous and linear *Neumann trace* as

$$\gamma_{\cdot n} : \begin{cases} H^1(\Omega)^d & \to & L_2(\partial\Omega) \\ U & \mapsto & (\gamma U) \cdot n \end{cases},$$

where n denotes the outer unit normal. Writing $H^{-1/2}(\partial\Omega) := (H^{1/2}(\partial\Omega))'$ and using (cf. [7, p. 3743])

$$\|\gamma_{\cdot n}(U)\|_{H^{-1/2}(\partial\Omega)} \leq C \,\|U\|_{H(\mathrm{div}, \Omega)}$$

for U from the dense subset $H^1(\Omega)^d \subseteq H(\mathrm{div}, \Omega)$, we can uniquely extend $\gamma_{\cdot n}$ to the continuous and linear function

$$\gamma_{\cdot n} : H(\mathrm{div}, \Omega) \to H^{-1/2}(\partial\Omega)$$

with (integration by parts)

$$\gamma_{\cdot n}(U)(\gamma u) = \langle \mathrm{div}\, U, u \rangle_{L_2(\Omega)} + \langle U, \mathrm{grad}\, u \rangle_{L_2(\Omega)^d} \tag{4}$$

for $U \in H(\mathrm{div}, \Omega)$ and $u \in H^1(\Omega)$. For every $F \in H^{-1/2}(\partial\Omega)$, the Riesz Representation Theorem yields a unique $u \in H^1(\Omega)$ with

$$\langle u, v \rangle_{L_2(\Omega)} + \langle \mathrm{grad}\, u, \mathrm{grad}\, v \rangle_{L_2(\Omega)^d} = F(\gamma v) \tag{5}$$

for all $v \in H^1(\Omega)$. Choosing $v \in C_c^\infty(\Omega)$, we obtain $\mathrm{grad}\, u \in H(\mathrm{div}, \Omega)$ with $\mathrm{div}(\mathrm{grad}\, u) = u$. With (4), we obtain $\gamma_{\cdot n}(\mathrm{grad}\, u) = F$, which shows that the Neumann trace is in fact onto.

For our purposes, we will need special higher-dimensional versions of these differential operators and of their traces. With (the index sym here stands for symmetric matrices)

$$\begin{cases} C_c^\infty(\Omega)^d \subseteq L_2(\Omega)^d & \to & L_2(\Omega)_{\text{sym}}^{d\times d} \\ (\varphi_j)_{j=1}^d & \mapsto & \frac{1}{2}(\partial_k\varphi_j + \partial_j\varphi_k)_{k,j=1}^d \end{cases} \tag{6}$$

and

$$\begin{cases} C_c^\infty(\Omega)_{\text{sym}}^{d\times d} \subseteq L_2(\Omega)_{\text{sym}}^{d\times d} & \to & L_2(\Omega)^d \\ (\varphi_{jk})_{j,k=1}^d & \mapsto & \left(\sum_{k=1}^d \partial_k\varphi_{jk}\right)_{j=1}^d \end{cases}, \tag{7}$$

we define the weak *symmetrized gradient*, Grad, as the negative adjoint of (7) and the weak *symmetrized divergence*, Div, as the negative adjoint of (6). Once again, we write $H(\text{Grad}, \Omega)$ and $H(\text{Div}, \Omega)$ for the respective domains endowed with the respective graph inner products that make them Hilbert spaces. Using methods based on Korn's second inequality (cf. Remark 2.4), we obtain $H(\text{Grad}, \Omega) \simeq H^1(\Omega)^d$ in the sense that the sets coincide and that the norms are equivalent. Hence, applying the Dirchlet trace to every component yields the linear, continuous and onto (d-dimensional) Dirchlet trace

$$\gamma : H(\text{Grad}, \Omega) \to H^{1/2}(\partial\Omega)^d.$$

For Div, we easily obtain

$$H(\text{Div}, \Omega) \simeq H(\text{div}, \Omega)^d \cap L_2(\Omega)_{\text{sym}}^{d\times d} \subseteq H(\text{div}, \Omega)^d$$

and

$$\text{Div}\left((f_{jk})_{j,k=1}^d\right) = \left(\text{div}\left((f_{jk})_{k=1}^d\right)\right)_{j=1}^d.$$

Hence, applying the Neumann trace to every component yields the continuous and linear (d-dimensional) Neumann trace

$$\gamma_{\cdot n} : H(\text{Div}, \Omega) \to H^{-1/2}(\partial\Omega)^d \simeq \left(H^{1/2}(\partial\Omega)^d\right)',$$

and thus (4) turns into

$$\sum_{j=1}^d \gamma_{\cdot n}(U_j)(\gamma u_j) = \langle \text{Div}\, U, u\rangle_{L_2(\Omega)^d} + \langle U, \text{Grad}\, u\rangle_{L_2(\Omega)_{\text{sym}}^{d\times d}} \tag{8}$$

for $U \in H(\text{Div}, \Omega)$ and $u \in H(\text{Grad}, \Omega)$. An argument similar to (5) proves that the Neumann trace is even onto.

In the case $d = 3$, we define the weak curl: $\mathrm{dom}(\mathrm{curl}) \subseteq L_2(\Omega)^3 \to L_2(\Omega)^3$ as

$$(f, g) \in \mathrm{curl} :\Longleftrightarrow \forall \varphi \in C_c^\infty(\Omega)^3 : \int_\Omega f(x) \cdot \mathrm{curl}\, \varphi(x)\, \mathrm{d}x = \int_\Omega g(x) \cdot \varphi(x)\, \mathrm{d}x,$$

and write $H(\mathrm{curl}, \Omega)$ for the Hilbert space $\mathrm{dom}(\mathrm{curl})$ endowed with the graph inner product. We recall the following (classical) traces and associated spaces for $H(\mathrm{curl}, \Omega)$. These were originally discussed in [8] and later considered in [7, Section 4] and [9, Definition 2.15 and Remark 2.16]. The closed subspace

$$L_2^t(\partial\Omega) := \{ f \in L_2(\partial\Omega)^3 : f \cdot n = 0 \}$$

of $L_2(\partial\Omega)^3$ is called the space of *tangential vector-fields* on the boundary. We define the continuous and linear *tangential-components trace*

$$\pi_t : \begin{cases} H^1(\Omega)^3 & \to & L_2^t(\partial\Omega) \\ U & \mapsto & -n \times (n \times \gamma U) \end{cases}$$

and the continuous and linear *tangential trace*

$$\gamma_t : \begin{cases} H^1(\Omega)^3 & \to & L_2^t(\partial\Omega) \\ U & \mapsto & \gamma U \times n \end{cases}.$$

The image spaces V_{π_t} of π_t and V_{γ_t} of γ_t are Hilbert spaces with respective norms given by

$$\|v\|_{V_{\pi_t}} := \inf \left\{ \|\gamma U\|_{H^{1/2}(\partial\Omega)^3} : U \in H^1(\Omega)^3, \pi_t U = v \right\}$$

and

$$\|v\|_{V_{\gamma_t}} := \inf \left\{ \|\gamma U\|_{H^{1/2}(\partial\Omega)^3} : U \in H^1(\Omega)^3, \gamma_t U = v \right\}.$$

Integration by parts yields

$$\langle \pi_t U_1, \gamma_t U_2 \rangle_{L_2^t(\partial\Omega)} = \langle \mathrm{curl}\, U_1, U_2 \rangle_{L_2(\Omega)^3} - \langle U_1, \mathrm{curl}\, U_2 \rangle_{L_2(\Omega)^3} \tag{9}$$

for $U_1, U_2 \in H^1(\Omega)^3$. This implies that (cf. [7, Proposition 4.3])

$$\pi_t : H^1(\Omega)^3 \subseteq H(\mathrm{curl}, \Omega) \to V_{\gamma_t}'$$

and

$$\gamma_t \colon H^1(\Omega)^3 \subseteq H(\mathrm{curl},\, \Omega) \to V'_{\pi_t}$$

are both continuous. Since $H^1(\Omega)^3$ is a dense subset, we can uniquely extend π_t and γ_t to continuous and linear functions from $H(\mathrm{curl},\, \Omega)$ to V'_{γ_t} and V'_{π_t}, respectively.

Remark 2.4 In [10], it is claimed that the weak and strong definition of any linear first-order differential operator with Lipschitz continuous coefficients on any open set Ω coincide. That is to say [10] extends the original Meyers-Serrin Theorem [11] to a vast class of differential operators. This class obviously includes Grad. Hence, $C^\infty(\Omega)^d$ could be shown to be dense both in $H(\mathrm{Grad},\, \Omega)$ and in $H^1(\Omega)^d$. Thus, Korn's second inequality (e.g. [12]) would immediately show $H(\mathrm{Grad},\, \Omega) \simeq H^1(\Omega)^d$ for any bounded Lipschitz domain Ω. Unfortunately, the authors were not able to fathom how the method of [13] was applied in the argumentation of [10].

Another direct approach, which treats the operator Grad (cf. e.g. [14] or [15, Chapter 7]), is to prove that the linear, bounded and one-to-one canonical embedding $\iota \colon H^1(\Omega)^d \to H(\mathrm{Grad},\, \Omega)$ is onto for any bounded Lipschitz domain Ω using[1]

$$f \in H^{-1}(\Omega) \wedge \partial_i f \in H^{-1}(\Omega) \text{ for } i = 1, \ldots, d \implies f \in L_2(\Omega). \tag{10}$$

For $u \in H(\mathrm{Grad},\, \Omega)$ and $i, j, k = 1, \ldots, d$, we have $\partial_j u_k \in H^{-1}(\Omega)$ and

$$\partial_i \partial_j u_k = \partial_i \underbrace{\frac{1}{2}(\partial_k u_j + \partial_j u_k)}_{\in L_2(\Omega)} + \partial_j \underbrace{\frac{1}{2}(\partial_k u_i + \partial_i u_k)}_{\in L_2(\Omega)} - \partial_k \underbrace{\frac{1}{2}(\partial_i u_j + \partial_j u_i)}_{\in L_2(\Omega)} \in H^{-1}(\Omega).$$

Therefore, (10) yields $\partial_j u_k \in L_2(\Omega)$ for $j, k = 1, \ldots, d$, i.e. $u \in H^1(\Omega)^d$. $\qquad \triangledown$

2.3 Abstract Boundary Data Spaces

Armed with the above classical traces and spaces, we now recall abstract boundary data spaces. The importance of these spaces for us cannot be understated. When formulating our own model, we will work with these abstract means instead of using the typical classical tools. In the following, let $\Omega \subseteq \mathbb{R}^3$ be an arbitrary open set and let the differential operators be defined in the same way as before. Setting

$$H_0^1(\Omega) := \mathrm{dom}(\mathrm{div}^*) \subseteq H^1(\Omega),$$

$$H_0(\mathrm{div},\, \Omega) := \mathrm{dom}(\mathrm{grad}^*) \subseteq H(\mathrm{div},\, \Omega),$$

---

[1] Here, the partial derivatives have to be understood in the distributional sense.

$$H_0(\mathrm{curl},\,\Omega) := \mathrm{dom}(\mathrm{curl}^*) \subseteq H(\mathrm{curl},\,\Omega),$$

$$H_0(\mathrm{Grad},\,\Omega) := \mathrm{dom}(\mathrm{Div}^*) \subseteq H(\mathrm{Grad},\,\Omega) \text{ and}$$

$$H_0(\mathrm{Div},\,\Omega) := \mathrm{dom}(\mathrm{Grad}^*) \subseteq H(\mathrm{Div},\,\Omega),$$

we define the following *abstract boundary data spaces* (cf. [5, Chapter 12], [9] or [2, Section 4.1]):

Lemma 2.5 *We have*

$$\mathrm{BD}(\mathrm{grad}) := H_0^1(\Omega)^{\perp_{H^1(\Omega)}}$$

$$= \left\{ u \in H^1(\Omega) : \mathrm{grad}\, u \in \mathrm{dom}(\mathrm{div}),\, \mathrm{div}\,\mathrm{grad}\, u = u \right\},$$

$$\mathrm{BD}(\mathrm{div}) := H_0(\mathrm{div},\,\Omega)^{\perp_{H(\mathrm{div},\Omega)}}$$

$$= \{ U \in H(\mathrm{div},\,\Omega) : \mathrm{div}\, U \in \mathrm{dom}(\mathrm{grad}),\, \mathrm{grad}\,\mathrm{div}\, U = U \},$$

$$\mathrm{BD}(\mathrm{curl}) := H_0(\mathrm{curl},\,\Omega)^{\perp_{H(\mathrm{curl},\Omega)}}$$

$$= \{ U \in H(\mathrm{curl},\,\Omega) : \mathrm{curl}\, U \in \mathrm{dom}(\mathrm{curl}),\, \mathrm{curl}\,\mathrm{curl}\, U = -U \},$$

$$\mathrm{BD}(\mathrm{Grad}) := H_0(\mathrm{Grad},\,\Omega)^{\perp_{H(\mathrm{Grad},\Omega)}}$$

$$= \{ u \in H(\mathrm{Grad},\,\Omega) : \mathrm{Grad}\, u \in \mathrm{dom}(\mathrm{Div}),\, \mathrm{Div}\,\mathrm{Grad}\, u = u \} \text{ and}$$

$$\mathrm{BD}(\mathrm{Div}) := H_0(\mathrm{Div},\,\Omega)^{\perp_{H(\mathrm{Div},\Omega)}}$$

$$= \{ U \in H(\mathrm{Div},\,\Omega) : \mathrm{Div}\, U \in \mathrm{dom}(\mathrm{Grad}),\, \mathrm{Grad}\,\mathrm{Div}\, U = U \}.$$

Proof The proofs of these identities follow immediately from the definitions of the respective orthogonal complements and adjoints, e.g. $(\mathrm{grad}\,\lceil_{H_0^1(\Omega)})^* = -\,\mathrm{div}$. $\square$

For these spaces, Lemma 2.5 immediately yields:

Lemma 2.6 *The mappings*

$$\mathrm{grad}_{\mathrm{BD}} : \begin{cases} \mathrm{BD}(\mathrm{grad}) & \to & \mathrm{BD}(\mathrm{div}) \\ u & \mapsto & \mathrm{grad}\, u \end{cases},$$

$$\mathrm{div}_{\mathrm{BD}} : \begin{cases} \mathrm{BD}(\mathrm{div}) & \to & \mathrm{BD}(\mathrm{grad}) \\ U & \mapsto & \mathrm{div}\, U \end{cases},$$

$$\mathrm{curl}_{\mathrm{BD}} : \begin{cases} \mathrm{BD}(\mathrm{curl}) & \to & \mathrm{BD}(\mathrm{curl}) \\ U & \mapsto & \mathrm{curl}\, U \end{cases},$$

$$\mathrm{Grad}_{\mathrm{BD}} : \begin{cases} \mathrm{BD}(\mathrm{Grad}) & \to & \mathrm{BD}(\mathrm{Div}) \\ u & \mapsto & \mathrm{Grad}\, u \end{cases}$$

and

$$\mathrm{Div_{BD}}: \begin{cases} \mathrm{BD(Div)} & \to & \mathrm{BD(Grad)} \\ U & \mapsto & \mathrm{Div}\, U \end{cases}$$

are unitary with $\mathrm{grad}^*_{\mathrm{BD}} = \mathrm{div_{BD}}$, $\mathrm{curl}^*_{\mathrm{BD}} = -\mathrm{curl_{BD}}$ *and* $\mathrm{Grad}^*_{\mathrm{BD}} = \mathrm{Div_{BD}}$.

We can even obtain integration-by-parts formulae for these operators (cf. [5, Proposition 12.4.2]).

Lemma 2.7 *For $u \in H^1(\Omega)$ and $U \in H(\mathrm{div}, \Omega)$, we have*

$$\langle \mathrm{div}\, U, u \rangle_{L_2(\Omega)} + \langle U, \mathrm{grad}\, u \rangle_{L_2(\Omega)^3} = \langle \mathrm{div_{BD}}\, \iota^*_{\mathrm{BD(div)}} U, \iota^*_{\mathrm{BD(grad)}} u \rangle_{\mathrm{BD(grad)}}$$

$$= \langle \iota^*_{\mathrm{BD(div)}} U, \mathrm{grad_{BD}}\, \iota^*_{\mathrm{BD(grad)}} u \rangle_{\mathrm{BD(div)}}.$$

Additionally, for $U, V \in H(\mathrm{curl}, \Omega)$, we have

$$\langle \mathrm{curl}\, U, V \rangle_{L_2(\Omega)^3} - \langle U, \mathrm{curl}\, V \rangle_{L_2(\Omega)^3} = \langle \mathrm{curl_{BD}}\, \iota^*_{\mathrm{BD(curl)}} U, \iota^*_{\mathrm{BD(curl)}} V \rangle_{\mathrm{BD(curl)}}$$

$$= -\langle \iota^*_{\mathrm{BD(curl)}} U, \mathrm{curl_{BD}}\, \iota^*_{\mathrm{BD(curl)}} V \rangle_{\mathrm{BD(curl)}}.$$

Finally, for $u \in H(\mathrm{Grad}, \Omega)$ and $U \in H(\mathrm{Div}, \Omega)$, we have

$$\langle \mathrm{Div}\, U, u \rangle_{L_2(\Omega)^3} + \langle U, \mathrm{Grad}\, u \rangle_{L_2(\Omega)^{3\times3}_{\mathrm{sym}}} = \langle \mathrm{Div_{BD}}\, \iota^*_{\mathrm{BD(Div)}} U, \iota^*_{\mathrm{BD(Grad)}} u \rangle_{\mathrm{BD(Grad)}}$$

$$= \langle \iota^*_{\mathrm{BD(Div)}} U, \mathrm{Grad_{BD}}\, \iota^*_{\mathrm{BD(Grad)}} u \rangle_{\mathrm{BD(Div)}}.$$

Proof Consider the first case with $u \in H^1(\Omega)$ and $U \in H(\mathrm{div}, \Omega)$. We can write $u = u_0 + \iota^*_{\mathrm{BD(grad)}} u$, where $u_0 \in H^1_0(\Omega)$, and $U = U_0 + \iota^*_{\mathrm{BD(div)}} U$, with $U_0 \in H_0(\mathrm{div}, \Omega)$. Using this decomposition for U together with the fact that $\langle \mathrm{div}\, U_0, u \rangle_{L_2(\Omega)} = -\langle U_0, \mathrm{grad}\, u \rangle_{L_2(\Omega)^3}$, we obtain

$$\langle \mathrm{div}\, U, u \rangle_{L_2(\Omega)} + \langle U, \mathrm{grad}\, u \rangle_{L_2(\Omega)^3}$$

$$= \langle \mathrm{div}\, \iota^*_{\mathrm{BD(div)}} U, u \rangle_{L_2(\Omega)} + \langle \iota^*_{\mathrm{BD(div)}} U, \mathrm{grad}\, u \rangle_{L_2(\Omega)^3}. \tag{11}$$

From here, we use the decomposition for u and

$$\langle \mathrm{div}\, \iota^*_{\mathrm{BD(div)}} U, u_0 \rangle_{L_2(\Omega)} = -\langle \iota^*_{\mathrm{BD(div)}} U, \mathrm{grad}\, u_0 \rangle_{L_2(\Omega)^3},$$

so that (11) becomes

$$\langle \mathrm{div}\, U, u \rangle_{L_2(\Omega)} + \langle U, \mathrm{grad}\, u \rangle_{L_2(\Omega)^3}$$

$$= \langle \mathrm{div}\, \iota^*_{\mathrm{BD(div)}} U, \iota^*_{\mathrm{BD(grad)}} u \rangle_{L_2(\Omega)} + \langle \iota^*_{\mathrm{BD(div)}} U, \mathrm{grad}\, \iota^*_{\mathrm{BD(grad)}} u \rangle_{L_2(\Omega)^3}. \tag{12}$$

On account of Lemma 2.5, we have $\operatorname{grad}\operatorname{div} \iota^{*}_{\mathrm{BD(div)}} U = \iota^{*}_{\mathrm{BD(div)}} U$ so that (12) now reads

$$\langle \operatorname{div} U, u \rangle_{L_2(\Omega)} + \langle U, \operatorname{grad} u \rangle_{L_2(\Omega)^3} = \langle \operatorname{div}_{\mathrm{BD}} \iota^{*}_{\mathrm{BD(div)}} U, \iota^{*}_{\mathrm{BD(grad)}} u \rangle_{\mathrm{BD(grad)}}.$$

Finally, Lemma 2.6 ($\operatorname{grad}_{\mathrm{BD}} = \operatorname{div}^{*}_{\mathrm{BD}}$) implies the second identity. The remaining two cases can be proven analogously. $\qquad\square$

The next theorem (cf. [5, Corollary 12.2.3]) explains in which sense these abstract boundary data spaces are an abstract version of the classical traces discussed in Sect. 2.2.

Theorem 2.8 *Let $\Omega \subseteq \mathbb{R}^3$ be a bounded Lipschitz domain. Then, the restricted traces*

$$\gamma \lceil_{\mathrm{BD(grad)}} : \ \mathrm{BD(grad)} \to H^{1/2}(\partial\Omega),$$

$$\gamma_{\cdot n} \lceil_{\mathrm{BD(div)}} : \ \mathrm{BD(div)} \to H^{-1/2}(\partial\Omega),$$

$$\gamma \lceil_{\mathrm{BD(Grad)}} : \ \mathrm{BD(Grad)} \to H^{1/2}(\partial\Omega)^3 \ \textit{and}$$

$$\gamma_{\cdot n} \lceil_{\mathrm{BD(Div)}} : \ \mathrm{BD(Div)} \to H^{-1/2}(\partial\Omega)^3$$

are continuous and bijective, and the restricted traces

$$\pi_t \lceil_{\mathrm{BD(curl)}} : \ \mathrm{BD(curl)} \to V'_{\gamma_t} \ \textit{and}$$

$$\gamma_t \lceil_{\mathrm{BD(curl)}} : \ \mathrm{BD(curl)} \to V'_{\pi_t}$$

are continuous and one-to-one.

Proof In view of Sect. 2.2 and the definition of the BD-spaces, it suffices to show that the H_0-spaces are the kernels of the respective operators. Since div^{*} is the closure of the operator $\operatorname{grad} \lceil_{C_c^{\infty}(\Omega)}$, and similar statements hold true for the other differential operators, the continuity of the traces shows that the H_0-spaces are subsets of the respective kernels. The other inclusions easily follow from the integration-by-parts formulae (4), (8) and (9). $\qquad\square$

Remark 2.9 We mention that we could also make $\pi_t|_{\mathrm{BD(curl)}}$ and $\gamma_t|_{\mathrm{BD(curl)}}$ onto by replacing their image spaces with suitable smaller Hilbert spaces (cf. [8, Theorem 4.1]). $\qquad\triangledown$

Remark 2.10 We also stress that these abstract boundary data spaces cannot be considered as proper generalizations of the classical traces. On the one hand (cf. [5, Proposition 12.4.2]), it turns out from an integration-by-parts point of view that the suitable analogue to $\gamma_{\cdot n}$ is $\operatorname{div}_{\mathrm{BD}} \iota^{*}_{\mathrm{BD(div)}}$ and not only $\iota^{*}_{\mathrm{BD(div)}}$. An analogous statement holds true for curl. For a more in-depth view, see the discussion in [2, Section 4.3.1]. On the other hand (cf. [5, Proposition 12.5.3]), the Robin boundary condition $\gamma_n H = -i\gamma u$, for $H \in H(\mathrm{div}, \Omega)$ and $u \in H^1(\Omega)$, is not equivalent to $\mathrm{div}_{\mathrm{BD}} \, \iota^*_{\mathrm{BD(div)}} H = -i\iota^*_{\mathrm{BD(grad)}} u$ (in the case of a bounded Lipschitz domain). $\qquad \triangledown$

3 Boundary Conditions and Model System

In the following, let $\Omega \subseteq \mathbb{R}^3$ be open and non-empty and let $\mathbb{K}$ stand for either $\mathbb{R}$ or $\mathbb{C}$.

3.1 The Underlying System Equations

We recall the underlying system equations and material relations of thermo-piezo-electromagnetism (cf. [3, Section 2]). The basic system is made up of the equation of elasticity, Maxwell's equations and the heat equation. We have the equation of elasticity

$$\partial_t^2 \rho_* u - \mathrm{Div}\, T = F_0,$$

where $u : \mathbb{R} \times \Omega \to \mathbb{R}^3$ denotes the displacement of the elastic body, Ω, and $T : \mathbb{R} \times \Omega \to \mathbb{K}^{3 \times 3}_{\mathrm{sym}}$ the stress tensor. The function $\rho_* \in L_\infty(\Omega; \mathbb{R})$ describes the density of Ω, and $F_0 : \mathbb{R} \times \Omega \to \mathbb{R}^3$ is an external balancing force. Assuming that *Ohm's Law* holds, Maxwell's equations read

$$\partial_t B + \mathrm{curl}\, E = F_2,$$

$$\partial_t D - \mathrm{curl}\, H = F_1 - \sigma E,$$

where E, H, B and $D : \mathbb{R} \times \Omega \to \mathbb{R}^3$ are, respectively, the electric field, the magnetic field, the magnetic flux density and the electric displacement field. The functions $F_1, F_2 : \mathbb{R} \times \Omega \to \mathbb{R}^3$ denote given current sources, whereas $\sigma \in L_\infty(\Omega; \mathbb{R}^{3 \times 3})$ describes the electrical conductivity. The heat equation is

$$\partial_t \Theta_0 \eta + \mathrm{div}\, q = F_3,$$

where $\eta : \mathbb{R} \times \Omega \to \mathbb{R}$ is the entropy density, $q : \mathbb{R} \times \Omega \to \mathbb{R}^3$ describes the heat flux, $F_3 : \mathbb{R} \times \Omega \to \mathbb{R}$ denotes a given external heat source and $\Theta_0 : \Omega \to \mathbb{R}$, with $\Theta_0, \Theta_0^{-1} \in L_\infty(\Omega)$, is the reference temperature. It is assumed that the *Maxwell-Cattaneo-Vernotte modification* holds, which relates the temperature $\theta : \Omega \to \mathbb{R}$ and the heat flux via

$$\partial_t \kappa_1 q + \kappa_0^{-1} q + \mathrm{grad}\, \Theta_0^{-1} \theta = 0$$

for bijective operators $\kappa_0, \kappa_1 \in L(L_2(\Omega)^3)$.

3.2 *Formulating New Boundary Conditions*

Having recalled the underlying system equations and unknowns involved in our problem, we can now present our new boundary conditions. We first recall the generalization of the boundary conditions (1) to the setting of abstract boundary data spaces indicated in the introduction. In this setting, the boundary conditions (1) take the form (cf. Remark 2.10)

$$\mathrm{curl}_{\mathrm{BD}}\, \iota^{*}_{\mathrm{BD(curl)}} H - \mathrm{curl}_{\mathrm{BD}}\, Q^{*} \iota^{*}_{\mathrm{BD(Grad)}} v + \iota^{*}_{\mathrm{BD(curl)}} E = 0,$$

$$\mathrm{Div}_{\mathrm{BD}}\, \iota^{*}_{\mathrm{BD(Div)}} T - Q\, \mathrm{curl}_{\mathrm{BD}}\, \iota^{*}_{\mathrm{BD(curl)}} E + \left(1 + \alpha \partial_t^{-1}\right) \iota^{*}_{\mathrm{BD(Grad)}} v = 0. \tag{13}$$

Here the given boundary mappings $\widetilde{Q}$ and $\widetilde{\alpha}$ have been replaced by arbitrary (bounded) boundary operators

$$Q\colon \mathrm{BD(curl)} \to \mathrm{BD(Grad)} \quad \text{and} \quad \alpha\colon \mathrm{BD(Grad)} \to \mathrm{BD(Grad)}$$

respectively. In the case of a bounded Lipschitz domain, $\widetilde{Q}$ and $\widetilde{\alpha}$ could be recovered via

$$Q\colon \begin{cases} \mathrm{BD(curl)} & \to & \mathrm{BD(Grad)} \\ H & \mapsto & \gamma^{-1} \widetilde{Q} \gamma_t H \end{cases}$$

and

$$\alpha\colon \begin{cases} \mathrm{BD(Grad)} & \to & \mathrm{BD(Grad)} \\ v & \mapsto & \gamma^{-1} \widetilde{\alpha} \gamma v \end{cases}.$$

An appropriate extension of (13) will yield a novel set of impedance boundary conditions suitable for full thermo-piezo-electromagnetic data. This is achieved in part by supplementing the above two equations with a new boundary equation for thermal data. In addition, the existing two equations for piezo and electromagnetic boundary data are extended by adding in thermal boundary terms. These observations are explored further in Remark 3.1. Using (13) as the starting point, we arrive at the following set of novel boundary conditions. We present

$$\mathrm{curl}_{\mathrm{BD}}\, \iota^{*}_{\mathrm{BD(curl)}} H - \mathrm{curl}_{\mathrm{BD}}\, Q^{*} \iota^{*}_{\mathrm{BD(Grad)}} v + \iota^{*}_{\mathrm{BD(curl)}} E$$
$$+ \beta \iota^{*}_{\mathrm{BD(grad)}} \Theta_0^{-1} \theta = 0,$$

$$\mathrm{Div}_{\mathrm{BD}}\, \iota^{*}_{\mathrm{BD(Div)}} T - Q\, \mathrm{curl}_{\mathrm{BD}}\, \iota^{*}_{\mathrm{BD(curl)}} E + \left(1 + \alpha \partial_t^{-1}\right) \iota^{*}_{\mathrm{BD(Grad)}} v$$
$$+ Q \beta \iota^{*}_{\mathrm{BD(grad)}} \Theta_0^{-1} \theta = 0,$$

$$- \mathrm{div}_{\mathrm{BD}}\, \iota^{*}_{\mathrm{BD(div)}} q - \beta^{*} Q^{*} \iota^{*}_{\mathrm{BD(Grad)}} v - \beta^{*} \iota^{*}_{\mathrm{BD(curl)}} E + \iota^{*}_{\mathrm{BD(grad)}} \Theta_0^{-1} \theta = 0,$$

$$\tag{14}$$

where the arbitrary (bounded) boundary operator $\beta \colon \mathrm{BD(grad)} \to \mathrm{BD(curl)}$ has been introduced that once again could be traced back to an underlying (bounded and linear) boundary mapping $\widetilde{\beta} \colon H^{1/2}(\partial\Omega) \to V_{\gamma_t}$ via

$$\beta \colon \begin{cases} \mathrm{BD(grad)} & \to & \mathrm{BD(curl)} \\ u & \mapsto & \gamma_t^{-1}\widetilde{\beta}\gamma u \end{cases}$$

in the case of a bounded Lipschitz domain.

Before coming to consider the full model with combined boundary dynamics, the following remark is offered to contextualize the modelling choices behind the abstract boundary conditions formulated above.

Remark 3.1 There are several key observations justifying this extension which we now outline. Notice first of all that each of the original piezo-electromagnetic boundary conditions in (13) are, respectively, posed on BD (curl) and BD (Grad). To see this, recall Lemmas 2.6 and 2.2, noting the action of the orthogonal projectors involved. As such, there needs to be an entirely new equation formed for boundary data pertaining to the thermal part of the system. This new equation needs then to be framed on BD (grad). Indeed the last, and entirely new, equation in (14) is posed there.

Secondly, notice that the boundary equations in (13) each involve *both* of the respective unknowns for the corresponding part of the system. Thus, the new equation for the thermal part of the system needs to expressly involve the heat flux, q, and relative temperature, $\Theta_0^{-1}\theta$.

Thirdly, and finally, the original boundary conditions (13) need to be modified to accommodate for, and couple with, the new thermal boundary data. Notice that a similar coupling already exists in (13) between the piezo and electromagnetic boundary data. This is on account of the underlying boundary operators Q, α, and the boundary spaces they map between (again, consider the action of the orthogonal projectors involved). The introduction of the new boundary operator β allows us to achieve this with the relative temperature, $\Theta_0^{-1}\theta$. In the first two equations of (14), notice how β is used to translate thermal boundary data to the realms of electromagnetic and piezo boundary data. In the latter of these cases, one also needs to make use of Q to properly realize this translation. ∇

Setting

$$\tau_q := -\operatorname{div}_{\mathrm{BD}} \iota^*_{\mathrm{BD(div)}} q,$$

$$\tau_H := \operatorname{curl}_{\mathrm{BD}} \iota^*_{\mathrm{BD(curl)}} H \text{ and} \tag{15}$$

$$\tau_T := \operatorname{Div}_{\mathrm{BD}} \iota^*_{\mathrm{BD(Div)}} T,$$

and introducing the weight $\nu \in \mathbb{R}_{>0}$, we can encode the new set of boundary conditions (14) as the block operator equation

$$\begin{pmatrix} \tau_q \\ \tau_H \\ \tau_T \end{pmatrix} + \begin{pmatrix} 1 & -\beta^* & -\beta^* Q^* \\ \beta & 1 & -\operatorname{curl}_{\mathrm{BD}} Q^* \\ Q\beta & -Q\operatorname{curl}_{\mathrm{BD}} & \left(1 + \alpha \partial_{t,\nu}^{-1}\right) \end{pmatrix} \begin{pmatrix} \iota^*_{\mathrm{BD(grad)}} \left(\Theta_0^{-1}\theta\right) \\ \iota^*_{\mathrm{BD(curl)}} E \\ \iota^*_{\mathrm{BD(Grad)}} \nu \end{pmatrix} = 0, \qquad (16)$$

which we will have recourse to use in the sequel.

Remark 3.2 As something of an aside, we conclude this subsection by pointing out that in the classical setting, these new boundary conditions correspond formally to

$$n \times H - n \times \widetilde{Q}^* \nu + n \times (E \times n) + \widetilde{\beta}\left(\Theta_0^{-1}\theta\right) = 0 \text{ on } \partial\Omega,$$

$$T \cdot n - \widetilde{Q}(n \times E) + \left(1 + \widetilde{\alpha}\partial_t^{-1}\right)\nu + \widetilde{Q}\widetilde{\beta}\left(\Theta_0^{-1}\theta\right) = 0 \text{ on } \partial\Omega,$$

$$-q \cdot n + \widetilde{\beta}^* \widetilde{Q}^* \nu + \widetilde{\beta}^*(n \times (E \times n)) + \Theta_0^{-1}\theta = 0 \text{ on } \partial\Omega.$$

$$\triangledown$$

3.3 The Model for Thermo-Piezo-Electromagnetism with Boundary Dynamics

Armed with the novel boundary conditions of interest, we turn our attention back to the formulation of the thermo-piezo-electromagnetic model with boundary dynamics.

In order to enable material coupling to occur between the underlying system equations recalled in Sect. 3.1, they need to be complemented by the additional material relations (cf. [3, Section 3] or [16])

$$T = C \operatorname{Grad} u - eE - \lambda\theta,$$

$$D = e^* \operatorname{Grad} u + \varepsilon E + p\theta,$$

$$B = \mu H,$$

$$\eta = \lambda^* \operatorname{Grad} u + p^* E + \alpha \Theta_0^{-1}\theta.$$

$$(17)$$

These material relations will also allow us to determine the form of the material law operators required in our formulation of the system as an evolutionary equation. Here, $C \in L(L_2(\Omega)^{3\times3}_{sym})$ is continuously invertible and denotes the elasticity tensor. Moreover, $\varepsilon, \mu \in L(L_2(\Omega))$ are the permittivity and permeability, respectively. We introduce $\alpha := \rho_* c \in L(L_2(\Omega))$ where $c \in L(L_2(\Omega))$ is the specific heat capacity. Here, the operators $e \in L(L_2(\Omega)^3; L_2(\Omega)^{3\times3}_{\mathrm{sym}})$, $\lambda \in L(L_2(\Omega); L_2(\Omega)^{3\times3}_{\mathrm{sym}})$ and $p \in L(L_2(\Omega); L_2(\Omega)^3)$ act as coupling parameters. The form of the material law operators in our model is also influenced by an additional factor, which we discuss next.

Following the methodology used in [2], we use abstract boundary data spaces in order to formulate any boundary dynamics *within* the model itself. This is done by introducing auxiliary Hilbert spaces on which to form our boundary dynamics (cf. [9]). As we have three parts to our system (a thermo, a piezo and an electromagnetic part), we introduce a corresponding auxiliary Hilbert space for each of them. This point will become clear once we look at the constituent elements of our model in greater detail. To this end, consider the following lemma (cf. [9] or [2, Section 4.3.2]).

Lemma 3.3 *We have the inclusion*

$$\begin{pmatrix} \mathrm{grad} \\ \iota^*_{\mathrm{BD(grad)}} \end{pmatrix}^* \subseteq (-\,\mathrm{div}\ 0) \tag{18}$$

as well as

$$\begin{pmatrix} \mathrm{curl} \\ \iota^*_{\mathrm{BD(curl)}} \end{pmatrix}^* \subseteq (\mathrm{curl}\ 0) \ \text{and} \ \begin{pmatrix} -\,\mathrm{Grad} \\ \iota^*_{\mathrm{BD(Grad)}} \end{pmatrix}^* \subseteq (\mathrm{Div}\ 0). \tag{19}$$

Moreover, $\begin{pmatrix} \mathrm{grad} \\ \iota^*_{\mathrm{BD(grad)}} \end{pmatrix}^*$ *has as its domain*

$$\{(q, \tau_q) \in H(\mathrm{div}, \Omega) \oplus \mathrm{BD(grad)} : \tau_q = -\,\mathrm{div}_{\mathrm{BD}}\, \iota^*_{\mathrm{BD(div)}} q\}.$$

Similarly, $\begin{pmatrix} \mathrm{curl} \\ \iota^*_{\mathrm{BD(curl)}} \end{pmatrix}^*$ *has as its domain*

$$\{(H, \tau_H) \in H(\mathrm{curl}, \Omega) \oplus \mathrm{BD(curl)} : \tau_H = \mathrm{curl}_{\mathrm{BD}}\, \iota^*_{\mathrm{BD(curl)}} H\},$$

and $\begin{pmatrix} -\,\mathrm{Grad} \\ \iota^*_{\mathrm{BD(Grad)}} \end{pmatrix}^*$ *has as its domain*

$$\{(T, \tau_T) \in H(\mathrm{Div}, \Omega) \oplus \mathrm{BD(Grad)} : \tau_T = \mathrm{Div}_{\mathrm{BD}}\, \iota^*_{\mathrm{BD(Div)}} T\}.$$

Proof Adjoining the inclusion

$$\begin{pmatrix} \mathrm{grad}\ \lceil_{H^1_0(\Omega)} \\ 0 \end{pmatrix} \subseteq \begin{pmatrix} \mathrm{grad} \\ \iota^*_{\mathrm{BD(grad)}} \end{pmatrix}$$

allows us to obtain (18). By definition, $(q, \tau_q) \in H(\mathrm{div}, \Omega) \oplus \mathrm{BD(grad)}$ is in the domain of $\begin{pmatrix} \mathrm{grad} \\ \iota^*_{\mathrm{BD(grad)}} \end{pmatrix}^*$ iff

$$\langle q, \mathrm{grad}\,u \rangle_{L_2(\Omega)^3} + \langle \tau_q, \iota^*_{\mathrm{BD(grad)}} u \rangle_{\mathrm{BD(grad)}} = -\langle \mathrm{div}\,q, u \rangle_{L_2(\Omega)^3}$$

holds for all $u \in H^1(\Omega)$. Using integration by parts (Lemma 2.7), this is equivalent to

$$\langle \tau_q, \iota^*_{BD(grad)} u\rangle_{BD(grad)} = \langle \mathrm{div}_{BD}\, \iota^*_{BD(div)} q, \iota^*_{BD(grad)} u\rangle_{BD(grad)}$$

for all $u \in H^1(\Omega)$. Clearly, this yields the desired domain. The remaining two cases follow by an analogous means. $\qquad\square$

The boundary data spaces appearing in Lemma 3.3 are precisely the auxiliary Hilbert spaces we alluded to above. Thus, as an evolutionary equation on $L_{2,\nu}\,(\mathbb{R}; \mathcal{H})$, the model for thermo-piezo-electromagnetism with boundary dynamics is

$$\left(\partial_{t,\nu} M_0 + M_1\left(\partial_{t,\nu}\right) + A\right)
\begin{pmatrix} v \\ \begin{pmatrix} T \\ \tau_T \end{pmatrix} \\ E \\ \begin{pmatrix} H \\ \tau_H \end{pmatrix} \\ \Theta_0^{-1}\theta \\ \begin{pmatrix} q \\ \tau_q \end{pmatrix} \end{pmatrix}
=
\begin{pmatrix} F_0 \\ \begin{pmatrix} 0 \\ 0 \end{pmatrix} \\ F_1 \\ \begin{pmatrix} F_2 \\ 0 \end{pmatrix} \\ F_3 \\ \begin{pmatrix} 0 \\ 0 \end{pmatrix} \end{pmatrix}, \tag{20}$$

(where v denotes the first-time-derivative of u and, as in [3], the temperature, θ, is replaced by the relative temperature, $\Theta_0^{-1}\theta$, as the unknown temperature function) with $\mathcal{H}$, A, M_0 and $M_1\left(\partial_{t,\nu}\right)$ to be specified. We are on the Hilbert space

$$\mathcal{H} := L_2\,(\Omega)^3 \oplus L_2\,(\Omega)^{3\times 3}_{\mathrm{sym}} \oplus BD\,(\mathrm{Grad}) \oplus$$
$$L_2\,(\Omega)^3 \oplus L_2\,(\Omega)^3 \oplus BD\,(\mathrm{curl}) \oplus \qquad . \tag{21}$$
$$L_2\,(\Omega) \oplus L_2\,(\Omega)^3 \oplus BD\,(\mathrm{grad})$$

The operator A is

$$A :=$$

$$\begin{pmatrix}
0 & -\begin{pmatrix} -\mathrm{Grad} \\ \iota^*_{\mathrm{Grad}} \end{pmatrix}^* & 0 & 0_{1\times 2} & 0 & 0_{1\times 2} \\
\begin{pmatrix} -\mathrm{Grad} \\ \iota^*_{\mathrm{Grad}} \end{pmatrix} & 0_{2\times 2} & 0_{2\times 1} & 0_{2\times 2} & 0_{2\times 1} & 0_{2\times 2} \\
0 & 0_{1\times 2} & 0 & -\begin{pmatrix} \mathrm{curl} \\ \iota^*_{\mathrm{curl}} \end{pmatrix}^* & 0 & 0_{1\times 2} \\
0_{2\times 1} & 0_{2\times 2} & \begin{pmatrix} \mathrm{curl} \\ \iota^*_{\mathrm{curl}} \end{pmatrix} & 0_{2\times 2} & 0_{2\times 1} & 0_{2\times 2} \\
0 & 0_{1\times 2} & 0 & 0_{1\times 2} & 0 & -\begin{pmatrix} \mathrm{grad} \\ \iota^*_{\mathrm{grad}} \end{pmatrix}^* \\
0_{2\times 1} & 0_{2\times 2} & 0_{2\times 1} & 0_{2\times 2} & \begin{pmatrix} \mathrm{grad} \\ \iota^*_{\mathrm{grad}} \end{pmatrix} & 0_{2\times 2}
\end{pmatrix} . \tag{22}$$

The operator A encodes the purely spatial derivatives of our model. On account of the extended operators recalled in (18) and (19), A also encodes the orthogonal projectors needed to isolate boundary data. It is important to note that the inherent boundary conditions (15) are present in our system implicitly via Lemma 3.3. As for the material operator M_0, we have

$$M_0 :=$$

$$\begin{pmatrix} \rho_* & 0_{1\times2} & 0 & 0_{1\times2} & 0 & 0_{1\times2} \\ 0_{2\times1} & M_{0,33} & 0_{2\times1} & M_{0,36} & \begin{pmatrix} C^{-1}\lambda\Theta_0 \\ 0 \end{pmatrix} & 0_{2\times2} \\ 0 & 0_{1\times2} & \varepsilon + e^*C^{-1}e & 0_{1\times2} & p\Theta_0 + e^*C^{-1}\lambda\Theta_0 & 0_{1\times2} \\ 0_{2\times1} & M_{0,36}{}^* & 0_{2\times1} & M_{0,66} & 0_{2\times1} & 0_{2\times2} \\ 0 & \begin{pmatrix} \Theta_0\lambda^*C^{-1} & 0 \end{pmatrix} & \Theta_0 p^* + \Theta_0\lambda^*C^{-1}e & 0_{1\times2} & \gamma_0 + \Theta_0\lambda^*C^{-1}\lambda\Theta_0 & 0_{1\times2} \\ 0_{2\times1} & 0_{2\times2} & 0_{2\times1} & 0_{2\times2} & 0_{2\times1} & M_{0,99} \end{pmatrix}, \tag{23}$$

(notice the introduction of the shorthand $\gamma_0 := \Theta_0\alpha$—again see [3]) where for notational ease, we have introduced the blocks

$$M_{0,33} := \begin{pmatrix} C^{-1} & 0 \\ 0 & 0 \end{pmatrix}, \; M_{0,36} := \begin{pmatrix} C^{-1}e & 0 \\ 0 & 0 \end{pmatrix},$$

$$M_{0,66} := \begin{pmatrix} \mu & 0 \\ 0 & 0 \end{pmatrix}, \; M_{0,99} := \begin{pmatrix} \kappa_1 & 0 \\ 0 & 0 \end{pmatrix}.$$

It is clear that M_0 is self-adjoint by construction. The remaining material operator, $M_1\left(\partial_{t,v}\right)$, is given by

$$M_1\left(\partial_{t,v}\right) :=$$

$$\begin{pmatrix} 0 & 0_{1\times2} & 0 & 0_{1\times2} & 0 & 0_{1\times2} \\ 0_{2\times1} & M_{1,33}\left(\partial_{t,v}\right) & 0_{2\times1} & M_{1,36}\left(\partial_{t,v}\right) & 0_{2\times1} & M_{1,39}\left(\partial_{t,v}\right) \\ 0 & 0_{1\times2} & \sigma & 0_{1\times2} & 0 & 0_{1\times2} \\ 0_{2\times1} & M_{1,63}\left(\partial_{t,v}\right) & 0_{2\times1} & M_{1,66}\left(\partial_{t,v}\right) & 0_{2\times1} & M_{1,69}\left(\partial_{t,v}\right) \\ 0 & 0_{1\times2} & 0 & 0_{1\times2} & 0 & 0_{1\times2} \\ 0_{2\times1} & M_{1,93}\left(\partial_{t,v}\right) & 0_{2\times1} & M_{1,96}\left(\partial_{t,v}\right) & 0_{2\times1} & M_{1,99}\left(\partial_{t,v}\right) \end{pmatrix}, \tag{24}$$

where for $i, j \in \{3, 6, 9\}$, $(i, j) \neq 9$, we have introduced the block operators

$$M_{1,ij}\left(\partial_{t,v}\right) := \begin{pmatrix} 0 & 0 \\ 0 & K_{ij}\left(\partial_{t,v}\right) \end{pmatrix}, \tag{25}$$

and for the case $i = j = 9$,

$$M_{1,99}\left(\partial_{t,v}\right) := \begin{pmatrix} \kappa_0^{-1} & 0 \\ 0 & K_{99}\left(\partial_{t,v}\right) \end{pmatrix}, \tag{26}$$

with the specific operator coefficients $K_{ij}(\partial_{t,\nu})$ to be specified shortly. We first point out that in our PDE system (20), M_0 and $M_1(\partial_{t,\nu})$ encode the underlying constitutive relations behind the physics of our problem. This is done with the material coupling of (17).

Upon recalling the block operator formulation of our boundary equations, (16), we can first compute and then apply the inverse[2] to instead equivalently consider

$$
\begin{pmatrix}
1 & -\beta^* & -\beta^* Q^* \\
\beta & 1 & -\operatorname{curl}_{\mathrm{BD}} Q^* \\
Q\beta & -Q\operatorname{curl}_{\mathrm{BD}} & \left(1+\alpha\partial_{t,\nu}^{-1}\right)
\end{pmatrix}^{-1}
\begin{pmatrix}
\tau_T \\ \tau_H \\ \tau_q
\end{pmatrix}
+
\begin{pmatrix}
\iota^*_{\mathrm{BD(Grad)}}\, v \\
\iota^*_{\mathrm{BD(curl)}}\, E \\
\iota^*_{\mathrm{BD(grad)}}\left(\Theta_0^{-1}\theta\right)
\end{pmatrix}
= 0.
$$

The computed inverse

$$
\begin{pmatrix}
1 & -\beta^* & -\beta^* Q^* \\
\beta & 1 & -\operatorname{curl}_{\mathrm{BD}} Q^* \\
Q\beta & -Q\operatorname{curl}_{\mathrm{BD}} & \left(1+\alpha\partial_{t,\nu}^{-1}\right)
\end{pmatrix}^{-1}
=
\begin{pmatrix}
K_{99}(\partial_{t,\nu}) & K_{96}(\partial_{t,\nu}) & K_{93}(\partial_{t,\nu}) \\
K_{69}(\partial_{t,\nu}) & K_{66}(\partial_{t,\nu}) & K_{63}(\partial_{t,\nu}) \\
K_{39}(\partial_{t,\nu}) & K_{36}(\partial_{t,\nu}) & K_{33}(\partial_{t,\nu})
\end{pmatrix}
$$

has for diagonal coefficients

$$
K_{33}(\partial_{t,\nu}) = \left(1 + Q\beta\,(Q\beta)^* + \alpha\partial_{t,\nu}^{-1} - \left[Q\beta\beta^* - Q\operatorname{curl}_{\mathrm{BD}}\right]\left(1+\beta\beta^*\right)^{-1}\right.
$$

$$
\left.\cdot\left[\beta\,(Q\beta)^* - Q\operatorname{curl}_{\mathrm{BD}}\right]\right)^{-1},
$$

$$
K_{66}(\partial_{t,\nu}) = \left(1+\beta\beta^*\right)^{-1} + \left(1+\beta\beta^*\right)^{-1}\left[\beta\,(Q\beta)^* - \operatorname{curl}_{\mathrm{BD}} Q^*\right] K_{33}(\partial_{t,\nu})
$$

$$
\cdot\left[Q\beta\beta^* - Q\operatorname{curl}_{\mathrm{BD}}\right]\left(1+\beta\beta^*\right)^{-1},
$$

$$
K_{99}(\partial_{t,\nu}) = 1 + \left[-\beta^*\left(1+\beta\beta^*\right)^{-1}\beta + \left[\beta^*\left(1+\beta\beta^*\right)^{-1}\left[\beta\,(Q\beta)^* - \operatorname{curl}_{\mathrm{BD}} Q^*\right]\right.\right.
$$

$$
\left.\left. - (Q\beta)^*\right] K_{33}(\partial_{t,\nu})\left[Q\beta - \left[Q\beta\beta^* - Q\operatorname{curl}_{\mathrm{BD}}\right]\left(1+\beta\beta^*\right)^{-1}\beta\right]\right],
$$

and for off-diagonal coefficients

$$
K_{96}(\partial_{t,\nu}) = -\left[\left[(Q\beta)^* - \beta^*\left(1+\beta\beta^*\right)^{-1}\left[\beta\,(Q\beta)^* - \operatorname{curl}_{\mathrm{BD}} Q^*\right] - \beta^*\right]\right.
$$

$$
\left.\cdot K_{33}(\partial_{t,\nu})\left[Q\beta\beta^* - Q\operatorname{curl}_{\mathrm{BD}}\right]\right]\left(1+\beta\beta^*\right)^{-1},
$$

$$
K_{69}(\partial_{t,\nu}) = -\left(1+\beta\beta^*\right)^{-1}\left[\beta - \left[\beta\,(Q\beta)^* - \operatorname{curl}_{\mathrm{BD}} Q^*\right]\right.
$$

$$
\left.\cdot K_{33}(\partial_{t,\nu})\left[Q\beta - \left[Q\beta\beta^* - Q\operatorname{curl}_{\mathrm{BD}}\right]\left(1+\beta\beta^*\right)^{-1}\beta\right]\right],
$$

---

[2] Using Lemma 2.1, we prove the invertibility for large enough ν in (27).

and

$$K_{93}\left(\partial_{t,\nu}\right) = -\left[\beta^*\left(1+\beta\beta^*\right)^{-1}\left[\beta\left(Q\beta\right)^* - \mathrm{curl}_{\mathrm{BD}}\, Q^*\right] - \left(Q\beta\right)^*\right] K_{33}\left(\partial_{t,\nu}\right),$$

$$K_{39}\left(\partial_{t,\nu}\right) = -K_{33}\left(\partial_{t,\nu}\right)\left[Q\beta - \left[Q\beta\beta^* - Q\,\mathrm{curl}_{\mathrm{BD}}\right]\left(1+\beta\beta^*\right)^{-1}\beta\right],$$

as well as

$$K_{63}\left(\partial_{t,\nu}\right) = -\left(1+\beta\beta^*\right)^{-1}\left[\beta\left(Q\beta\right)^* - \mathrm{curl}_{\mathrm{BD}}\, Q^*\right] K_{33}\left(\partial_{t,\nu}\right),$$

$$K_{36}\left(\partial_{t,\nu}\right) = -K_{33}\left(\partial_{t,\nu}\right)\left[Q\beta\beta^* - Q\,\mathrm{curl}_{\mathrm{BD}}\right]\left(1+\beta\beta^*\right)^{-1},$$

where we have used the skew-symmetry of $\mathrm{curl}_{\mathrm{BD}}$ (see Lemma 2.6). With these entries computed, the actual form of $M_1\left(\partial_{t,\nu}\right)$ is fully realized.

Remark 3.4 The additional zeroes appearing in the block operators (22), (23) and (24) arise on account of encoding the boundary dynamics within the system itself. In particular, the increase in dimension is incurred by the construction (15) by Lemma 3.3. $\qquad\triangledown$

Remark 3.5 Formally replacing z by $\partial_{t,\nu}$ in the definition of $M_1(\partial_{t,\nu})$, we get the material law $M_1(z)$ with $s_b(M_1)$ being bounded above by $\|\alpha\|$ (cf. (27)). Using the definition of material operators and Lemma 2.3, we easily get that $M_1(\partial_{t,\nu})$ indeed is the material operator stemming from the material law $M_1(z)$. $\qquad\triangledown$

3.4 Evolutionary Well-Posedness of the Model

With the above preparations to hand, the main well-posedness result of this chapter can now be presented and proven.

Theorem 3.6 *Let $\nu \in \mathbb{R}_{>0}$ and $z \in \mathbb{C}_{\mathrm{Re}>\nu}$. Let $\Omega \subseteq \mathbb{R}^d$ be open and $\mathcal{H}$ as in (21). Additionally, let M_0, $M_1(z) \in L(\mathcal{H})$ be as in (23) and (24), respectively, and A as in (22). Furthermore, introduce the notation*

$$m_{0,55} := \gamma_0 - \Theta_0\lambda^*C^{-1}e\left(\mu - e^*C^{-1}e\right)^{-1}e^*C^{-1}\lambda\Theta_0 \text{ and}$$

$$m_{0,44} := \varepsilon + e^*C^{-1}e - \left(p\Theta_0 + e^*C^{-1}\lambda\Theta_0\right)^*\left(m_{0,55}\right)^{-1}\left(p\Theta_0 + e^*C^{-1}\lambda\Theta_0\right).$$

Assume ρ_*, ε, μ, C and γ_0 are each self-adjoint and non-negative. Moreover, assume ρ_*, C, $m_{0,55} \gg 0$, as well as

$$\mu - e^*C^{-1}e, \ \nu\, m_{0,44} + \sigma, \ \nu\kappa_1 + \kappa_0^{-1} \gg 0,$$

for large enough $v \in \mathbb{R}_{>0}$. Then, for all $v \in \mathbb{R}_{>0}$ sufficiently large, the operator

$$\partial_{t,v} M_0 + M_1(\partial_{t,v}) + A$$

is densely defined and closable in $L_{2,v}(\mathbb{R}; \mathcal{H})$. The respective closure is continuously invertible with causal inverse being eventually independent of v.

Proof The assertion will follow from applying Picard's Theorem to the material law (cf. Remark 3.5)

$$M(z) := M_0 + z^{-1} M_1(z)$$

and spatial operator A. As already noted in Sect. 3.3, it is clear that A is skew-self-adjoint and M_0 self-adjoint by construction. As such, we need only establish

$$z M_0 + \operatorname{Re} M_1(z) \gg 0$$

uniformly in $z \in \mathbb{C}_{\operatorname{Re} \geq v}$ for large enough $v \in \mathbb{R}_{>0}$. An elementary first permutation yields the congruence

$$v M_0 + \operatorname{Re} M_1(z) \sim v \mathcal{N} + \operatorname{Re} \mathcal{M}(z)$$

where

$$\mathcal{N} := \begin{pmatrix} \rho_* & 0 & 0 & 0 \\ 0 & \mathcal{N}' & 0 & 0 \\ 0 & 0 & 0_{3\times 3} & 0 \\ 0 & 0 & 0 & \kappa_1 \end{pmatrix}, \quad \mathcal{M}(z) := \begin{pmatrix} 0 & 0 & 0 & 0 \\ 0 & \mathcal{M}' & 0 & 0 \\ 0 & 0 & \mathcal{K}(z) & 0 \\ 0 & 0 & 0 & \kappa_0^{-1} \end{pmatrix}$$

and where

$$\mathcal{N}' := \begin{pmatrix} \varepsilon + e^* C^{-1} e & 0 & 0 & p\Theta_0 + e^* C^{-1} \lambda \Theta_0 \\ 0 & C^{-1} & C^{-1} e & C^{-1} \lambda \Theta_0 \\ 0 & e^* C^{-1} & \mu & 0 \\ \Theta_0 p^* + \Theta_0 \lambda^* C^{-1} e & \Theta_0 \lambda^* C^{-1} & 0 & \gamma_0 + \Theta_0 \lambda^* C^{-1} \lambda \Theta_0 \end{pmatrix}$$

together with

$$\mathcal{M}' := \begin{pmatrix} \sigma & 0_{3\times 1} \\ 0_{1\times 3} & 0_{3\times 3} \end{pmatrix} \quad \text{and} \quad \mathcal{K}(z) := \begin{pmatrix} K_{33}(z) & K_{36}(z) & K_{39}(z) \\ K_{63}(z) & K_{66}(z) & K_{69}(z) \\ K_{93}(z) & K_{96}(z) & K_{99}(z) \end{pmatrix}.$$

It suffices to check the positive-definiteness condition for the block operators $v\mathcal{N}' + \operatorname{Re} \mathcal{M}'$ and $\mathcal{K}(z)$ alone. Starting with the former of these blocks, on account of a second permutation and a subsequent symmetric Gauss step (which isolates C^{-1} on the leading diagonal), we need only consider the sub-block operator

$$\nu \begin{pmatrix} \varepsilon + e^* C^{-1} e & 0 & p\Theta_0 + e^* C^{-1} \lambda \Theta_0 \\ 0 & \mu - e^* C^{-1} e & -e^* C^{-1} \lambda \Theta_0 \\ \Theta_0 p^* + \Theta_0 \lambda^* C^{-1} e & -\Theta_0 \lambda^* C^{-1} e & \gamma_0 \end{pmatrix} + \begin{pmatrix} \sigma & 0 & 0 \\ 0 & 0 & 0 \\ 0 & 0 & 0 \end{pmatrix}.$$

A third permutation yields the congruent operator

$$\nu \begin{pmatrix} \mu - e^* C^{-1} e & 0 & -e^* C^{-1} \lambda \Theta_0 \\ 0 & \varepsilon + e^* C^{-1} e & p\Theta_0 + e^* C^{-1} \lambda \Theta_0 \\ -\Theta_0 \lambda^* C^{-1} e & \Theta_0 p^* + \Theta_0 \lambda^* C^{-1} e & \gamma_0 \end{pmatrix} + \begin{pmatrix} 0 & 0 & 0 \\ 0 & \sigma & 0 \\ 0 & 0 & 0 \end{pmatrix}$$

which, under a subsequent pair of symmetric Gauss steps, is itself congruent to the operator

$$\nu \begin{pmatrix} \mu - e^* C^{-1} e & 0 & 0 \\ 0 & m_{0,44} & 0 \\ 0 & 0 & m_{0,55} \end{pmatrix} + \begin{pmatrix} 0 & 0 & 0 \\ 0 & \sigma & 0 \\ 0 & 0 & 0 \end{pmatrix}$$

which is positive-definite by assumption. As for the remaining block operator, $\mathcal{K}(z)$, we will use Lemma 2.1 to indirectly establish the desired property. First of all, for $x \in \mathrm{BD}(\mathrm{Grad})$, compute

$$\left\langle x, 1 + \mathrm{Re}\left(\alpha z^{-1}\right) x \right\rangle_{\mathrm{BD(Grad)}} = \|x\|_{\mathrm{BD(Grad)}}^2 + \left\langle x, \mathrm{Re}\left(\alpha z^{-1}\right) x \right\rangle_{\mathrm{BD(Grad)}}$$

$$= \|x\|_{\mathrm{BD(Grad)}}^2 + \mathrm{Re}\left\langle x, \alpha z^{-1} x \right\rangle_{\mathrm{BD(Grad)}}$$

$$\geq \|x\|_{\mathrm{BD(Grad)}}^2 - \|\alpha\| \, |z^{-1}| \|x\|_{\mathrm{BD(Grad)}}^2$$

$$\geq \left(1 - \frac{\|\alpha\|}{\nu}\right) \|x\|_{\mathrm{BD(Grad)}}^2.$$

We then compute

$$\mathrm{Re}\begin{pmatrix} 1 & -\beta^* & -\beta^* Q^* \\ \beta & 1 & -\mathrm{curl}_{\mathrm{BD}} Q^* \\ Q\beta & -Q\,\mathrm{curl}_{\mathrm{BD}} & (1 + \alpha z^{-1}) \end{pmatrix} = \begin{pmatrix} 1 & 0 & 0 \\ 0 & 1 & 0 \\ 0 & 0 & \mathrm{Re}\,(1 + \alpha z^{-1}) \end{pmatrix}$$

$$\geq \min\left\{1, 1 - \frac{\|\alpha\|}{\nu}\right\} \tag{27}$$

$$= 1 - \frac{\|\alpha\|}{\nu}.$$

By Lemma 2.1, we can use this to estimate the real-part of the remaining block operator occurring in the congruent form above. Indeed, we then have

$$
\mathrm{Re}\begin{pmatrix} 1 & -\beta^* & -\beta^* Q^* \\ \beta & 1 & -\operatorname{curl}_{\mathrm{BD}} Q^* \\ Q\beta & -Q\operatorname{curl}_{\mathrm{BD}} & (1+\alpha z^{-1}) \end{pmatrix}^{-1}
$$

$$
\geq \left(1 - \frac{\|\alpha\|}{\nu}\right)\left\|\begin{pmatrix} 1 & -\beta^* & -\beta^* Q^* \\ \beta & 1 & -\operatorname{curl}_{\mathrm{BD}} Q^* \\ Q\beta & -Q\operatorname{curl}_{\mathrm{BD}} & (1+\alpha z^{-1}) \end{pmatrix}\right\|^{-2}
$$

yielding the desired positive-definiteness of the system. $\qquad\square$

Remark 3.7 The application of the indicated permutations as congruence transforms in the proof above is necessary to retain the possibility of an *eddy-current approximation* (see Remark 2.1 in [3]). Put succinctly, the eddy-current approximation allows us to accommodate for the limit case

$$
\varepsilon = \left(p\Theta_0 + e^* C^{-1}\lambda\Theta_0\right)^* (m_{0,55})^{-1}\left(p\Theta_0 + e^* C^{-1}\lambda\Theta_0\right) - e^* C^{-1} e,
$$

provided that σ is large enough to compensate. Were one not to intermittently permute the system as done in the above proof—and instead solely apply sequential symmetric Gauss steps as congruence transforms—then one might arrive at a sub-block operator of the form

$$
\nu\begin{pmatrix} \varepsilon + e^* C^{-1} e & 0 & 0 \\ 0 & \mu - e^* C^{-1} e & 0 \\ 0 & 0 & \gamma_0' \end{pmatrix} + \begin{pmatrix} \sigma & 0 & 0 \\ 0 & 0 & 0 \\ 0 & 0 & 0 \end{pmatrix}
$$

where, besides needing to additionally assume $\varepsilon + e^* C^{-1} e$ invertible, arises the term

$$
\gamma_0' := \gamma_0 - \left(e^* C^{-1}\lambda\Theta_0\right)^*\left(\mu - e^* C^{-1} e\right)^{-1} e^* C^{-1}\lambda\Theta_0
$$

$$
- \left(p\Theta_0 + e^* C^{-1}\lambda\Theta_0\right)^*\left(\varepsilon + e^* C^{-1} e\right)^{-1}\left(p\Theta_0 + e^* C^{-1}\lambda\Theta_0\right).
$$

In this alternative formulation, it is still possible to choose the operator ε to be close to $-e^* C^{-1} e$; however, the eddy-current approximation $\varepsilon = -e^* C^{-1} e$ is excluded. $\qquad\triangledown$

Acknowledgments Author b would like to acknowledge the support provided by the Engineering and Physical Sciences Research Council [grant number EP/S515632/1] in preparing this work.

References

1. K. Ammari, S. Nicaise, Stabilization of a piezoelectric system. Asymptot. Anal. **73**(3), 125–146 (2011). ISSN: 0921-7134. https://doi.org/10.3233/ASY-2011-1033
2. R. Picard, On well-posedness for a piezo-electromagnetic coupling model with boundary dynamics. Comput. Methods Appl. Math. **17**(3), 499–513 (2017). ISSN: 1609-4840. https://doi.org/10.1515/cmam-2017-0005
3. A.J. Mulholland, R. Picard, S. Trostorff, M. Waurick, On well-posedness for some thermo-piezoelectric coupling models. Math. Methods Appl. Sci. **39**(15), 4375–4384 (2016). https://doi.org/10.1002/mma.3866
4. R. Picard, A structural observation for linear material laws in classical mathematical physics. Math. Methods Appl. Sci. **32**(14), 1768–1803 (2009). ISSN: 0170-4214. https://doi.org/10.1002/mma.1110
5. C. Seifert, S. Trostorff, M. Waurick, *Evolutionary Equations. Picard's Theorem for Partial Differential Equations, and Applications*, vol. 287. Operator Theory: Advances and Applications (Birkhäuser/Springer, Cham, 2022), pp. xii+317. ISBN: 978-3-030-89396-5. https://doi.org/10.1007/978-3-030-89397-2
6. R. Picard, S. Trostorff, M. Waurick, On evolutionary equations with material laws containing fractional integrals. Math. Methods Appl. Sci. **38**(15), 3141–3154 (2015). ISSN: 0170-4214. https://doi.org/10.1002/mma.3286
7. G. Weiss, O.J. Staffans, Maxwell's equations as a scattering passive linear system. SIAM J. Control Optim. **51**(5), 3722–3756 (2013). ISSN: 0363-0129. https://doi.org/10.1137/120869444
8. A. Buffa, M. Costabel, D. Sheen, On traces for $\mathbf{H}(\mathbf{curl}, \Omega)$ in Lipschitz domains. J. Math. Anal. Appl. **276**(2), 845–867 (2002). ISSN: 0022-247X. https://doi.org/10.1016/S0022-247X(02)00455-9
9. R. Picard, S. Seidler, S. Trostorff, M. Waurick, On abstract grad-div systems. J. Differ. Equ. **260**(6), 4888–4917 (2016). ISSN: 0022-0396. https://doi.org/10.1016/j.jde.2015.11.033
10. B. Franchi, R. Serapioni, F. Serra Cassano, Meyers-Serrin type theorems and relaxation of variational integrals depending on vector fields. Houst. J. Math. **22**(4), 859–890 (1996). ISSN: 0362-1588
11. N.G. Meyersm J. Serrin, $H = W$. Proc. Natl. Acad. Sci. USA **51**, 1055–1056 (1964). ISSN: 0027-8424. https://doi.org/10.1073/pnas.51.6.1055
12. J.A. Nitsche, On Korn's second inequality. RAIRO, Anal. Numér. **15**, 237–248 (1981). ISSN: 0399-0516. https://doi.org/10.1051/m2an/1981150302371
13. K.O. Friedrichs, The identity of weak and strong extensions of differential operators. Trans. Am. Math. Soc. **55**, 132–151 (1944). ISSN: 0002-9947. https://doi.org/10.2307/1990143
14. G. Geymonat, P. Suquet, Functional spaces for Norton-Hoff materials. Math. Meth. Appl. Sci. **8**(1), 206–222 (1986). ISSN: 0170-4214. https://doi.org/10.1002/mma.1670080113
15. F. Demengel, G. Demengel, *Functional Spaces for the Theory of Elliptic Partial Differential Equations*. Trans. French by R. Erné. Universitext (Springer, London, 2012), pp. xviii+465. ISBN: 978-1-4471-2806-9. https://doi.org/10.1007/978-1-4471-2807-6
16. R.D. Mindlin, Equations of high frequency vibrations of thermopiezoelectric crystal plates. Int. J. Solids Struct. **10**(6), 625–637 (1974). ISSN: 0020-7683. https://doi.org/10.1016/0020-7683(74)90047-X

A Note on Some Non-Local Boundary Conditions and Their Use in Connection with Beltrami Fields

Rainer Picard and Sascha Trostorff

1 Introduction

A typical non-local boundary condition we have in mind is the condition of periodicity—say—on the unit interval $]-1/2, 1/2[$ for the standard one-dimensional derivative ∂. In an $L_2\left(]-1/2, 1/2[\right)$ setting, this condition can conveniently be described (see Example 3.3) by

$$\partial u \perp 1, \tag{1}$$

resulting in a skew-self-adjoint operator $\partial^{\#}$. If we denote by ∂_0 the derivative with vanishing boundary data, we have

$$\partial = -\partial_0^*.$$

We clearly have the orthogonal decomposition

$$L_2\left(]-1/2, 1/2[\,,\mathbb{R}\right) = \overline{\mathrm{ran}}\left(\partial_0\right) \oplus \ker\left(\partial\right),$$
$$= \overline{\mathrm{ran}}\left(\partial_0\right) \oplus \mathbb{R},$$

R. Picard (✉)
Institut für Analysis, TU Dresden, Dresden, Germany
e-mail: rainer.picard@tu-dresden.de

S. Trostorff
Mathematisches Seminar, CAU Kiel, Kiel, Germany
e-mail: trostorff@math.uni-kiel.de

© The Author(s), under exclusive license to Springer Nature Switzerland AG 2024   25
F. L. Schwenninger, M. Waurick (eds.), *Systems Theory and PDEs*,
Trends in Mathematics, https://doi.org/10.1007/978-3-031-64991-2_2

so that the non-local boundary condition (1) defining $\partial^{\#}$ can be rephrased as

$$\partial u \in \overline{\operatorname{ran}}\,(\partial_0)\,. \tag{2}$$

Remarkably, this is essentially the same situation as in the case of the vector-analytical operator curl with boundary conditions associated with the topic of force-free magnetic field, the eigensolutions of which are frequently referred to as Beltrami fields. In classical terms, this boundary condition is $n \cdot \operatorname{curl} H = 0$ on the boundary of a domain Ω, n denoting the unit normal vector field. In an $L_2\left(\Omega; \mathbb{R}^3\right)$ setting, this boundary condition can be formulated as an orthogonality condition

$$\operatorname{curl} H \perp \operatorname{ran}(\operatorname{grad})\,.$$

In topologically more complex domains, one needs to require additional constraints to obtain a reasonably small spectrum, which leads to the analogue to (2)

$$\operatorname{curl} H \in \overline{\operatorname{ran}}\,(\operatorname{curl}_0)\,.$$

The resulting operator $\operatorname{curl}^{\#}$ has been extensively studied; see the discussion in [9, 10]. After a long pre-history based on assumptions warranting that $\operatorname{ran}(\operatorname{curl}_0)$ is actually closed, N. Filonov, [3], has found, based on potential theoretical considerations, that it suffices to assume that Ω is merely of finite measure. This is in sharp contrast to the situation of $\operatorname{ran}(\operatorname{curl}_0)$ closed, for which at least some boundary regularity appears to be required.

Filonov's result suggests that there actually might be an abstract functional analytical mechanism in the background. In this chapter, we shall indeed present such an approach, which allows to cover the case of bounded domains. Exterior domains clearly require adjustments, and it appears that to control the asymptotics of unbounded domains of finite measure, more concrete methods, such as potential theory, are required.

In the following, we shall develop an abstract setting covering these examples and expanding the reach of the concept to a larger class of applications. In Sect. 2, we first collect some basic facts useful for our framework. With the periodic boundary condition case as a root example in mind, we shall speak of 'abstract periodicity', which we introduce in Sect. 3. In Sect. 4, we deal with the closed range property of $A^{\#}$, which is a crucial ingredient of linear solution theory. The last section, Sect. 5, is dedicated to some illustrative examples.

2 Preliminaries

In this section, we collect some probably well-known results, which will be useful in further sections of this note (see e.g. [7, Lemmas 4.1, 4.3]). For the readers' convenience, we include the proofs. Throughout, let H_0 and H_1 be Hilbert spaces and $A\colon \operatorname{dom}(A) \subseteq H_0 \to H_1$ a closed, densely defined, linear operator. We begin to introduce the reduced operator for A.

Definition We define the *reduced operator* A_{red} by

$$A_{\mathrm{red}}\colon \operatorname{dom}(A) \cap \ker(A)^\perp \subseteq \ker(A)^\perp \to \overline{\operatorname{ran}}(A), \quad x \mapsto Ax.$$

Moreover, we set $D_{A_{\mathrm{red}}} := \operatorname{dom}(A) \cap \ker(A)^\perp$ and equip it with the graph norm of A_{red}.

Clearly, A_{red} is one-to-one and has dense range. Moreover, A_{red} is closed, and hence, $D_{A_{\mathrm{red}}}$ is complete.

Proposition 2.1 *We consider A and the reduced operator A_{red}. Then*

(a) $\operatorname{ran}(A)$ *is closed if and only if A_{red} is boundedly invertible.*
(b) $D_{A_{\mathrm{red}}} \hookrightarrow\hookrightarrow H_0$ *if and only if A_{red} is compactly invertible.*

Proof

(a) If $\operatorname{ran}(A)$ is closed, then A_{red} is onto and hence bijective. Moreover, since A_{red} is closed, the bounded invertibility is a consequence of the closed graph theorem. If, on the other hand, A_{red} is boundedly invertible, then it is onto, which yields $\overline{\operatorname{ran}}(A) = \operatorname{ran}(A_{\mathrm{red}}) = \operatorname{ran}(A)$, hence A has closed range.

(b) Assume that $D_{A_{\mathrm{red}}} \hookrightarrow\hookrightarrow H_0$. Since $D_{A_{\mathrm{red}}} \subseteq \ker(A)^\perp$ and $\ker(A)^\perp$ is closed in H_0, we infer that $D_{A_{\mathrm{red}}} \hookrightarrow\hookrightarrow \ker(A)^\perp$. We claim that A_{red} is boundedly invertible. Indeed, if A_{red} is not boundedly invertible, we find a sequence $(x_n)_{n\in\mathbb{N}}$ in $\operatorname{dom}(A_{\mathrm{red}})$ such that $A_{\mathrm{red}}x_n \to 0$ and $\|x_n\|_{H_0} = 1$ for each $n \in \mathbb{N}$. The latter gives that $(x_n)_n$ is a bounded sequence in $D_{A_{\mathrm{red}}}$, and thus, it possesses a convergent sub-sequence. So, we assume without loss of generality that $x_n \to x$ for some $x \in \ker(A)^\perp$. By the closedness of A_{red}, we obtain $x \in \operatorname{dom}(A_{\mathrm{red}})$ and $A_{\mathrm{red}}x = 0$. Thus, $x = 0$, which contradicts $\|x\| = \lim_{n\to\infty}\|x_n\| = 1$. Thus, A_{red} is boundedly invertible. Finally, $A_{\mathrm{red}}^{-1}\colon \operatorname{ran}(A) \to D_{A_{\mathrm{red}}}$ is bounded, and hence, $A_{\mathrm{red}}^{-1}\colon \operatorname{ran}(A) \to \ker(A)^\perp$ is compact.

Conversely, assume that A_{red} is compactly invertible and set $\iota\colon D_{A_{\mathrm{red}}} \to \ker(A)^\perp$ by $\iota x = x$. Consider $A_{\mathrm{red}}\colon D_{A_{\mathrm{red}}} \to \operatorname{ran}(A)$, which is a bounded operator, and hence, $\iota = A_{\mathrm{red}}^{-1}A_{\mathrm{red}}\colon D_{A_{\mathrm{red}}} \to \ker(A)^\perp$ is compact, which in turn is equivalent to $D_{A_{\mathrm{red}}} \hookrightarrow\hookrightarrow H_0$.

$\square$

The following proposition can also be found in [16, Lemma 2.4].

Proposition 2.2 *We have*

$$(A_{\mathrm{red}})^* = \left(A^*\right)_{\mathrm{red}}.$$

Proof Let $y \in \mathrm{dom}\left(A_{\mathrm{red}}^{*}\right)$. Then for each $x \in \mathrm{dom}(A_{\mathrm{red}}) = \mathrm{dom}(A) \cap \ker(A)^{\perp}$, we have

$$\langle A_{\mathrm{red}}x, y \rangle = \langle x, (A_{\mathrm{red}})^{*} y \rangle.$$

Denote by $P : H_0 \to H_0$ the orthogonal projection onto $\ker(A)^{\perp} = \overline{\mathrm{ran}}(A^{*})$. Then we obtain for each $x \in \mathrm{dom}(A)$

$$\langle Ax, y \rangle = \langle A_{\mathrm{red}} Px, y \rangle = \langle Px, (A_{\mathrm{red}})^{*} y \rangle = \langle x, (A_{\mathrm{red}})^{*} y \rangle,$$

where we have used $(A_{\mathrm{red}})^{*} y \in \ker(A)^{\perp}$ in the last equality. Hence, $y \in \mathrm{dom}(A^{*})$ with $A^{*} y = (A_{\mathrm{red}})^{*} y$. Since $y \in \overline{\mathrm{ran}}(A) = \ker(A^{*})^{\perp}$ by definition, we infer that $y \in \mathrm{dom}\left((A^{*})_{\mathrm{red}}\right)$, and thus, $(A^{*})_{\mathrm{red}} \, y = A^{*} y = (A_{\mathrm{red}})^{*} y$, which shows $(A_{\mathrm{red}})^{*} \subseteq (A^{*})_{\mathrm{red}}$. For the reverse inclusion, let $y \in \mathrm{dom}((A^{*})_{\mathrm{red}})$; i.e. $y \in \mathrm{dom}(A^{*}) \cap \ker(A^{*})^{\perp}$. Then for each $x \in \mathrm{dom}(A_{\mathrm{red}}) = \mathrm{dom}(A) \cap \ker(A)^{\perp}$, we compute

$$\langle A_{\mathrm{red}}x, y \rangle = \langle Ax, y \rangle = \langle x, A^{*} y \rangle,$$

and since $A^{*} y \in \mathrm{ran}(A^{*}) \subseteq \ker(A)^{\perp}$, we infer $y \in \mathrm{dom}((A_{\mathrm{red}})^{*})$, which shows the assertion. $\square$

Corollary 2.3

(a) $\mathrm{ran}(A)$ *is closed if and only if* $\mathrm{ran}(A^{*})$ *is closed.*
(b) $D_{A_{\mathrm{red}}} \hookrightarrow \hookrightarrow H_0$ *if and only if* $D_{A_{\mathrm{red}}^{*}} \hookrightarrow \hookrightarrow H_1$.

Proof This is a direct consequence of Proposition 2.1 and Proposition 2.2 and the fact that the adjoint of a compact operator is again compact. $\square$

We now focus on the case when A is self-adjoint. Note that then $H_0 = H_1$ and $\overline{\mathrm{ran}}(A) = \overline{\mathrm{ran}}(A^{*}) = \ker(A)^{\perp}$. Hence, A and A_{red} are operators acting on one Hilbert space.

Proposition 2.4 *Let A be self-adjoint. Then*

$$\sigma(A) \cup \{0\} = \sigma(A_{\mathrm{red}}) \cup \{0\}.$$

Proof Let $\lambda \in \rho(A_{\mathrm{red}})$ with $\lambda \neq 0$. We show that $\lambda \in \rho(A)$; that is, we show the bijectivity of $\lambda - A$. First, $\lambda - A$ is one-to-one, since $\lambda x = Ax$ for $x \in \mathrm{dom}(A)$ implies $x \in \ker(A)$, since otherwise $\lambda x = A_{\mathrm{red}}x$. However, if $x \in \ker(A)$ then $\lambda x = Ax = 0$ and since $\lambda \neq 0$, we infer $x = 0$. Let $y \in H_1$ and denote by $P : H_0 \to H_0$ the orthogonal projector onto $\ker(A)^{\perp} = \overline{\mathrm{ran}}(A)$. We set $x_0 :=$

$(\lambda - A_{\mathrm{red}})^{-1} P f$ and $x_1 = \frac{1}{\lambda}(1 - P)f$. Then $x_0 \in \mathrm{dom}(A_{\mathrm{red}}) = \mathrm{dom}(A) \cap \ker(A)^{\perp}$ and $x_1 \in \ker(A) \subseteq \mathrm{dom}(A)$. Setting $x := x_0 + x_1 \in \mathrm{dom}(A)$, we obtain

$$(\lambda - A)x = (\lambda - A)x_0 + \lambda x_1 = Pf + (1 - P)f = f,$$

and hence $\lambda - A$ is onto.

Assume now conversely that $\lambda \in \rho(A)$ with $\lambda \neq 0$. Then clearly, $\lambda - A_{\mathrm{red}}$ is one-to-one and for $f \in \overline{\mathrm{ran}}(A)$, we set

$$x := (\lambda - A)^{-1} f.$$

Then

$$\lambda x = Ax + f \in \overline{\mathrm{ran}}(A) = \ker(A)^{\perp}$$

and since $\lambda \neq 0$, we infer $x \in \ker(A)^{\perp}$. Hence, $x \in \mathrm{dom}(A_{\mathrm{red}})$ and clearly, $(\lambda - A_{\mathrm{red}})x = f$, which proves $\lambda \in \rho(A_{\mathrm{red}})$. $\qquad\square$

Corollary 2.5 *Assume that A is self-adjoint. Then* $\mathrm{ran}(A)$ *is closed if and only if there exists $r > 0$ such that $B(0, r) \cap \sigma(A) \subseteq \{0\}$.*

Proof First note that A_{red} is self-adjoint and one-to-one. Hence, 0 is not an isolated value of $\sigma(A_{\mathrm{red}})$ (see e.g. [14, Corollary 5.11]). Thus, by Proposition 2.4, $B(0, r) \cap \sigma(A) \subseteq \{0\}$ is equivalent to $0 \in \rho(A_{\mathrm{red}})$, which by Proposition 2.1 is equivalent to the closedness of $\mathrm{ran}(A)$. $\qquad\square$

3   Abstract Periodicity

Let H_0 and H_1 be two Hilbert spaces and $A_c \colon \mathrm{dom}(A_c) \subseteq H_0 \to H_1$ as well as $B_c \colon \mathrm{dom}(B_c) \subseteq H_1 \to H_0$ be two densely defined linear operators satisfying

$$A_c \subseteq -B_c^*.$$

By taking adjoints, we obtain

$$B_c \subseteq -A_c^*$$

and we set

$$A := -B_c^* \quad B := -A_c^*.$$

As a consequence, $A_c \subseteq A$ and $B_c \subseteq B$ are closable and we set $A_0 := \overline{A_c}$ and $B_0 := \overline{B_c}$. Note that $A = -B_0^*$ and $B = -A_0^*$.

Remark 3.1 The abstract setting above reflects the classical definition of linear differential operators with and without boundary conditions. Indeed, if $H_0 = L_2(\Omega)$ and $H_1 = L_2(\Omega)^n$ for some open $\Omega \subseteq \mathbb{R}^n$, we can set $A_c \colon C_c^\infty(\Omega) \subseteq L_2(\Omega) \to L_2(\Omega)^n$ by $A_c\phi := \operatorname{grad}\phi$, where $C_c^\infty(\Omega)$ denotes the space of compactly supported infinitely often differentiable functions on Ω and $B_c \colon C_c^\infty(\Omega)^n \subseteq L_2(\Omega)^n \to L_2(\Omega)$ by $B_c\Psi := \operatorname{div}\Psi$. Then integration by parts yields $A_c \subseteq -B_c^*$ and by definition $A = -B_c^* = \operatorname{grad}$ with domain $H^1(\Omega)$. Likewise, $A_0 = \overline{A_c}$ is the gradient with domain $H_0^1(\Omega)$; that is, the elements satisfy an abstract homogeneous Dirichlet boundary condition. Similarly, $B = -A_c^* = \operatorname{div}$ with maximal domain; i.e.

$$\operatorname{dom}(B) = \{\Psi \in L_2(\Omega)^n \,;\, \operatorname{div}\Psi \in L_2(\Omega)\},$$

where the divergence is defined in the distributional sense and the elements in $\operatorname{dom}(B_0)$ satisfy a generalised homogeneous Neumann boundary condition (see [12, 15] for details).

The focus of this section is on the following restrictions of A and B.

Definition We define the restrictions $A^\#$ and $B^\#$ of A and B, respectively, by the domains

$$\operatorname{dom}(A^\#) := \{x \in \operatorname{dom}(A) \,;\, Ax \in \overline{\operatorname{ran}}(A_0)\}$$

$$\operatorname{dom}(B^\#) := \{y \in \operatorname{dom}(B) \,;\, Bx \in \overline{\operatorname{ran}}(B_0)\}.$$

Remark 3.2 Note that $\overline{\operatorname{ran}}(A_0) = \left(\ker A_0^*\right)^\perp = (\ker B)^\perp$, and thus,

$$\operatorname{dom}(A^\#) = \{x \in \operatorname{dom}(A) \,;\, Ax \perp \ker B\},$$

and likewise for $B^\#$. Moreover, by definition, $A_0 \subseteq A^\# \subseteq A$ and $B_0 \subseteq B^\# \subseteq B$. Finally, it is immediate that $A^\#$ and $B^\#$ are closed.

Example 3.3 The boundary conditions induced by the domain constraints of $A^\#$ can be interpreted as an abstract version of periodicity. Indeed, if $\Omega =\,]-1/2, 1/2[$, then as discussed already in the introduction, $A_0 = B_0 = \partial_0$ and $A = B = \partial$, where $\operatorname{dom}(A) = H^1(]-1/2, 1/2[)$ and $\operatorname{dom}(A_0) = H_0^1(]-1/2, 1/2[) = \{u \in H^1(]-1/2, 1/2[) \,;\, u(-1/2) = u(1/2) = 0\}$. Then $u \in \operatorname{dom}(A^\#)$ if and only if $u \in H^1(]-1/2, 1/2[)$ and

$$\partial u \perp \ker \partial.$$

Since $\ker \partial = \operatorname{span}\{1\}$, we infer that

$$\int_{-1/2}^{1/2} \partial u(t)\,dt = 0,$$

which is equivalent to $u(1/2) = u(-1/2)$. Hence,

$$\mathrm{dom}(A^{\#}) = \{u \in H^1(]-1/2, 1/2[) \, ; \, u(-1/2) = u(1/2)\}.$$

In this section, we aim to discuss when $A^{\#} = -\left(B^{\#}\right)^*$. First note that this cannot hold in general as the next example illustrates.

Example 3.4 We again consider the derivatives ∂_0 and ∂ but this time on the interval $[0, \infty[$. More precisely, we set $A_0 = B_0 := \partial_0$ with domain $H_0^1([0, \infty[) = \{u \in H^1([0, \infty[) \, ; \, u(0) = 0\}$ and $A = B = \partial$ with domain $H^1([0, \infty[)$. Then $\ker \partial = \{0\}$, and thus, $u \in \mathrm{dom}(\partial^{\#})$ is equivalent to $u \in \mathrm{dom}(\partial)$, and thus, $\partial^{\#} = \partial$. Since $\partial^* = -\partial_0 \neq -\partial$, the desired relation for the operator $\partial^{\#}$ cannot hold.

In order to characterise when $A^{\#} = -(B^{\#})^*$ actually holds, we start with the following observation:

Proposition 3.5 *We have that*

$$\overline{A|_{\ker(A)+\mathrm{dom}(A_0)}} = -\left(B^{\#}\right)^*.$$

Proof For notational convenience, we set $\widetilde{A} := A|_{\ker(A)+\mathrm{dom}(A_0)}$. Since $\widetilde{A} \subseteq A$, we infer that $\widetilde{A}$ is closable. Thus, the statement we want to show is equivalent to

$$-B^{\#} = \widetilde{A}^*.$$

Let $y \in \mathrm{dom}(\widetilde{A}^*)$. Then, for all $x \in \mathrm{dom}(\widetilde{A}) = \mathrm{dom}(A_0) + \ker A$, we have

$$\langle \widetilde{A}x, y \rangle = \langle x, \widetilde{A}^*y \rangle.$$

Choosing $x \in \mathrm{dom}(A_0)$, we infer that

$$\langle A_0 x, y \rangle = \langle x, \widetilde{A}^*y \rangle$$

and hence, $y \in \mathrm{dom}(A_0^*) = \mathrm{dom}(B)$ and $\widetilde{A}^*y = -By$. Moreover, choosing $x \in \ker A$, we infer

$$\langle x, \widetilde{A}^*y \rangle = 0,$$

that is, $-By = \widetilde{A}^*y \in (\ker A)^{\perp} = \overline{\mathrm{ran}}(B_0)$, which yields $y \in \mathrm{dom}(B^{\#})$, and hence, $\widetilde{A}^* \subseteq -B^{\#}$. For the other inclusion, let $y \in \mathrm{dom}(B^{\#})$. Then for $x \in \mathrm{dom}(\widetilde{A})$, that is, $x = x_0 + x_1$ with $x_0 \in \mathrm{dom}(A_0)$ and $x_1 \in \ker(A)$, we compute

$$\langle y, \widetilde{A}x \rangle = \langle y, A_0 x_0 \rangle = \langle -B^{\#}y, x_0 \rangle = \langle -B^{\#}y, x \rangle,$$

where we have used $B^{\#}y \in (\ker A)^{\perp}$ in the last equality. This shows $-B^{\#} \subseteq \widetilde{A}^*$.

$\square$

The latter result shows that always $-(B^{\#})^{*} \subseteq A^{\#}$, since $A|_{\ker(A)+\mathrm{dom}(A_0)} \subseteq A^{\#}$ and equality holds, if and only if $\ker(A) + \mathrm{dom}(A_0)$ is a core for $A^{\#}$. We inspect this property a bit closer.

Proposition 3.6 *The set* $\ker(A) + \mathrm{dom}(A_0)$ *is a core for* $A^{\#}$ *if and only if* $1 \notin P\sigma(B^{\#}A^{\#})$, *i.e.,* 1 *in no eigenvalue of* $B^{\#}A^{\#}$.

Proof We remark that $\ker(A) + \mathrm{dom}(A_0)$ is a core for $A^{\#}$ if and only if $\ker(A) + \mathrm{dom}(A_0)$ is dense in $\mathrm{dom}(A^{\#})$ with respect to the graph norm of $A^{\#}$, which in turn is equivalent to

$$(\ker(A) + \mathrm{dom}(A_0))^{\perp_{A^{\#}}} = \{0\},$$

where the orthogonal complement is taken in the graph inner product of $A^{\#}$. We characterise the set on the left-hand side as follows:

$$x \in (\ker(A) + \mathrm{dom}(A_0))^{\perp_{A^{\#}}}$$

$$\Leftrightarrow \forall y \in \ker(A) + \mathrm{dom}(A_0) : \langle x, y \rangle + \langle A^{\#}x, A^{\#}y \rangle = 0,$$

$$\Leftrightarrow x \in \ker(A)^{\perp} \wedge \forall y \in \mathrm{dom}(A_0) : \langle x, y \rangle + \langle A^{\#}x, A_0 y \rangle = 0,$$

$$\Leftrightarrow x \in \overline{\mathrm{ran}}(B_0) \wedge A^{\#}x \in \mathrm{dom}(B) \wedge BA^{\#}x = x,$$

$$\Leftrightarrow A^{\#}x \in \mathrm{dom}(B^{\#}) \wedge B^{\#}A^{\#}x = x,$$

$$\Leftrightarrow x \in \ker(1 - B^{\#}A^{\#}).$$

Thus, $\ker(A) + \mathrm{dom}(A_0)$ is a core for $A^{\#}$ if and only if $\ker(1 - B^{\#}A^{\#}) = \{0\}$, which means $1 \notin P\sigma(B^{\#}A^{\#})$. $\qquad\square$

Next, we provide a sufficient condition for $\ker(A) + \mathrm{dom}(A_0)$ being a core for $A^{\#}$.

Proposition 3.7 *Assume that* $\mathrm{ran}(A^{\#})$ *is closed. Then the space* $\mathrm{dom}(A_0) + \ker A \subseteq \mathrm{dom}(A^{\#})$ *is a core for* $A^{\#}$.

To prove the latter proposition, we show a more general theorem, which can be applied in the above situation.

Theorem 3.8 *Let* $S \colon \mathrm{dom}(S) \subseteq H_0 \to H_1$ *and* $T \colon \mathrm{dom}(T) \subseteq H_0 \to H_1$ *be two densely defined linear operators with* $S \subseteq T$. *Assume further that*

- *T is closed with closed range.*
- *$\mathrm{ran}(S)$ lies dense in* $\mathrm{ran}(T)$.
- *$\ker(S) = \ker(T)$.*

Then $T = \overline{S}$.

Proof The inclusion $\overline{S} \subseteq T$ is obvious. For the other inclusion, we recall that T_{red} is boundedly invertible by Proposition 2.1. Let $x \in \mathrm{dom}(T)$ and decompose $x = x_0 + x_1$ with $x_0 \in \ker(T)$ and $x_1 \in \ker(T)^{\perp}$. Since $x_0 \in \ker(S) \subseteq \mathrm{dom}(S)$, it suffices to show $x_1 \in \mathrm{dom}(\overline{S})$. For this, we choose a sequence $(u_n)_{n \in \mathbb{N}}$ in $\mathrm{dom}(S)$ with $Su_n \to Tx_1$. Let P denote the orthogonal projection onto $\ker(S)^\perp = \ker(T)^\perp$. Then

$$T_{\mathrm{red}} P u_n = T u_n = S u_n \to T x_1 = T_{\mathrm{red}} x_1$$

and hence, by the bounded invertibility of T_{red}, we infer $P u_n \to x_1$. Since $P u_n \in \ker(S)^\perp$, we infer that $P u_n \in \mathrm{dom}(S)$ and hence $x_1 \in \mathrm{dom}(\overline{S})$, since $S P u_n = S u_n \to T x_1$. $\square$

Proof of Proposition 3.7 We apply Theorem 3.8 to $S = A|_{\ker(A)+\mathrm{dom}(A_0)}$ and $T = A^{\#}$. Then clearly, $S \subseteq T$ and T has a closed range by assumption. Moreover, $\mathrm{ran}(S) = \mathrm{ran}(A_0)$ and since $\mathrm{ran}(T) = \mathrm{ran}(A^{\#}) \subseteq \overline{\mathrm{ran}(A_0)}$, the range of S lies dense in the range of T. Finally, we have that $A^{\#} x = 0$ for some $x \in \mathrm{dom}(A^{\#})$ implies $x \in \ker(A)$, and thus, $x \in \mathrm{dom}(S)$ with $Sx = 0$. Thus, $\ker S = \ker T$ and Theorem 3.8 yields the assertion. $\square$

We summarise the result of this section in the next theorem.

Theorem 3.9 *We have*

$$-\left(B^{\#}\right)^{*} \subseteq A^{\#}.$$

Moreover, the following statements are equivalent:

(i) $-(B^{\#})^* = A^{\#}$.
(ii) $\ker(A) + \mathrm{dom}(A_0)$ *is a core for* $A^{\#}$.
(iii) $1 \notin P\sigma(B^{\#} A^{\#})$.

Moreover, the statements (i) to (iii) hold, if $\mathrm{ran}(A^{\#})$ *is closed.*

Remark 3.10 We note that statement (i) in Theorem 3.9 is symmetric in A and B, so we can replace the statements (ii) and (iii) by the analogous statements involving B instead of A. By the same reason, the closedness of $\mathrm{ran}(B^{\#})$ would also be sufficient for statement (i) to be true.

Example 3.11 Let $S \subseteq \mathrm{dom}(S) \subseteq H \to H$ be a symmetric operator on some Hilbert space H with $S \geq cI$ for some $c > 0$. Then we set $A_0 := \overline{S}$ and $B_0 := -\overline{S}$ and by the symmetry of S, we infer $A_0 \subseteq -B_0^* =: A$, where $A = S^*$. Hence, we are in the setting of this section. Moreover, since $A_0 \geq cI$ for some $c > 0$, we infer that A_0 has closed range and hence, so has $A^{\#}$ (see also Proposition 4.1 below). Thus, by Proposition 3.7

$$A^{\#} = \overline{S^*|_{\mathrm{dom}(\overline{S})+\ker(S^*)}} = \overline{S^*|_{\mathrm{dom}(S)+\ker(S^*)}}$$

and by Theorem 3.9

$$A^{\#} = -(B^{\#})^{*} = \left(A^{\#}\right)^{*}.$$

Thus, $A^{\#}$ is a self-adjoint extension of S known as the Krein–von Neumann extension of S (see e.g. [14, Section 13.3] or [4] and the references therein).

We emphasise that the closedness of $\mathrm{ran}(A^{\#})$ is just a sufficient condition as the following example shows.

Example 3.12 Let $\Omega = \mathbb{R}^3 \setminus B[0, 1] = \{x \in \mathbb{R}^3 \, ; \, \|x\| > 1\}$ and consider the operators curl_0 and curl on $L_2(\Omega)^3$ given as the usual vector-analytical operator with the domains

$$\mathrm{dom}(\mathrm{curl}) := \{u \in L_2(\Omega)^3 \, ; \, \mathrm{curl}\, u \in L_2(\Omega)^3\},$$

$$\mathrm{dom}(\mathrm{curl}_0) := \overline{C_c^{\infty}(\Omega)}^{\,\mathrm{dom}(\mathrm{curl})},$$

where in the first domain $\mathrm{curl}\, u$ is defined in the sense of distributions and in the second domain the closure is taken with respect to the graph norm of curl. Then $\mathrm{curl}_0 \subseteq \mathrm{curl}$ are both closed densely defined operators and $\mathrm{curl}_0^{*} = \mathrm{curl}$. This situation fits in the abstract setting considered in this section by choosing $A_0 := \mathrm{curl}_0, B_0 := -\mathrm{curl}_0$ as well as $A := \mathrm{curl}, B := -\mathrm{curl}$. In contrast to the bounded domain case, where the spectrum is discrete (see Theorem 5.2), we have that $\sigma(\mathrm{curl}^{\#}) = \mathbb{R}$ according to [10, p.333-334]. Moreover, by [9] and [10, Theorem 2.6] $\mathrm{curl}^{\#}$ is indeed self-adjoint, and hence, $\mathrm{ran}(\mathrm{curl}^{\#})$ cannot be closed by Corollary 2.5.

Remark 3.13 The construction in this example extends to the exterior derivative d on q-forms as an operator in $L_2^q(M)$ with M a bounded open subset of an N-dimensional Riemannian Lipschitz manifold with $N = 2q + 1$.

With the above mechanism we get

$$* \, \mathrm{d}^{\#} \text{ is } \begin{cases} \text{self-adjoint for} & q \text{ odd,} \\ \text{skew-self-adjoint for} & q \text{ even,} \end{cases}$$

where d denotes the exterior derivative on q-forms and $*$ is the Hodge-star-operator. These cases correspond to $N = 4k + 3$ and $N = 4k + 1$ with $k = \lfloor \frac{q}{2} \rfloor$, respectively. In particular, for $q = 1$, we recover the force-free magnetic field case. For example, if $q = 0, 2$, i.e. $N = 1$ or $N = 5$, respectively, we have that $* \mathrm{d}^{\#}$ is skew-self-adjoint. The case $N = 1$ recovers the standard one-dimensional periodic boundary case (on finite intervals). In both these cases (and in contrast to $N = 3$), we have well-posedness of the *real* (i.e. commuting with conjugation) evolutionary (in the sense of [12]) problem

$$\left(\partial_t + *d^{\#}\right) u = f.$$

In fact, we can insert any suitable material law operator as a coefficient of ∂_t (see [11, 15]). The main problem lies in the proof of the closedness of $\mathrm{ran}(d^{\#})$, which can be shown with the help of compact embedding results for the exterior derivative, and we refer to [13, 17] for sufficient conditions.

4 Conditions for a Closed Range of $A^{\#}$

We recall the setting from the previous section. Let H_0 and H_1 be two Hilbert spaces and A_0 and A densely defined closed linear operators with $A_0 \subseteq A$ and $B_0 := -A^*$ and $B := -A_0^*$. Moreover, we set $A^{\#} \subseteq A$ with domain

$$\mathrm{dom}(A^{\#}) := \{x \in \mathrm{dom}(A) ;\ Ax \in \overline{\mathrm{ran}}(A_0)\}$$

and analogously, we define $B^{\#}$. By Theorem 3.9, we know that $-\left(B^{\#}\right)^* = A^{\#}$ if we can ensure that $\mathrm{ran}(A^{\#})$ is closed. This section is devoted to some conditions ensuring the closedness of $\mathrm{ran}(A^{\#})$. Moreover, we will address the question, whether $D_{A^{\#}_{\mathrm{red}}}$ is compactly embedded into H_0, which by Proposition 2.1 would also imply the closedness of $\mathrm{ran}(A^{\#})$.

We start with a simple observation.

Proposition 4.1 *If* $\mathrm{ran}(A_0)$ *or* $\mathrm{ran}(A)$ *is closed, then* $\mathrm{ran}(A^{\#})$ *is closed as well.*

Proof Note that $\mathrm{ran}(A_0) \subseteq \mathrm{ran}(A^{\#}) \subseteq \overline{\mathrm{ran}}(A_0)$ by definition, and hence, if $\mathrm{ran}(A_0)$ is closed, we infer that $\mathrm{ran}(A^{\#}) = \mathrm{ran}(A_0)$ is closed. If on the other hand $\mathrm{ran}(A)$ is closed and we have a sequence $(x_n)_{n \in \mathbb{N}}$ in $\mathrm{dom}(A^{\#})$ such that $A^{\#}x_n \to y$ for some $y \in H$, we derive $y = Ax$ for some $x \in \mathrm{dom}(A)$, since $A^{\#} \subseteq A$. Since $Ax = y = \lim_{n \to \infty} A^{\#}x_n \in \overline{\mathrm{ran}}(A_0)$, we obtain $x \in \mathrm{dom}(A^{\#})$ and $y = A^{\#}x \in \mathrm{ran}(A^{\#})$, yielding the claim. $\square$

An analogous result holds for the compact embedding of $\mathrm{dom}(A^{\#}) \cap \ker(A^{\#})^{\perp}$ into H_0.

Proposition 4.2 *We have* $D_{A^{\#}_{\mathrm{red}}} \hookrightarrow\hookrightarrow H_0$ *if*

- $D_{(A_0)_{\mathrm{red}}} \hookrightarrow\hookrightarrow H_0$ *or*
- $D_{A_{\mathrm{red}}} \hookrightarrow\hookrightarrow H_0$

Proof Assume first that $D_{A_{\mathrm{red}}} \hookrightarrow\hookrightarrow H_0$. Since $\ker(A^{\#}) = \ker(A)$, we infer that $D_{A^{\#}_{\mathrm{red}}}$ is a closed subspace of $D_{A_{\mathrm{red}}}$, and hence, it is also compactly embedded into H_0.

Assume now that $D_{(A_0)_{\mathrm{red}}} \hookrightarrow\hookrightarrow H_0$. By Corollary 2.3 (b), we know that $D_{B_{\mathrm{red}}} \hookrightarrow\hookrightarrow H_1$, and hence, $D_{B_{\mathrm{red}}^{\#}} \hookrightarrow\hookrightarrow H_1$ by the first part of the proof. However, since $\mathrm{ran}(A_0)$ is closed by Proposition 2.1 (b), we infer that $\mathrm{ran}(A^{\#})$ is closed by Proposition 4.1, and thus, $A^{\#} = (-B^{\#})^*$ according to Theorem 3.9. Employing Corollary 2.3 (b) again, we infer that $D_{A_{\mathrm{red}}^{\#}} \hookrightarrow\hookrightarrow H_0$. $\qquad\square$

For applications, it will be useful to study the case, where the operator A_0 can be extended by an operator $\mathcal{A}$ on a larger space, which has closed range.

Proposition 4.3 *Let $\mathcal{A} : \mathrm{dom}\,(\mathcal{A}) \subseteq X \to Y$ be a densely defined closed linear operator between two Hilbert spaces X, Y. Moreover, assume that* $\mathrm{ran}(\mathcal{A})$ *is closed and there exist isometries $\iota_X : H_0 \to X, \iota_Y : H_1 \to Y$ such that*

$$\iota_Y A_0 \subseteq \mathcal{A}\iota_X \text{ and } \iota_X B_0 \subseteq -\mathcal{A}^*\iota_Y.$$

Then, $A^{\#}$ has closed range. Moreover, if $D_{\mathcal{A}_{\mathrm{red}}} \hookrightarrow\hookrightarrow X$, then $D_{A_{\mathrm{red}}^{\#}} \hookrightarrow\hookrightarrow H_0$.

Proof We need to show that $\mathrm{ran}(A^{\#}) = \overline{\mathrm{ran}}(A_0)$. So let $f \in \overline{\mathrm{ran}}(A_0)$. Then there exists a sequence $(\phi_n)_n$ in $\mathrm{dom}(A_0)$ such that $A_0\phi_n \to f$. Consequently, $\iota_Y A_0\phi_n \to \iota_Y f$, and thus, by assumption $\mathcal{A}\iota_X\phi_n \to \iota_Y f$. Since $\mathcal{A}$ has closed range, we infer

$$\exists v \in \mathrm{dom}(\mathcal{A}) : \ \mathcal{A}v = \iota_Y f. \tag{3}$$

Next, we show that $\iota_X^* v \in \mathrm{dom}(A)$. For this, let $\psi \in \mathrm{dom}(A^*) = \mathrm{dom}(B_0)$ and compute

$$
\begin{aligned}
\langle -B_0\psi, \iota_X^* v \rangle_{H_0} &= -\langle \iota_X B_0\psi, v \rangle_X \\
&= \langle \mathcal{A}^*\iota_Y\psi, v \rangle_X \\
&= \langle \iota_Y\psi, \mathcal{A}v \rangle_Y \\
&= \langle \psi, \iota_Y^*\mathcal{A}v \rangle_{H_1} \\
&= \langle \psi, f \rangle_{H_1}.
\end{aligned}
$$

Hence, indeed $\iota_X^* v \in \mathrm{dom}(A)$ and $A\iota_X^* v = f \in \overline{\mathrm{ran}}(A_0)$. Thus, we have

$$\iota_X^* v \in \mathrm{dom}(A^{\#}) \text{ with } A^{\#}\iota_X^* v = f. \tag{4}$$

Assume now that $D_{\mathcal{A}_{\mathrm{red}}} \hookrightarrow\hookrightarrow X$. Then $\mathcal{A}_{\mathrm{red}}$ is compactly invertible by Proposition 2.1. For $f \in \mathrm{ran}(A^{\#})$, we have by (3) and (4) $A^{\#}v = f$ for $v := \iota_X^* \mathcal{A}_{\mathrm{red}}^{-1}\iota_Y f$. Hence,

$$\left(A_{\mathrm{red}}^{\#}\right)^{-1} = P\iota_X^* \mathcal{A}_{\mathrm{red}}^{-1}\iota_Y,$$

where P denotes the orthogonal projector onto $\ker(A)^{\perp} = \ker(A^{\#})^{\perp}$. Since $\mathcal{A}_{\mathrm{red}}^{-1}$ is compact, so is $(A_{\mathrm{red}}^{\#})^{-1}$, and thus $D_{A_{\mathrm{red}}^{\#}} \hookrightarrow\hookrightarrow H_0$ by Proposition 2.1. $\square$

Remark 4.4 It is remarkable that the latter theorem just yields the closedness of $\mathrm{ran}(A^{\#})$ and neither of $\mathrm{ran}(A_0)$ nor of $\mathrm{ran}(A)$. Indeed, the arguments used in the proof suggest that the closedness of the range can only be achieved for the operator $A^{\#}$ and not for any other extension of A_0 by such an extension technique. However, we were not able to construct a concrete example for an operator A such that Proposition 4.3 is applicable but $\mathrm{ran}(A_0)$ and $\mathrm{ran}(A)$ are not closed.

5 Applications

Besides the already mentioned application to the transport equation with periodic boundary condition, we will consider Maxwell's equations as well as the heat and wave equation with abstract periodic boundary conditions.

5.1 The Operators $\mathbf{grad}^{\#}$ and $\mathbf{div}^{\#}$

Let $\Omega \subseteq \mathbb{R}^n$ be open. As in Remark 3.1, we define the operators

$$\mathrm{grad}_c \colon C_c^{\infty}(\Omega) \subseteq L_2(\Omega) \to L_2(\Omega)^n, \quad \phi \mapsto (\partial_j \phi)_{j \in \{1,\dots,n\}}$$

and

$$\mathrm{div}_c \colon C_c^{\infty}(\Omega)^n \subseteq L_2(\Omega)^n \to L_2(\Omega), \quad \Psi \mapsto \sum_{j=1}^{n} \partial_j \Psi_j.$$

Clearly, both operators are densely defined and linear, and by integration by parts, we infer $\mathrm{grad}_c \subseteq -(\mathrm{div}_c)^*$. Hence, we are in the framework of Sect. 3 with $A_c := \mathrm{grad}_c$ and $B_c := \mathrm{div}_c$. As in the abstract setting, we define

$$\mathrm{grad}_0 := \overline{\mathrm{grad}_c}, \ \mathrm{div}_0 := \overline{\mathrm{div}_c}, \quad \mathrm{grad} := -\mathrm{div}_0^*, \ \mathrm{div} := -\mathrm{grad}_0^*.$$

The domains are then given by

$$\mathrm{dom}(\mathrm{grad}) = H^1(\Omega), \quad \mathrm{dom}(\mathrm{div}) = \{\Psi \in L_2(\Omega)^n ; \sum_{j=1}^{n} \partial_j \Psi_j \in L_2(\Omega)\}$$

and likewise for grad_0 and div_0, where the classical boundary conditions $u = 0$ on $\partial\Omega$ and $\Psi \cdot n = 0$ on $\partial\Omega$ are implemented in a generalised sense (if $\partial\Omega$ is smooth enough, one can make sense of these boundary traces, see e.g. [6, Section 2.4]). Now consider the operators $\mathrm{grad}^{\#}$ and $\mathrm{div}^{\#}$ with the domains

$$\mathrm{dom}(\mathrm{grad}^{\#}) = \{u \in \mathrm{dom}(\mathrm{grad}) \,;\, \mathrm{grad}\, u \in \overline{\mathrm{ran}}(\mathrm{grad}_0)\},$$

$$\mathrm{dom}(\mathrm{div}^{\#}) = \{\Psi \in \mathrm{dom}(\mathrm{div}) \,;\, \mathrm{div}\, \Psi \in \overline{\mathrm{ran}}(\mathrm{div}_0)\}.$$

Theorem 5.1 *Assume that Ω is bounded in one dimension; that is, there exists a vector $x \in \mathbb{R}^n$ with $\|x\| = 1$ and a number $m > 0$ such that*

$$\Omega \subseteq \{y \in \mathbb{R}^n \,;\, |\langle y, x \rangle| \leq m\}.$$

Then $\mathrm{ran}(\mathrm{grad}^{\#})$ *is closed and* $\mathrm{grad}^{\#} = -\left(\mathrm{div}^{\#}\right)^{*}$ *. Moreover, if Ω is bounded, then $D_{\mathrm{grad}^{\#}_{\mathrm{red}}} \hookrightarrow\hookrightarrow L_2(\Omega)$ and $D_{\mathrm{div}^{\#}_{\mathrm{red}}} \hookrightarrow\hookrightarrow L_2(\Omega)^n$.*

Proof Since Ω is bounded in one dimension, we find a constant $c > 0$ such that

$$\|u\|_{L_2(\Omega)} \leq c\|\, \mathrm{grad}_0\, u\|_{L_2(\Omega)^n} \quad (u \in \mathrm{dom}(\mathrm{grad}_0)).$$

This is Poincare's inequality (see e.g. [15, Proposition 11.3.1]). Hence, grad_0 has a closed range, and thus, $\mathrm{ran}(\mathrm{grad}^{\#})$ is closed by Proposition 4.1. Thus, $\mathrm{grad}^{\#} = -\left(\mathrm{div}^{\#}\right)^{*}$ follows from Theorem 3.9.

Moreover, if Ω is bounded, we have $D_{\mathrm{grad}_0} \hookrightarrow\hookrightarrow L_2(\Omega)$ by Rellich's selection theorem (see e.g. [2, Chapter 5.7] or [15, Theorem 14.2.5]), and thus, the compactness of $D_{\mathrm{grad}^{\#}_{\mathrm{red}}} \hookrightarrow\hookrightarrow L_2(\Omega)$ and $D_{\mathrm{div}^{\#}_{\mathrm{red}}} \hookrightarrow\hookrightarrow L_2(\Omega)^n$ follows from Corollary 2.3 (b) and Proposition 4.2. $\square$

It is remarkable that the closedness of $\mathrm{ran}(\mathrm{grad}^{\#})$ and the compact embedding does not depend on the regularity of the boundary of Ω, which is needed for analogous results for grad. The above result can be used to study diffusion or wave phenomena with abstract periodic boundary conditions. For instance, the heat or the wave equation

$$\partial_t u - \mathrm{div}^{\#}\, \mathrm{grad}^{\#}\, u = f,$$

$$\partial_t^2 u - \mathrm{div}^{\#}\, \mathrm{grad}^{\#}\, u = f,$$

are well-posed, due to the self-adjointness of the operator $\mathrm{div}^{\#}\, \mathrm{grad}^{\#}$. Moreover, if Ω is bounded, the solution can be computed explicitly by using eigenvalue expansions for $\mathrm{div}^{\#}\, \mathrm{grad}^{\#}$.

5.2   The Operator $\mathrm{curl}^{\#}$

Let $\Omega \subseteq \mathbb{R}^3$ open. Similar to grad and div, we define the operator

$$\mathrm{curl}_c \colon C_c^\infty(\Omega)^3 \subseteq L_2(\Omega)^3 \to L_2(\Omega)^3,$$

$$\Psi \mapsto (\partial_2 \Psi_3 - \partial_3 \Psi_2,\ \partial_3 \Psi_1 - \partial_1 \Psi_3,\ \partial_1 \Psi_2 - \partial_2 \Psi_1).$$

Again, this is clearly a densely defined linear operator, and integration by parts gives $\mathrm{curl}_c \subseteq \mathrm{curl}_c^*$. Hence, we are in the framework of Sect. 3 with $A_c = -B_c = \mathrm{curl}_c$. Thus, we may define

$$\mathrm{curl}_0 := \overline{\mathrm{curl}_c}, \quad \mathrm{curl} := \mathrm{curl}_0^*$$

where

$$\mathrm{dom}(\mathrm{curl}) := \{\Psi \in L_2(\Omega)^3\,;\ \mathrm{curl}\,\Psi \in L_2(\Omega)^3\}$$

and the elements in $\mathrm{dom}(\mathrm{curl}_0)$ satisfy a generalised electric boundary condition $n \times \Psi = 0$ on $\partial\Omega$ (see [1] for trace spaces associated with curl). We define the operator $\mathrm{curl}^{\#}$ by

$$\mathrm{dom}(\mathrm{curl}^{\#}) := \{\Psi \in \mathrm{dom}(\mathrm{curl})\,;\ \mathrm{curl}\,\Psi \in \overline{\mathrm{ran}}(\mathrm{curl}_0)\}.$$

Theorem 5.2 *Assume that Ω is bounded. Then $D_{\mathrm{curl}^{\#}_{\mathrm{red}}} \hookrightarrow\hookrightarrow L_2(\Omega)^3$ and in particular $\mathrm{curl}^{\#}$ is self-adjoint and $\sigma(\mathrm{curl}^{\#})$ is discrete.*

Proof We want to apply Proposition 4.3. For this, let $R > 0$ be such that $\Omega \subseteq B(0, R)$ and let $\mathcal{A}$ be the operator curl_0 established on $L_2(B(0, R))^3$. Then $D_{\mathcal{A}_{\mathrm{red}}} \hookrightarrow\hookrightarrow L_2(B(0, R))^3$ (see e.g. [5] or [8]). We denote by $\iota\colon L_2(\Omega)^3 \to L_2(B(0, R))^3$ the canonical injection, that is,

$$(\iota\Psi)(x) := \begin{cases} \Psi(x) & \text{if } x \in \Omega, \\ 0 & \text{otherwise.} \end{cases}$$

By Proposition 4.3, it suffices to check

$$\iota\,\mathrm{curl}_0 \subseteq \mathcal{A}\iota, \quad \iota\,\mathrm{curl}_0 \subseteq \mathcal{A}^*\iota.$$

Since $\mathcal{A}$ is symmetric, it suffices to show the first inclusion. So let $\Psi \in \mathrm{dom}(\mathrm{curl}_0)$. Then we find a sequence $(\Psi_n)_{n\in\mathbb{N}}$ in $C_c^\infty(\Omega)^3$ with $\Psi_n \to \Psi$ and $\mathrm{curl}\,\Psi_n \to \mathrm{curl}_0\,\Psi$ in $L_2(\Omega)^3$. Thus, $\iota\Psi_n \to \iota\Psi$ and $\mathrm{curl}\,\iota\Psi_n = \iota\,\mathrm{curl}\,\Psi_n \to \iota\,\mathrm{curl}_0\,\Psi$ in $L_2(B(0, R))^3$. Since $\iota\Psi_n \in C_c^\infty(B(0, R))^3$ for each $n \in \mathbb{N}$, we infer that $\iota\Psi \in \mathrm{dom}(\mathcal{A})$ and $\mathcal{A}\iota\Psi = \iota\,\mathrm{curl}_0\,\Psi$, which shows the claim. $\qquad\square$

Remark 5.3 The latter theorem provides an abstract functional analytical proof of a variant of the main result in [3], where the author shows the self-adjointness of $\mathrm{curl}^{\#}$ and the discreteness of its spectrum by using a similar extension procedure and potential theory. Moreover, the author provides a Weyl asymptotic of the eigenvalues. It should be noted that in [3], Ω is assumed to be of finite measure and not necessarily bounded.

Remark 5.4 We again emphasise that the compact embedding of $D_{\mathrm{curl}^{\#}_{\mathrm{red}}}$ does not require any regularity of the boundary of Ω, while for a corresponding result for curl, such regularity assumptions are needed (see e.g. [13, 17]).

References

1. A. Buffa, M. Costabel, D. Sheen, On traces for $H(curl, \Omega)$ in Lipschitz domains. J. Math. Anal. Appl. **276**(2), 845–867 (2002)
2. L.C. Evans, *Partial Differential Equations*, volume 19 of Graduate Studies in Mathematics (American Mathematical Society, Providence, 1998)
3. N.D. Filonov, The operator rot in domains of finite measure. Zap. Nauchn. Sem. S.-Peterburg. Otdel. Mat. Inst. Steklov. (POMI) **262**(Issled. po Lineĭn. Oper. i Teor. Funkts. 27), 227–230, 236 (1999)
4. G. Fucci, F. Gesztesy, K. Kirsten, L.L. Littlejohn, R. Nichols, J. Stanfill, The Krein–von Neumann extension revisited. Appl. Anal. **101**(5), 1593–1616 (2022)
5. R. Leis, Rand- und Eigenwertaufgaben in der Theorie elektromagnetischer Schwingungen. Z. Angew. Math. Mech. **54**, T36–T40 (1974)
6. J. Nečas, *Direct Methods in the Theory of Elliptic Equations*. Springer Monographs in Mathematics. Springer, Heidelberg, 2012. Translated from the 1967 French original by Gerard Tronel and Alois Kufner, Editorial coordination and preface by Šárka Nečasová and a contribution by Christian G. Simader
7. D. Pauly, A global div-curl-lemma for mixed boundary conditions in weak Lipschitz domains and a corresponding generalized A_0^*-A_1-lemma in Hilbert spaces. Anal. München **39**(2), 33–58 (2019)
8. R. Picard, An elementary proof for a compact imbedding result in generalized electromagnetic theory. Math. Z. **187**(2), 151–164 (1984)
9. R. Picard, On a self-adjoint realization of curl and some of its applications. Ricerche Mat. **47**(1), 153–180 (1998)
10. R. Picard, On a selfadjoint realization of curl in exterior domains. Math. Z. **229**(2), 319–338 (1998)
11. R. Picard, A structural observation for linear material laws in classical mathematical physics. Math. Methods Appl. Sci. **32**(14), 1768–1803 (2009)
12. R. Picard, D. McGhee, *Partial Differential Equations*, volume 55 of De Gruyter Expositions in Mathematics (Walter de Gruyter GmbH & Co. KG, Berlin, 2011). A unified Hilbert space approach
13. R. Picard, N. Weck, K.-J. Witsch, Time-harmonic Maxwell equations in the exterior of perfectly conducting, irregular obstacles. Analysis (Munich) **21**(3), 231–263 (2001)
14. K. Schmüdgen, *Unbounded Self-Adjoint Operators on Hilbert Space*, volume 265 of Graduate Texts in Mathematics (Springer, Dordrecht, 2012)
15. C. Seifert, S. Trostorff, M. Waurick, *Evolutionary Equations*, volume 287 of Operator Theory: Advances and Applications (Birkhäuser/Springer, Cham, 2022). Picard's theorem for partial differential equations, and applications

16. S. Trostorff, M. Waurick, A note on elliptic type boundary value problems with maximal monotone relations. Math. Nachr. **287**(13), 1545–1558 (2014)
17. N. Weck, Maxwell's boundary value problem on Riemannian manifolds with nonsmooth boundaries. J. Math. Anal. Appl. **46**, 410–437 (1974)

Spectral Theory for Schrödinger Operators on Compact Metric Graphs with δ and δ' Couplings: A Survey

Jonathan Rohleder and Christian Seifert

1 Introduction

In the last three decades, differential operators on metric graphs, so-called quantum graphs, have been studied extensively. Metric graphs yield effective one-dimensional models for networks of quasi-one-dimensional structures such as nanotubes or waveguides, see, e.g., the survey in [13, Section 7.6]. Moreover, they provide easily accessible models due to its one-dimensional nature on the edges, whereas metric graphs exhibit a nontrivial behavior due to the coupling in the vertices.

By now a vast body of literature studying especially self-adjoint Schrödinger operators on metric graphs exists, see, e.g., [13, 44–48, 57, 65] for a non-exhaustive list, but also other types of operators such as first-order operators [16, 69, 73], operators related to diffusion processes on metric graphs [30, 31, 39, 70], and higher order operators have been investigated, e.g., in [5, 9, 23, 24, 53, 58, 71]. Furthermore, descriptions of certain non-self-adjoint operators on metric graphs and spectral estimates for them appear in the literature [6, 29]. Besides these continuum models

This chapter is based upon work from COST Action 18232 MAT-DYN-NET, supported by COST (European Cooperation in Science and Technology), www.cost.eu. J.R. acknowledges financial support by the Swedish Research Council (VR), grant no. 2022-03342. The authors are grateful to James B. Kennedy for some useful remarks.

J. Rohleder
Matematiska Institutionen, Stockholms Universitet, Stockholm, Sweden
e-mail: jonathan.rohleder@math.su.se

C. Seifert (✉)
Technische Universität Hamburg, Institut für Mathematik, Hamburg, Germany
e-mail: christian.seifert@tuhh.de

© The Author(s), under exclusive license to Springer Nature Switzerland AG 2024
F. L. Schwenninger, M. Waurick (eds.), *Systems Theory and PDEs*,
Trends in Mathematics, https://doi.org/10.1007/978-3-031-64991-2_3

on metric graphs, difference operators acting on the vertices of discrete graphs have a long history, and there is an intimate relation between operators on metric graphs and operators on discrete graphs, see [19, 21, 40, 41, 72], as well as [34], for an extensive overview on operators on discrete graphs.

In this chapter our focus is on Schrödinger operators

$$-\frac{d^2}{dx^2} + q$$

acting on the edges of a metric graph, equipped with a real-valued potential q and suitable vertex conditions; cf. Sect. 2 for details. We will admit general self-adjoint vertex conditions, but often we focus particularly on δ and δ' vertex conditions; the most prominent continuity-Kirchhoff (also called standard or Neumann) vertex conditions are a special case of the first-mentioned class, and their dual counterparts, anti-Kirchhoff conditions, belong to the second class.

Many properties of the nontrivial behavior of Schrödinger operators on metric graphs can be investigated in terms of spectral theory, which has been studied in recent years in various perspectives such as spectral estimates, properties of eigenfunctions, inverse problems, or questions of isospectrality, see [1, 2, 4, 11, 14, 15, 17, 21, 25–27, 37, 38, 42, 43, 55, 60, 61, 63, 64] for a few of the most recent developments.

In recent years, the effect of geometric manipulations of the metric graph onto the eigenvalues of Schrödinger operators has received much attention. This is mainly due to the fact that such so-called surgery principles have turned out to be powerful tools for obtaining eigenvalue bounds and prove isoperimetric inequalities [3, 10, 11, 17, 18, 36, 52, 54, 56, 59, 61, 66, 67].

In this survey we review different aspects of the spectral theory for self-adjoint Schrödinger operators on metric graphs. While some of the results presented here are valid for general self-adjoint vertex conditions, in many cases we put emphasis on the comparison between δ and δ' vertex conditions. Among other things, we focus on the effect of surgery principles. These may be divided basically into two groups: on the one hand manipulations that change the total length of the graph such as:

- Increasing the length of one or all edges
- Adding a new edge or attaching another graph
- Inserting another graph into a vertex
- Shrinking some edge lengths to zero

and on the other hand surgical operations that preserve the total length of the graph, for instance:

- Changing the coupling condition or its strength at a vertex
- Joining vertices
- Unfolding parallel or pendant edges

In addition, we study properties of the ground state eigenvalue and the corresponding eigenfunction, which especially are of some use to the above-named surgery principles. Moreover, we discuss bounds for the lowest eigenvalue of the Laplacian with δ and δ' vertex conditions, where some of the above surgery principles can be of use.

This chapter has mostly the character of a survey. Many of the results come from recent literature in this or a similar form, though we in some cases provide slight generalizations and modified proofs. However, we complement these results by some modest additions such as Theorem 4.13 or Theorem 5.4 and a bunch of illuminating examples and counterexamples. In addition, we formulate some open questions.

Let us outline the content of this chapter. In Sect. 2 we describe the setup of metric graphs and Schrödinger operators on them. Furthermore, we collect the basic facts on spectral theory we need. In Sect. 3 we focus on the ground state eigenvalue and eigenfunction. The subsequent sections are devoted to surgery principles, and they are divided into Sect. 4 on graph modifications which may change the total length and Sect. 5 on those which preserve it. In Sect. 6 we provide Hadamard-type variational formulas that describe the variation of spectral data (we focus on simple eigenvalues here) under variation of the model data; specifically, we consider variation of the coupling conditions. In the final section, Sect. 7 we discuss, as an application, some bounds for the lowest eigenvalue for δ and δ' vertex conditions.

2 Schrödinger Operators on Metric Graphs

2.1 Metric Graphs

Let Γ be a metric graph constituted by a finite set of vertices $\mathcal{V} := \mathcal{V}(\Gamma)$, a finite set of edges $\mathcal{E} := \mathcal{E}(\Gamma)$, and a length function $L \colon \mathcal{E} \to (0, \infty)$. Upon parametrizing each edge $e \in \mathcal{E}$ by the interval $[0, L(e)] \subseteq \mathbb{R}$ and considering Γ as the one-dimensional simplicial complex consisting of the intervals $[0, L(e)]$, $e \in \mathcal{E}$, with corresponding coupling at the vertices, Γ becomes a compact metric space, and we will refer to such Γ as a compact metric graph. For each vertex $v \in \mathcal{V}$, we denote by $\mathcal{E}_v \subseteq \mathcal{E}$ the set of edges incident to v and by $\deg(v)$ the degree of v. We write $v = o(e)$ or $v = t(e)$ and say that the edge e originates from or terminates at the vertex v, respectively, if v corresponds to the end point 0 (respectively, $L(e)$) according to our parametrization. We will denote by

$$L^2(\Gamma) := \bigoplus_{e \in \mathcal{E}} L^2(0, L(e))$$

the usual L^2 space on Γ and by

$$\tilde{H}^k(\Gamma) := \bigoplus_{e \in \mathcal{E}} H^k(0, L(e))$$

the L^2-based Sobolev space of order $k \in \mathbb{N}$. For $f \in L^2(\Gamma)$ and $e \in \mathcal{E}$, we write f_e for the restriction of f to the edge $e \in \mathcal{E}$ (i.e., the interval $(0, L(e))$ corresponding to e).

2.2 *Schrödinger Operators on Metric Graphs*

In this chapter we focus on self-adjoint Schrödinger operators in $L^2(\Gamma)$ subject to self-adjoint coupling conditions on the vertices; that is, we consider Schrödinger operators in $L^2(\Gamma)$ acting as

$$(\mathcal{L}f)_e := -f_e'' + q_e f_e, \quad e \in \mathcal{E}, \tag{2.1}$$

with a real-valued potential $q = (q_e)_{e \in \mathcal{E}}$, where we assume that $q \in L^\infty(\Gamma)$. This assumption is made for reasons of simplicity; many of the results discussed in this survey extend naturally to larger classes of potentials such as $L^1(\Gamma)$.

Next we specify vertex conditions with which $\mathcal{L}$ will constitute a self-adjoint operator. For $v \in \mathcal{V}$, let $\{e_1, \ldots, e_l\} := \{e \in \mathcal{E}_v : v = o(e)\}$ and $\{e_{l+1}, \ldots e_m\} := \{e \in \mathcal{E}_v : v = t(e)\}$ be enumerations of the sets of edges originating from and terminating at v, respectively. Note that, in the special case that an edge e is a loop, i.e., $o(e) = t(e)$, e will belong to both sets. For $f \in \widetilde{H}^1(\Gamma)$, we write

$$F(v) := \begin{pmatrix} f_{e_1}(0) \\ \vdots \\ f_{e_l}(0) \\ f_{e_{l+1}}(L(e_{l+1})) \\ \vdots \\ f_{e_m}(L(e_m)) \end{pmatrix}$$

for the collection of boundary values of the restriction of f to the adjacent edges at v. Moreover, for $f \in \widetilde{H}^2(\Gamma)$, we write

$$F'(v) := \begin{pmatrix} f_{e_1}'(0) \\ \vdots \\ f_{e_l}'(0) \\ -f_{e_{l+1}}'(L(e_{l+1})) \\ \vdots \\ -f_{e_m}'(L(e_m)) \end{pmatrix},$$

for the collection of derivatives of f at v in the direction pointing out of v into the edges. The following description of all self-adjoint incarnations of $\mathcal{L}$ in $L^2(\Gamma)$ with local coupling conditions is standard.

Proposition 2.1 ([13, Theorem 1.4.4]) *Let Γ be a compact metric graph, $q \in L^{\infty}(\Gamma)$ be real-valued, and $\mathcal{L}$ be the Schrödinger differential expression in (2.1). For each vertex $v \in \mathcal{V}$, let $P_{v,\mathrm{D}}$, $P_{v,\mathrm{N}}$, and $P_{v,\mathrm{R}}$ be orthogonal projections in $\mathbb{C}^{\deg(v)}$ with mutually orthogonal ranges such that $P_{v,\mathrm{D}} + P_{v,\mathrm{N}} + P_{v,\mathrm{R}} = I$, and let Λ_v be a self-adjoint, invertible operator in $\mathrm{ran}\, P_{v,\mathrm{R}}$. Then the operator H in $L^2(\Gamma)$ given by*

$$Hf := \mathcal{L}f,$$

$$\mathrm{dom}\, H := \Big\{ f \in \tilde{H}^2(\Gamma) : P_{v,\mathrm{D}}F(v) = 0,\ P_{v,\mathrm{N}}F'(v) = 0,$$

$$P_{v,\mathrm{R}}F'(v) = \Lambda_v P_{v,\mathrm{R}}F(v)\ \textit{for each } v \in \mathcal{V} \Big\}$$

is self-adjoint (and each self-adjoint realization of $\mathcal{L}$ in $L^2(\Gamma)$ subject to local coupling conditions can be written in this form). Furthermore, the closed quadratic form h corresponding to the operator H in the sense of, e.g., [33, Chapter VI, Theorem 2.1] is given by

$$h(f) := \int_{\Gamma} |f'|^2 + \int_{\Gamma} q|f|^2 + \sum_{v \in \mathcal{V}} \langle \Lambda_v P_{v,\mathrm{R}}F(v), P_{v,\mathrm{R}}F(v) \rangle,$$

$$\mathrm{dom}\, h := \Big\{ f \in \tilde{H}^1(\Gamma) : P_{v,\mathrm{D}}F(v) = 0\ \textit{for each } v \in \mathcal{V} \Big\}.$$

By a standard compact embedding argument, the spectrum of the Hamiltonian H on the compact metric graph Γ is always purely discrete and bounded from below, see, e.g., [47, Corollary 10 and Theorem 18]. We denote by

$$\lambda_1(H) \leq \lambda_2(H) \leq \dots$$

the eigenvalues of H in nondecreasing order and counted with multiplicities. If $q \geq 0$ and Λ_v is nonnegative for each vertex v, then all eigenvalues are nonnegative. We will frequently make use of the fact that these eigenvalues may be expressed via the min-max principle

$$\lambda_k(H) = \min_{\substack{F \subseteq \mathrm{dom}\, h \\ \dim F = k}} \max_{\substack{f \in F \\ f \neq 0}} \frac{h(f)}{\int_{\Gamma} |f|^2}$$

for $k = 1, 2, \dots$.

We will put special emphasis on so-called δ and δ' coupling conditions and their special variants, namely continuity-Kirchhoff and anti-Kirchhoff conditions.

To specify those within the framework of Proposition 2.1, for $d \in \mathbb{N}$ we denote by

$$\mathcal{P} := \mathcal{P}_d = \begin{pmatrix} \frac{1}{d} & \cdots & \frac{1}{d} \\ \vdots & & \vdots \\ \frac{1}{d} & \cdots & \frac{1}{d} \end{pmatrix} \quad \text{and} \quad \mathcal{Q} := \mathcal{Q}_d = \begin{pmatrix} \frac{d-1}{d} & -\frac{1}{d} & \cdots & -\frac{1}{d} \\ -\frac{1}{d} & \ddots & \ddots & \vdots \\ \vdots & \ddots & \ddots & -\frac{1}{d} \\ -\frac{1}{d} & \cdots & -\frac{1}{d} & \frac{d-1}{d} \end{pmatrix}$$

the orthogonal projections onto $\mathrm{span}\{(1, 1, \ldots, 1)^\top\}$ and onto its orthogonal complement $(\mathrm{span}\{(1, 1, \ldots, 1)^\top\})^\perp$, respectively.

Definition 2.2 Let $v \in \mathcal{V}$ and $d := \deg(v) > 0$. Then the vertex conditions at v are called:

(a) *Continuity-Kirchhoff conditions* if $P_{v,\mathrm{D}} = \mathcal{Q}$, $P_{v,\mathrm{N}} = \mathcal{P}$, and $P_{v,\mathrm{R}} = 0$.
(b) δ *coupling conditions* with *strength* $\alpha_v \in \mathbb{R}$ if $P_{v,\mathrm{D}} = \mathcal{Q}$, $P_{v,\mathrm{N}} = 0$, $P_{v,\mathrm{R}} = \mathcal{P}$, and Λ_v is the multiplication by $\frac{\alpha_v}{d}$.
(c) *Anti-Kirchhoff conditions* if $P_{v,\mathrm{N}} = \mathcal{Q}$, $P_{v,\mathrm{D}} = \mathcal{P}$, and $P_{v,\mathrm{R}} = 0$.
(d) δ' *coupling conditions* with *strength* $\beta_v \in \mathbb{R} \setminus \{0\}$ if $P_{v,\mathrm{N}} = \mathcal{Q}$, $P_{v,\mathrm{D}} = 0$, $P_{v,\mathrm{R}} = \mathcal{P}$, and Λ_v is the multiplication by $\frac{d}{\beta_v}$.

The vertex conditions given in Definition 2.2 can be written more explicitly. In fact, f satisfies δ vertex conditions with strength α_v at a vertex v if and only if $F(v)$ is equal to a constant vector, whose value we denote $f(v)$, and

$$\sum_{j=1}^{\deg(v)} F_j'(v) = \alpha_v f(v).$$

In particular, continuity-Kirchhoff coupling conditions equal δ coupling conditions with strength $\alpha_v = 0$. On the other hand, formally setting $\alpha_v = \infty$ results in a Dirichlet condition at v, i.e., $f(v) = 0$.

Furthermore, f satisfies δ' coupling conditions with strength β_v at v if and only if $F'(v)$ is equal to a constant vector, whose value we denote $f'(v)$, and

$$\sum_{j=1}^{d} F_j(v) = \beta_v f'(v).$$

In particular, an anti-Kirchhoff coupling condition can be interpreted as a δ' coupling condition with strength $\beta_v = 0$. Formally setting $\beta_v = \infty$ results in a Neumann condition at v, i.e., $f'(v) = 0$.

For δ and δ' couplings, the vertex term in the quadratic form h in Proposition 2.1 looks as follows.

Lemma 2.3 ([68, Lemma 2.5]) *Let Γ be a compact metric graph, let H be a self-adjoint Schrödinger operator in $L^2(\Gamma)$ as in Proposition 2.1, and let h be the* *corresponding quadratic form. Furthermore, let $v \in \mathcal{V}$, and let the vertex conditions for H at v be given in terms of $P_{v,\mathrm{D}}$, $P_{v,\mathrm{N}}$, $P_{v,\mathrm{R}}$, and Λ_v. Then, the following assertions hold for each $f \in \mathrm{dom}\, h$:*

(i) *If a continuity-Kirchhoff condition is imposed at v, then f is continuous at v and*

$$\left\langle \Lambda_v P_{v,\mathrm{R}} F(v),\, P_{v,\mathrm{R}} F(v) \right\rangle = 0.$$

(ii) *If a δ coupling condition with strength α_v is imposed at v, then f is continuous at v and*

$$\left\langle \Lambda_v P_{v,\mathrm{R}} F(v),\, P_{v,\mathrm{R}} F(v) \right\rangle = \alpha_v |f(v)|^2.$$

(iii) *If an anti-Kirchhoff condition is imposed at v, then* $\displaystyle\sum_{j=1}^{\deg(v)} F_j(v) = 0$ *and*

$$\left\langle \Lambda_v P_{v,\mathrm{R}} F(v),\, P_{v,\mathrm{R}} F(v) \right\rangle = 0.$$

(iv) *If a δ' coupling condition with strength β_v is imposed at v, then f does not satisfy any vertex conditions at v and*

$$\left\langle \Lambda_v P_{v,\mathrm{R}} F(v),\, P_{v,\mathrm{R}} F(v) \right\rangle = \frac{1}{\beta_v} \left| \sum_{j=1}^{\deg(v)} F_j(v) \right|^2.$$

3 Properties of the Ground State

In this section we review properties of the first eigenvalue and the corresponding eigenfunctions in case that the same coupling conditions are imposed at each vertex, either δ or δ'.

3.1 *Positivity of the First Eigenfunction*

Here we discuss Perron–Frobenius (or Courant) type properties of quantum graphs with δ and δ' vertex conditions. For, e.g., Schrödinger operators on Euclidean domains with suitable boundary conditions, it is well known that the first eigenvalue has multiplicity one, and the corresponding eigenfunction can be chosen positive. This turns out to be correct for δ coupling conditions but is no longer true for δ' conditions, as we will see below.

The following theorem corresponds to [20, Theorem 3.2], see also [51, Corollary 1]; for the convenience of the reader, we provide a proof.

Theorem 3.1 *Let Γ be a connected compact metric graph, let $q \in L^\infty(\Gamma)$ be real-valued, and let H be the Schrödinger operator on Γ defined in Proposition 2.1, and assume that δ coupling conditions are imposed at all vertices. Then the eigenspace of H corresponding to the first eigenvalue $\lambda_1(H)$ is one-dimensional. Moreover, the corresponding eigenfunction does not have any zero in Γ and, hence, can be chosen positive.*

Proof Let $f \in \ker(H - \lambda_1(H))$, $f \neq 0$, and let $g := |f|$. Then $|g'| \leq |f'|$, and therefore

$$h(g) = \int_\Gamma |g'|^2 + \int_\Gamma q|g|^2 + \sum_{v \in \mathcal{V}} \alpha_v |g(v)|^2$$

$$\leq \int_\Gamma |f'|^2 + \int_\Gamma q|f|^2 + \sum_{v \in \mathcal{V}} \alpha_v |f(v)|^2 = h(f),$$

and as f minimizes the Rayleigh quotient, the same is true for g. Hence, g is a nontrivial element of $\ker(H - \lambda_1(H))$; in particular, g satisfies δ coupling conditions. Note that g does not vanish on any vertex. Indeed, $g(0) = 0$, and the coupling conditions imply

$$\sum_{j=1}^{\deg(v)} G'_j(v) = 0,$$

and since g vanishes at v and is nonnegative everywhere, it follows $g'_e(v) = 0$ for each edge e incident to v. In particular, g vanishes on each edge incident to v, and it follows $g(w) = 0$ for each vertex w which is connected to v by an edge. Successively the same argument yields $g = 0$ identically on Γ as Γ is connected, a contradiction. As every nonvertex point on Γ can be interpreted as a vertex of degree 2 equipped with δ conditions of strength zero, it follows that g has no zero on Γ, and this implies that the arbitrarily chosen eigenfunction f has no zero.

Finally, having two linearly independent eigenfunctions f, k would imply that for any chosen $x_0 \in \Gamma$ the function

$$f - \frac{f(x_0)}{k(x_0)} k$$

belongs to $\ker(H - \lambda_1(H))$, is nontrivial, and vanishes at x_0, a contradiction. Thus $\ker(H - \lambda_1(H))$ is one-dimensional. $\qquad\square$

The case of δ' coupling conditions is fundamentally different, as the following example shows.

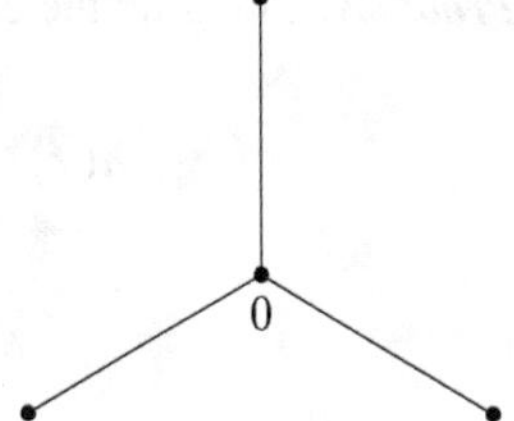

Fig. 1 The equilateral 3-star graph

Example 3.2 Consider a 3-star graph formed of three copies of the interval $[0, 1]$ coupled to one another at their zero end points, see Fig. 1. If we impose Neumann boundary conditions at the degree-1 vertices (which can be read as δ' conditions with strength ∞) and a δ' coupling condition with a strength $\beta \geq 0$ at the degree-3 vertex, then the corresponding quadratic form for the Laplacian H is nonnegative, i.e., $\lambda_1(H) \geq 0$. Moreover, the function f that is constantly 1 on one of the edges, -1 on another, and 0 on the remaining edge satisfies all imposed vertex conditions. As $f'' = 0$ on every edge, we have $f \in \ker(H - \lambda_1(H)) = \ker H$. But f is positive on one edge and negative on another, that is, there exist ground state eigenfunctions without fixed sign, and they can even be constantly zero on a whole edge. Furthermore, by interchanging the roles of the edges, one sees that the eigenspace is two-dimensional.

For the study of a larger class of vertex conditions that allow for uniqueness and positivity of the ground state eigenfunction, we refer the reader to [51].

3.2 Existence of Negative Eigenvalues

For a quantum graph with δ or δ' vertex conditions, it is not necessarily true that the existence of one negative coupling coefficient implies that the lowest eigenvalue is negative; cf. Example 3.6 below. In the following, sufficient conditions for existence of a negative eigenvalue are provided.

Theorem 3.3 *Let Γ be a compact metric graph and H a self-adjoint Schrödinger operator on Γ as defined in Proposition 2.1. Assume that at each vertex $v \in \mathcal{V}$, a δ coupling condition of strength α_v is imposed. If*

$$\int_\Gamma q + \sum_{v \in \mathcal{V}} \alpha_v < 0,$$

then $\lambda_1(H) < 0$.

Proof By plugging the constant function $f = 1$ into the quadratic form, we get

$$h(f) = \int_\Gamma |f'|^2 + \int_\Gamma q|f|^2 + \sum_{v \in V} \alpha_v |f(v)|^2$$

$$= \int_\Gamma q + \sum_{v \in V} \alpha_v < 0.$$

This implies $\lambda_1(H) < 0$. $\qquad\square$

For δ' couplings, a more localized condition for existence of negative eigenvalues can be given.

Theorem 3.4 *Let Γ be a compact metric graph and H a self-adjoint Schrödinger operator on Γ as defined in Proposition 2.1. Assume that at each vertex $v \in V$ a δ' coupling condition of strength β_v is imposed. If for some edge e_0,*

$$\int_{e_0} q + \frac{1}{\beta_{o(e_0)}} + \frac{1}{\beta_{t(e_0)}} < 0,$$

then $\lambda_1(H) < 0$.

Proof This time we may use the function f being constantly equal to 1 on e_0 and 0 otherwise. Then f belongs to dom h and

$$h(f) = \int_\Gamma |f'|^2 + \int_\Gamma q|f|^2 + \sum_{v \in V} \frac{1}{\beta_v} \left| \sum_{j=1}^{\deg(v)} F_j(v) \right|^2$$

$$= \int_{e_0} q + \frac{1}{\beta_{o(e_0)}} + \frac{1}{\beta_{t(e_0)}} < 0.$$

Thus $\lambda_1(H) < 0$. $\qquad\square$

Example 3.5 Let Γ be a connected compact metric graph and H the Laplacian (i.e., the Schrödinger operator with constant zero potential) on Γ with δ' coupling conditions at each vertex v, where $\beta_v = 1$ for all $v \neq v_0$ and $\beta_{v_0} = -1/2$, for a selected vertex v_0. If e_0 is any edge incident to v_0, then

$$\frac{1}{\beta_{o(e_0)}} + \frac{1}{\beta_{t(e_0)}} = -2 + 1 < 0.$$

Therefore, Theorem 3.4 yields $\lambda_1(H) < 0$.

The following example shows that the analogous condition for δ vertex conditions is not sufficient for existence of negative eigenvalues.

Example 3.6 Let Γ be any compact metric graph. Consider again the case $q = 0$ constantly, and suppose that at each vertex $v \in V$ a δ coupling condition of strength

α_v is imposed. Assume that for some $v_0 \in \mathcal{V}$, $\alpha_{v_0} < 0$, while the remaining α_v are all positive. Our aim is to show that $h(f) \geq 0$ for all $f \in \operatorname{dom} h = \widetilde{H}^1(\Gamma)$, supposed α_{v_0} is sufficiently close to zero, in relation to the lengths of its adjacent edges. Indeed, the Schrödinger operator $\widehat{H}$ obtained from H by replacing the coupling coefficient α_{v_0} at v_0 by zero satisfies $\lambda_1(\widehat{H}) > 0$. As the eigenvalue $\lambda_1(H)$ depends continuously on the coupling coefficients, $\lambda_1(H) > 0$ whenever $\alpha_{v_0} < 0$ is sufficiently close to zero.

We finally point out that the number of negative eigenvalues of Schrödinger operators on metric graphs with general self-adjoint vertex conditions was studied in [8]; cf. also [28].

4 Graph Manipulations Changing the Total Length of the Graph

In the following we review the effect of certain geometric manipulations of a given metric graph onto the spectrum of a Schrödinger operator on the graph. We start with manipulations that change the total length of the metric graph Γ. These include increasing or shrinking the length of one or all edges, attaching new edges or graphs, or inserting a graph at a vertex.

4.1 Increasing the Length of One Edge

Here, we consider the change of the eigenvalues in case we increase the length of one edge. For the Laplacian with continuity-Kirchhoff, δ, or Dirichlet vertex conditions, the following theorem can be found in [11, Corollary 3.12].

Theorem 4.1 *Let Γ be a compact metric graph, $q \in L^\infty(\Gamma)$ real-valued, and H the Schrödinger operator on Γ with potential q and arbitrary self-adjoint coupling conditions. Moreover, let $\widetilde{\Gamma}$ be the metric graph obtained from Γ by increasing the length of one edge, i.e., there exists $e_0 \in \mathcal{E}$ such that $\widetilde{L}(e) = L(e)$ for $e \neq e_0$ and $\widetilde{L}(e_0) > L(e_0)$, where $\widetilde{L}$ is the length function for $\widetilde{\Gamma}$. Set $\widetilde{q}_e := q_e$ for $e \neq e_0$ and $\widetilde{q}_{e_0} := q_{e_0}\left(\frac{L(e_0)}{\widetilde{L}(e_0)}\cdot\right)$. Denote by $\widetilde{H}$ the Schrödinger operator on $\widetilde{\Gamma}$ with potential $\widetilde{q}$ equipped with the same coupling conditions as for H. Then*

$$\lambda_k(\widetilde{H}) \leq \lambda_k(H)$$

holds for all $k \in \mathbb{N}$ such that $\lambda_k(H) \geq 0$. If all coupling conditions are of δ type and $\lambda_1(H) > 0$, i.e., the spectrum is positive, then

$$\lambda_1(\widetilde{H}) < \lambda_1(H).$$

Proof Let h be the quadratic form associated with H and $\widetilde{h}$ the quadratic form associated with $\widetilde{H}$; cf. Proposition 2.1. Let $k \in \mathbb{N}$ such that $\lambda_k(H) \geq 0$, and let $F \subseteq \mathrm{dom}\, h$ be a k-dimensional subspace such that

$$h(f) \leq \lambda_k(H) \int_\Gamma |f|^2 \quad \text{for all } f \in F.$$

Let $x_0 \in [0, L(e_0)]$. For each $f \in F$, define a function $\widetilde{f}$ on $\widetilde{\Gamma}$ by $\widetilde{f}_e := f_e$ for $e \neq e_0$ and

$$\widetilde{f}_{e_0}(x) := \begin{cases} f_{e_0}(x), & x \in [0, x_0], \\ f_{e_0}(x_0), & x \in (x_0, x_0 + \widetilde{L}(e_0) - L(e_0)], \\ f_{e_0}\big(x - (\widetilde{L}(e_0) - L(e_0))\big), & x \in (x_0 + \widetilde{L}(e_0) - L(e_0), \widetilde{L}(e_0)]. \end{cases}$$

Then $\widetilde{F}(v) = F(v)$ for each $v \in V$, in particular, $\widetilde{f} \in \mathrm{dom}\, \widetilde{h}$, $\widetilde{h}(\widetilde{f}) = h(f)$. Moreover,

$$\int_{\widetilde{\Gamma}} |\widetilde{f}|^2 = \int_\Gamma |f|^2 + (\widetilde{L}(e_0) - L(e_0))|f_{e_0}(x_0)|^2 \geq \int_\Gamma |f|^2, \quad f \in F. \tag{4.1}$$

For all $f \in F$ such that $h(f) < 0$, clearly $\frac{\widetilde{h}(\widetilde{f})}{\int_{\widetilde{\Gamma}} |\widetilde{f}|^2} < 0 \leq \lambda_k(H)$, and for all $f \in F$ with $h(f) \geq 0$, it follows

$$\frac{\widetilde{h}(\widetilde{f})}{\int_{\widetilde{\Gamma}} |\widetilde{f}|^2} \leq \frac{h(f)}{\int_\Gamma |f|^2} \leq \lambda_k.$$

Hence $\lambda_k(\widetilde{H}) \leq \lambda_k(H)$.

If each vertex is equipped with a δ coupling condition, then $\ker(H - \lambda_1(H))$ is one-dimensional, and the corresponding eigenfunction has no zeros on Γ, see Theorem 3.1. In particular, $f_{e_0}(x_0) \neq 0$, and (4.1) implies $\int_{\widetilde{\Gamma}} |\widetilde{f}|^2 > \int_\Gamma |f|^2$. Hence, if $\lambda_1(H) > 0$, then $\lambda_1(\widetilde{H}) < \lambda_1(H)$. $\qquad\square$

The restriction to nonnegative eigenvalues in the above theorem is necessary for some coupling conditions such as δ or δ' couplings, as the following examples show (see also [20, Theorem 4.1 and Section 5]).

Example 4.2 We consider a loop graph, i.e., a graph consisting of one vertex v and one edge e of length $L := L(e)$ originating from and terminating at v, see Fig. 2. At v we impose a δ coupling condition of strength $\alpha := \alpha_v < 0$. Then the only negative eigenvalue of H is given by the number $\lambda = -k^2$, where k is the positive solution of

$$2k - \alpha \sinh(kL) - 2k \cosh(kL) = 0,$$

Fig. 2 The loop graph

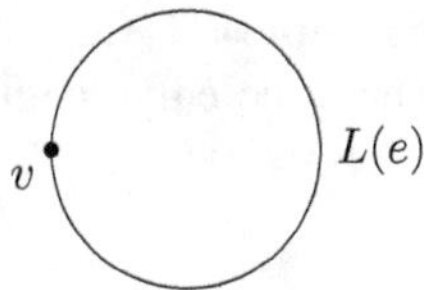

Fig. 3 The star graph

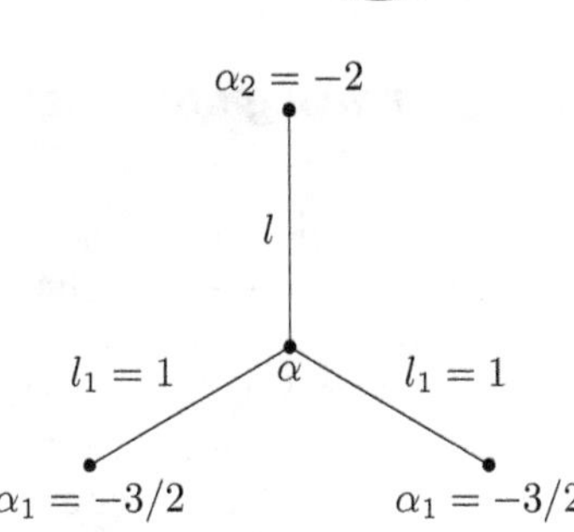

the corresponding eigenfunction being a linear combination of $x \mapsto \sinh(kx)$ and $x \mapsto \cosh(kx)$. It can be seen that k is decreasing when L increases, and thus λ increases.

Exactly the same behavior is displayed by δ' coupling conditions. In the same setting, for a δ' condition of strength $\beta := \beta_v < 0$, the negative eigenvalues are given by $\lambda = -k^2$, where k is the positive solution of

$$\beta k \sinh(kL) + 2\cosh(kL) + 2 = 0,$$

and again, λ increases when L is increased.

Example 4.3 The following example was studied in [20, Section 5]. Let Γ be a star graph, i.e., a graph consisting of a central vertex v_0 and a number of vertices $v_1, \ldots, v_r$ of degree 1, each of which is connected to v_0 by one edge. In this case, let us assume that Γ is a star with three edges, two of which have length $l_1 := 1$ and the third of which has length $l_2 := l > 0$. At the degree-1 vertices, we impose δ coupling conditions, with strength $\alpha_1 := -3/2$ for the edges of length 1 and $\alpha_2 := -2$ for the edge of length l. At the central vertex we impose a strength $\alpha < 0$, see Fig. 3. Let H be the Laplacian on Γ corresponding to these vertex conditions. Then $\lambda_1(H) < 0$ by Theorem 3.3. In [20, Section 5] it was shown that a constant $\alpha_c \approx -1.09$ with the following properties exists:

- If $\alpha_c < \alpha < 0$, then $\lambda_1(H)$ decreases is l increases.
- If $\alpha = \alpha_c$, then $\lambda_1(H)$ stays constant as l increases.
- If $\alpha < \alpha_c$, then $\lambda_1(H)$ increases as l increases.

Finally, we point out that strictness of the inequality in Theorem 4.1 cannot be guaranteed in the case of δ' coupling conditions, as the following example shows.

Example 4.4 Consider the metric graph and vertex conditions of Example 3.2. There the edge lengths do not play any role, and any changes of them will keep $\lambda_1(H) = 0$ fixed.

4.2 Increasing the Length of all Edges

Here, we consider the change of the eigenvalues in case we increase the length of all edges simultaneously. Iterating the results in Sect. 4.1, we easily obtain that the positive eigenvalues decrease if all edge lengths are increased. More can be said if all edge lengths are increased by the same factor and also the coupling conditions are adjusted according to the factor. At least for continuity-Kirchhoff and δ vertex conditions, the following theorem is folklore. However, for the sake of completeness, we present a proof.

Theorem 4.5 *Let Γ be a compact metric graph and $q \in L^\infty(\Gamma)$ real-valued, and let H be the Schrödinger operator on Γ with potential q and general self-adjoint vertex conditions as described in Proposition 2.1. Let $t > 0$, and obtain $\widetilde{\Gamma}$ by scaling each edge, the potential and the strengths, i.e., $\widetilde{\Gamma}$ has the length function $\widetilde{L}(e) := tL(e)$ and the potential $\widetilde{q}_e := \frac{1}{t^2}q_e(\frac{\cdot}{t})$ for $e \in \mathcal{E}$, as well as $\widetilde{\Lambda}_v = \frac{1}{t}\Lambda_v$ for $v \in \mathcal{V}$. Let $\widetilde{H}$ be the corresponding Schrödinger operator on $\widetilde{\Gamma}$. Then*

$$\lambda_k(\widetilde{H}) = \frac{1}{t^2}\lambda_k(H)$$

holds for all $k \in \mathbb{N}$.

Proof Let h and $\widetilde{h}$ be the quadratic forms associated with H and $\widetilde{H}$, respectively. For $f \in \widetilde{H}^1(\Gamma)$ and its restriction f_e to an arbitrary edge $e \in \mathcal{E}$, parameterized as $[0, L(e)]$, consider the function $\widetilde{f} \in \widetilde{H}^1(\widetilde{\Gamma})$ defined on each edge e by

$$\widetilde{f}_e(x) := f_e\left(\frac{x}{t}\right), \quad x \in [0, tL(e)].$$

Since the boundary values of $\widetilde{f}_e$ and f_e at each vertex are the same, i.e., $\widetilde{F}(v) = F(v)$ for each $v \in \mathcal{V}$, f belongs to dom h if and only if $\widetilde{f}$ belongs to dom $\widetilde{h}$. By the substitution $y = \frac{x}{t}$, we observe

$$\widetilde{h}(\widetilde{f}) = \sum_{e \in \mathcal{E}} \int_0^{tL(e)} |\widetilde{f}'|^2 + \sum_{e \in \mathcal{E}} \int_0^{tL(e)} \widetilde{q}_e|\widetilde{f}_e|^2 + \sum_{v \in \mathcal{V}} \langle \widetilde{\Lambda}_v P_{v,\mathrm{R}}\widetilde{F}(v), P_{v,\mathrm{R}}\widetilde{F}(v)\rangle$$

$$= \frac{1}{t}\sum_{e \in \mathcal{E}} \int_0^{L(e)} |f'|^2 + \frac{1}{t}\sum_{e \in \mathcal{E}} \int_0^{L(e)} q_e|f_e|^2 + \frac{1}{t}\sum_{v \in \mathcal{V}} \langle \Lambda_v P_{v,\mathrm{R}}F(v), P_{v,\mathrm{R}}F(v)\rangle$$

$$= \frac{1}{t}h(f).$$

Moreover, $\int_{\widetilde{\Gamma}} |\widetilde{f}|^2 = t \int_{\Gamma} |f|^2$, by the same substitution. Thus, the min-max principle yields

$$\lambda_k(\widetilde{H}) = \min_{\substack{\widetilde{F} \subseteq \operatorname{dom}\widetilde{h} \\ \dim \widetilde{F} = k}} \max_{\widetilde{f} \in \widetilde{F}} \frac{\widetilde{h}(\widetilde{f})}{\int_{\widetilde{\Gamma}} |\widetilde{f}|^2} = \min_{\substack{F \subseteq \operatorname{dom}h \\ \dim F = k}} \max_{f \in F} \frac{1}{t^2} \frac{h(f)}{\int_{\Gamma} |f|^2} = \frac{1}{t^2} \lambda_k(H)$$

for all $k \in \mathbb{N}$. $\qquad\square$

We emphasize the special cases of δ and δ' coupling conditions in the next corollary; observe that the coupling strengths for δ and δ' conditions have to be scaled differently.

Corollary 4.6 *Let Γ be a compact metric graph and $q \in L^\infty(\Gamma)$ real-valued, and let H be the Schrödinger operator on Γ with potential q such that at each vertex $v \in \mathcal{V}$ either a δ coupling condition with strength α_v or a δ' coupling condition with strength β_v is imposed. Let $t > 0$, and obtain $\widetilde{\Gamma}$ by scaling each edge, the potential and the strengths, i.e., $\widetilde{L}(e) := tL(e)$ and $\widetilde{q}_e := \frac{1}{t^2} q_e(\frac{\cdot}{t})$ for $e \in \mathcal{E}$, and $\widetilde{\alpha}_v := \frac{1}{t}\alpha_v$, respectively, $\widetilde{\beta}_v := t\beta_v$ for each $v \in \mathcal{V}$. Let $\widetilde{H}$ be the corresponding Schrödinger operator on $\widetilde{\Gamma}$. Then*

$$\lambda_k(\widetilde{H}) = \frac{1}{t^2} \lambda_k(H)$$

holds for all $k \in \mathbb{N}$.

Proof This follows directly from Theorem 4.5 and, at each vertex v, $\Lambda_v = \frac{\alpha_v}{\deg(v)}$ in the case of a δ coupling, respectively, and $\Lambda_v = \frac{\deg(v)}{\beta_v}$ in the case of a δ' coupling. $\qquad\square$

4.3 Attaching an Edge Between Two Vertices

Next we focus on the manipulation in which a new edge with a finite length between two vertices $v_1, v_2 \in \mathcal{V}$ is added, see Fig. 4. For the case of δ' coupling conditions at v_1 and v_2, this may only decrease the eigenvalues of H; this can be proven by simply extending test functions in the Rayleigh quotient by zero on the respective new edges.

Theorem 4.7 ([68, Theorem 3.2]) *Let Γ be a compact metric graph and $q \in L^\infty(\Gamma)$ real, and let $v_1, v_2 \in \mathcal{V}$ be two distinct vertices of Γ. Furthermore, let H be the Schrödinger operator on Γ with potential q subject to arbitrary self-adjoint coupling conditions at each vertex $v \in \mathcal{V} \setminus \{v_1, v_2\}$ and having δ' coupling conditions of strengths β_{v_1} and, respectively, β_{v_2} at v_1 and v_2; we allow the case of strengths zero, i.e., anti-Kirchhoff conditions. Let $\widetilde{\Gamma}$ be the graph obtained from Γ by adding an extra edge $\widetilde{e}$ of finite length connecting v_1 and v_2 with potential*

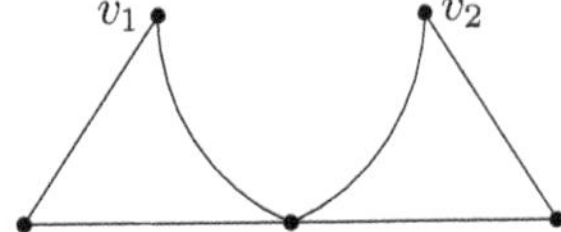 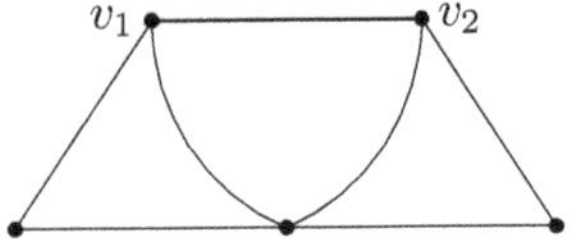

Fig. 4 Attaching a new edge between the vertices v_1 and v_2

$q_{\widetilde{e}} \in L^\infty((0, L(\widetilde{e})))$. *Moreover, let $\widetilde{H}$ be the Schrödinger operator in $L^2(\widetilde{\Gamma})$ having the same potentials on the edges and the same coupling conditions as H on all $v \in \mathcal{V} \setminus \{v_1, v_2\}$, as well as the same strengths for the δ' couplings at v_1 and v_2, respectively. Then*

$$\lambda_k(\widetilde{H}) \leq \lambda_k(H)$$

holds for all $k \in \mathbb{N}$.

Proof We provide a proof for the case $\beta_{v_1} \neq 0 \neq \beta_{v_2}$; the case of an anti-Kirchhoff condition at one or both of the vertices v_1, v_2 is analogous. Let h be the quadratic form associated with H and $\widetilde{h}$ be the quadratic form associated with $\widetilde{H}$. Furthermore, let F be a k-dimensional subspace of $\operatorname{dom} h$ such that

$$h(f) \leq \lambda_k(H) \int_\Gamma |f|^2 \qquad \text{for all } f \in F.$$

For $f \in F$, let $\widetilde{f}$ be the extension of f by zero on $\widetilde{e}$ and $\widetilde{F}$ the space of all these $\widetilde{f}$. Then $\widetilde{F}$ is k-dimensional and $\widetilde{F} \subseteq \operatorname{dom} \widetilde{h}$. For $\widetilde{f} \in \widetilde{F}$, we observe

$$\widetilde{h}(\widetilde{f}) = \int_\Gamma |f'|^2 + \int_\Gamma q|f|^2 + \sum_{v \in \mathcal{V} \setminus \{v_1, v_2\}} \langle \Lambda_v P_{v,\mathrm{R}} F(v), P_{v,\mathrm{R}} F(v) \rangle$$

$$+ \frac{1}{\beta_{v_1}} \left| \sum_{j=1}^{\deg(v_1)} F_j(v_1) \right|^2 + \frac{1}{\beta_{v_2}} \left| \sum_{j=1}^{\deg(v_2)} F_j(v_2) \right|^2 = h(f).$$

Thus,

$$\widetilde{h}(\widetilde{f}) = h(f) \leq \lambda_k(H) \int_\Gamma |f|^2 = \lambda_k(H) \int_{\widetilde{\Gamma}} |\widetilde{f}|^2 \quad \text{for all } \widetilde{f} \in \widetilde{F}.$$

The min-max principle yields the assertion. $\qquad\qquad\qquad\qquad\qquad\qquad\square$

The assertion of Theorem 4.7 is wrong for δ coupling conditions, and no definite statement is possible in this case, as the next example shows.

Example 4.8 ([52, Example 1] and [68, Example 3.3]) Let Γ be the metric graph with two vertices v_1, v_2 and one edge of length 1 connecting these two, see Fig. 5.

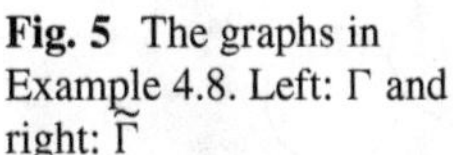 

Fig. 5 The graphs in Example 4.8. Left: Γ and right: $\widetilde{\Gamma}$

Let H be the Laplacian in $L^2(\Gamma)$ with a δ condition of strength $\alpha \in \mathbb{R}$ at v_1 and a Kirchhoff condition at v_2. Note that both vertices have degree 1; hence, the condition at v_1 is a Robin boundary condition and the one at v_2 is Neumann. Note further that for $\alpha = 0$ the condition at v_1 is Neumann, too. Moreover, let $\widetilde{\Gamma}$ be the graph obtained from Γ by adding another edge of length $\ell > 0$ that also connects v_1 to v_2, and let $\widetilde{H}$ be the Laplacian in $L^2(\widetilde{\Gamma})$ subject to the natural extensions of the conditions for H, namely a δ condition of strength α at v_1 and a Kirchhoff condition at v_2. We look at two different cases.

If $\alpha = 0$ (see [52, Example 1]), then $\lambda_k(H) = \pi^2(k-1)^2$ for all $k \in \mathbb{N}$, and $\widetilde{H}$ is unitarily equivalent to the Laplacian on an interval of length $1 + \ell$ with periodic boundary conditions, which implies $\lambda_1(\widetilde{H}) = 0$ and $\lambda_{2k}(\widetilde{H}) = \lambda_{2k+1}(\widetilde{H}) = \frac{4\pi^2}{(1+\ell)^2}k^2$ for all $k \in \mathbb{N}$. Hence, for $\ell < 1$, we observe $\lambda_2(\widetilde{H}) > \lambda_2(H)$, whereas for $\ell \geq 1$ we obtain $\lambda_k(\widetilde{H}) \leq \lambda_k(H)$ for all $k \in \mathbb{N}$ (and even a strict inequality for $k > 2$; for $\ell > 1$, the inequality is strict also for $k = 2$).

If $\alpha = 1$, then simple calculations yield $\lambda_1(H) \approx 0.74017$. Furthermore, letting $\ell = 0.1$, one obtains $\lambda_1(\widetilde{H}) \approx 0.83156$, that is, $\lambda_1(H) < \lambda_1(\widetilde{H})$. Hence, adding an edge may increase the eigenvalues also for $\alpha \neq 0$.

4.4   Attaching a Pendant Graph

In this section we study how the eigenvalues behave if we attach a whole new graph to one vertex of another given graph. In contrast to the above Theorem 4.7, extending test functions in the Rayleigh quotient constantly to the new edge is admissible for both δ and δ' couplings. The result includes, in particular, the case of attaching a single pendant edge.

Theorem 4.9 (cf. [11, Theorem 3.10] and [68, Theorem 3.5]) *Let Γ be a compact metric graph and $q \in L^\infty(\Gamma)$ real, and let $v \in V$ be a vertex of Γ. Furthermore, let H be the Schrödinger operator on Γ with potential q subject to arbitrary self-adjoint vertex conditions at each vertex $w \in V \setminus \{v\}$ and with a δ or δ' coupling condition at v. Let $\widehat{\Gamma}$ be a compact metric graph, $\widehat{q} \in L^\infty(\widehat{\Gamma})$, and $\widehat{v} \in \widehat{V}$ be a vertex of $\widehat{\Gamma}$, and choose arbitrary self-adjoint vertex conditions at each vertex $w \in \widehat{V} \setminus \{\widehat{v}\}$. Let $\widetilde{\Gamma}$ be the graph obtained from Γ by attaching the graph $\widehat{\Gamma}$ with vertex $\widehat{v}$ at v by identifying v and $\widehat{v}$, see Fig. 6. Let $\widetilde{H}$ be the Schrödinger operator in $L^2(\widetilde{\Gamma})$ having the same vertex conditions as H on all $w \in V \setminus \{v\}$ and the chosen coupling conditions at $w \in \widehat{V} \setminus \{\widehat{v}\}$ and with a δ or δ' coupling condition at v with the same strength as for H. If the condition at v is of δ type, we assume in addition that the condition of $\widetilde{H}$*

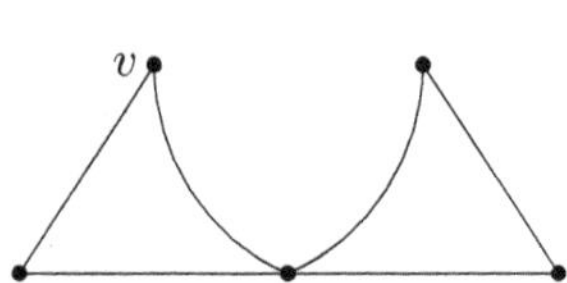 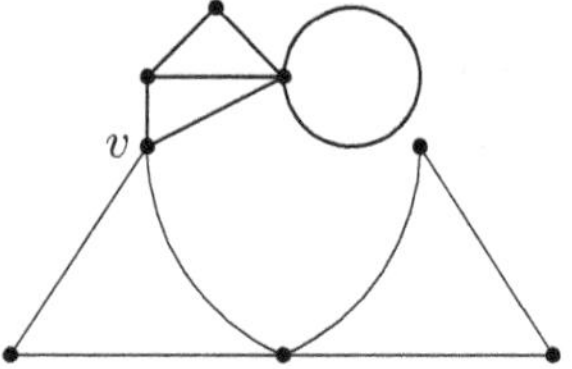

Fig. 6 The transformation of Theorem 4.9. Left: Γ and right: $\widetilde{\Gamma}$, where the graph $\widehat{\Gamma}$ (thick) is attached at v

at $w \in \widehat{\mathcal{V}} \setminus \{\widehat{v}\}$ is δ coupling conditions with strengths $\alpha_w \leq 0$ for all $w \in \widehat{\mathcal{V}} \setminus \{\widehat{v}\}$ and that $\int_{\widehat{\Gamma}} \widehat{q} \leq 0$. Then

$$\lambda_k(\widetilde{H}) \leq \lambda_k(H)$$

holds for all $k \in \mathbb{N}$.

Proof Let h be the quadratic form associated with H and $\widetilde{h}$ be the quadratic form associated with $\widetilde{H}$. Furthermore, let F be a k-dimensional subspace of $\mathrm{dom}\, h$ such that

$$h(f) \leq \lambda_k(H) \int_{\Gamma} |f|^2 \qquad \text{for all } f \in F.$$

For $f \in F$ let $\widetilde{f}$, be the extension of f by 0 to $\widetilde{\Gamma}$ (i.e., $\widetilde{f} = 0$ on $\widehat{\Gamma}$) in the case of a δ' coupling at v and by the constant value $f(v)$ to $\widetilde{\Gamma}$ (i.e., $\widetilde{f} = f(v)$ on $\widehat{\Gamma}$) in the case of a δ coupling at v. Moreover, let $\widetilde{F}$ be the space consisting of the extensions $\widetilde{f}$ of all $f \in F$. Then $\widetilde{F}$ is k-dimensional and $\widetilde{F} \subseteq \mathrm{dom}\, \widetilde{h}$. For $\widetilde{f} \in \widetilde{F}$, we observe

$$\widetilde{h}(\widetilde{f}) = \int_{\Gamma} |f'|^2 + \int_{\Gamma} q|f|^2 + \int_{\widehat{\Gamma}} \widehat{q}|\widetilde{f}_{\widehat{\Gamma}}|^2 + \sum_{w \in \mathcal{V} \setminus \{v\}} \langle \Lambda_w P_{w,\mathrm{R}} F(w), P_{w,\mathrm{R}} F(w) \rangle$$

$$+ \sum_{w \in \widehat{\mathcal{V}} \setminus \{\widehat{v}\}} \langle \widehat{\Lambda}_w \widehat{P}_{w,\mathrm{R}} \widehat{F}(w), \widehat{P}_{w,\mathrm{R}} F(w) \rangle$$

$$+ \begin{cases} \frac{1}{\beta_v} \sum_{j=1}^{\deg(v)} \widetilde{F}_j(v) & \text{for a } \delta' \text{ condition at } v, \\ \alpha_v |f(v)|^2 & \text{for a } \delta \text{ condition at } v. \end{cases}$$

Note that

$$\int_{\widehat{\Gamma}} \widehat{q}|\widetilde{f}|^2 \begin{cases} = 0 & \text{for a } \delta' \text{ condition at } v, \\ \leq 0 & \text{for a } \delta \text{ condition at } v. \end{cases}$$

$$
\begin{array}{cc}
\overset{\displaystyle v_0}{\underset{\displaystyle 0}{\bullet}} \quad \overset{\displaystyle v_1}{\underset{\displaystyle 1}{\bullet}}
\end{array}
\qquad\qquad
\begin{array}{ccc}
\overset{\displaystyle v_0}{\underset{\displaystyle 0}{\bullet}} \quad \overset{\displaystyle v_1}{\underset{\displaystyle 1}{\bullet}} \quad \overset{\displaystyle v_2}{\underset{\displaystyle 2}{\bullet}}
\end{array}
$$

Fig. 7 The graphs in Example 4.10. Left: Γ, right: $\widetilde{\Gamma}$

Moreover, note that the vertex terms at vertices $w \in \widehat{\mathcal{V}} \setminus \{\widehat{v}\}$ are

$$
\langle \widehat{\Lambda}_w \widehat{P}_{w,\mathrm{R}} \widehat{F}(w),\, \widehat{P}_{w,\mathrm{R}} F(w) \rangle
\begin{cases}
= 0 & \text{for a } \delta' \text{ condition at } v, \\
\leq 0 & \text{for a } \delta \text{ condition at } v,
\end{cases}
$$

according to the assumptions of the theorem. Thus,

$$
\widetilde{h}(\widetilde{f}) \leq h(f) \leq \lambda_k(H) \int_{\Gamma} |f|^2 \leq \lambda_k(H) \int_{\widetilde{\Gamma}} |\widetilde{f}|^2 \quad \text{for all } \widetilde{f} \in \widetilde{F}.
$$

The min-max principle yields the assertion. $\qquad\qquad\qquad\qquad\qquad\qquad\square$

If we have a δ type condition at the gluing vertex v, then a condition on the coupling coefficients at the further vertices of the newly attached graph is necessary, as the following example shows.

Example 4.10 ([68, Example 3.6]) Let Γ be the graph consisting of two vertices v_0 and v_1 and one edge connecting the two vertices, which is parameterized by the interval $[0, 1]$; cf. Fig. 7. Let us impose continuity-Kirchhoff conditions at v_0 and v_1. Since the degree is 1 in both cases, these conditions correspond to Neumann boundary conditions for H at 0 and at 1. Hence, $\lambda_k(H) = (k - 1)^2 \pi^2$ for all $k \in \mathbb{N}$. Now, let us add an edge of length 1, which connects v_1 to a new vertex v_2, see Fig. 7. At the vertex v_2 we impose either the condition $f'(2) + \alpha f(2) = 0$ for some $\alpha > 0$ or a Dirichlet condition. Then the spectrum of $\widetilde{H}$ is positive and, hence,

$$
\lambda_1(\widetilde{H}) > 0 = \lambda_1(H).
$$

4.5 Inserting a Graph at a Vertex

In this section we consider the graph manipulation of inserting a metric graph into a vertex of another given metric graph. This transformation was introduced in [11, Section 3], and its influence on the eigenvalues of the Laplacian with δ (and Dirichlet) vertex conditions was studied there. The operation may formally be defined as follows.

Definition 4.11 Let v_0 be one of the vertices of a compact metric graph Γ, and let Γ' be another compact metric graph. Denote by $\mathcal{E}_{v_0}$ the set of edges incident to v_0 in

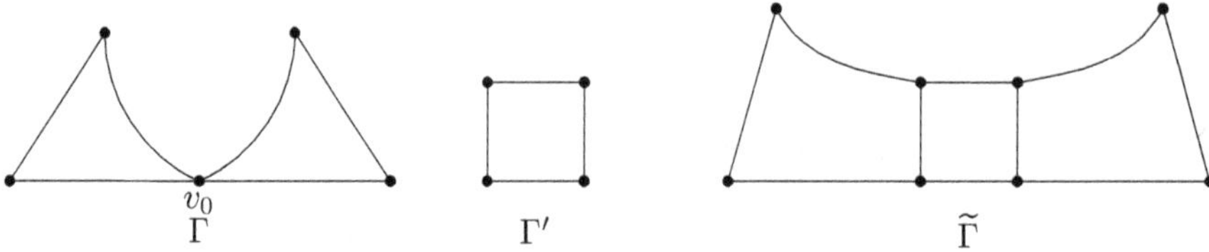

Fig. 8 Inserting the graph Γ' at the vertex $v_0 \in \Gamma$

Γ. We say that the graph $\widetilde{\Gamma}$ is obtained by *inserting* Γ' *into* Γ *at* v_0 if it is formed by removing v_0 from Γ and attaching each of the edges $e \in \mathcal{E}_{v_0}$ to one of the vertices of Γ'; cf. Fig. 8.

With this operation, the following theorem holds for δ vertex conditions. It was shown for the potential-free case of the Laplacian in [11, Theorem 3.10]. Here we present a slightly different proof and allow the presence of a potential q and more general vertex conditions at the vertices that are not affected by the manipulation.

Theorem 4.12 *Let Γ and Γ' be compact metric graphs, and let $\widetilde{\Gamma}$ be formed by inserting Γ' into Γ at a given vertex v_0 of Γ. Furthermore, let H be a Schrödinger operator on Γ with a real potential $q \in L^\infty(\Gamma)$, arbitrary self-adjoint vertex conditions at all vertices except v_0, and a δ vertex condition with strength $\alpha := \alpha_{v_0}$ at v_0. Consider the Schrödinger operator $\widetilde{H}$ on $\widetilde{\Gamma}$ with the same vertex conditions at all vertices of Γ except v_0 and the same potential on all edges of Γ; on the edges of Γ', potentials q_e are placed such that $\int_{\Gamma'} q \leq 0$, and on the remaining vertex set $\mathcal{V}' = \mathcal{V}(\widetilde{\Gamma}) \setminus (\mathcal{V}(\Gamma) \setminus \{v_0\})$, δ vertex conditions with coupling coefficients α_v are imposed such that*

$$\sum_{v \in \mathcal{V}'} \alpha_v \leq \alpha.$$

Then

$$\lambda_k(\widetilde{H}) \leq \lambda_k(H)$$

holds for all $k \in \mathbb{N}$.

Proof Let F denote the linear span of the spaces $\ker(H - \lambda_j(H))$ for $1 \leq j \leq k$. Then $\dim F \geq k$, and

$$h(f) \leq \lambda_k(H) \int_\Gamma |f|^2$$

holds for all $f \in F$. Moreover, let $\widetilde{F}$ be the space formed by extending all $f \in F$ constantly to all of $\widetilde{\Gamma}$ by identifying Γ as a subgraph of $\widetilde{\Gamma}$ in the natural way. Then also $\dim \widetilde{F} \geq k$, $\widetilde{F}$ belongs to the domain of the quadratic form $\widetilde{h}$ associated with

$\widetilde{H}$, and for each $\widetilde{f} \in \widetilde{F}$ we have

$$\widetilde{h}(\widetilde{f}) = \left(h(f) - \alpha |f(v_0)|^2\right) + |f(v_0)|^2 \left(\int_{\Gamma'} q + \sum_{v \in \mathcal{V}'} \alpha_v\right)$$

$$\leq h(f) \leq \lambda_k(H) \int_{\Gamma} |f|^2 \leq \lambda_k(H) \int_{\widetilde{\Gamma}} |f|^2.$$

This implies the assertion of the theorem. $\qquad\square$

The case of a δ' vertex condition at the vertex v_0 at which the additional graph Γ' is inserted is, on the one hand, less restrictive: Arbitrary self-adjoint vertex conditions and arbitrary potentials on the edges may be imposed on the inserted graph. On the other hand, a sign condition on the coupling conditions at the affected vertices is required.

Theorem 4.13 *Let Γ and Γ' be compact metric graphs, and let $\widetilde{\Gamma}$ be formed by inserting Γ' into Γ at a given vertex v_0 of Γ. Furthermore, let H be a Schrödinger operator on Γ with real potential $q \in L^\infty(\Gamma)$, arbitrary self-adjoint vertex conditions at all vertices except v_0, and a δ' vertex condition with strength $\beta := \beta_{v_0} < 0$ at v_0. Consider the Schrödinger operator $\widetilde{H}$ on $\widetilde{\Gamma}$ with the same vertex conditions at all vertices of Γ except v_0 and the same potential on all edges of Γ; on the edges of Γ', arbitrary potentials q_e are placed, on the vertex set $\widehat{V}$ obtained from splitting v_0, δ' vertex conditions with strengths $\beta_v < 0$ are imposed such that*

$$\sum_{v \in \widehat{V}} \beta_v = \beta, \tag{4.2}$$

and on the remaining vertex set $\mathcal{V}' = \mathcal{V}(\widetilde{\Gamma}) \setminus (\mathcal{V}(\Gamma) \setminus \{v_0\})$, arbitrary self-adjoint vertex conditions are imposed. Then

$$\lambda_k(\widetilde{H}) \leq \lambda_k(H)$$

holds for all $k \in \mathbb{N}$.

Proof Let $\widehat{\Gamma}$ be the graph obtained from Γ by splitting v_0 into a set of vertices $\widehat{V}$, and let $\widehat{H}$ be the Schrödinger operator on $\widehat{\Gamma}$ with the same potentials on the edges (after identification) and the same vertex conditions at all vertices in $\mathcal{V}(\widehat{\Gamma}) \setminus \widehat{V}$. Moreover, at the new vertices in $\widehat{V}$, δ' vertex conditions with strengths $\beta_v < 0$ are imposed which satisfy (4.2). Then, by Theorem 5.7 below,

$$\lambda_k(\widehat{H}) \leq \lambda_k(H)$$

for all $k \in \mathbb{N}$. Furthermore, attaching Γ' to $\widehat{\Gamma}$ in such a way that the resulting graph is $\widetilde{\Gamma}$, imposing the desired vertex conditions leads to the operator $\widetilde{H}$ defined in the theorem, and

$$\lambda_k(\widetilde{H}) \le \lambda_k(\widehat{H})$$

follows from a repeated application of Theorem 4.7 (respectively, Theorem 4.9). This completes the proof. □

4.6 Shrinking Edge Lengths to Zero

Here we review some results from [14], where shrinkage of edge lengths to zero has been considered. More precisely, in [14] the natural limit vertex conditions when some edge lengths shrink to zero were identified, and convergence of the corresponding Schrödinger operators was shown under certain conditions. For simplicity of presentation, we do restrict ourselves here to the potential-free case of the Laplacian, $q = 0$ identically.

We change now slightly the perspective in the following sense. Up to now, we viewed Schrödinger operators on metric graphs as having coupling conditions at the vertices, where the boundary values of functions came from all edges adjacent to a given vertex. Now, we focus on each edge separately such that functions in $\widetilde{H}^2$ have two boundary values on each edge for the function and for the (inward) derivative, one for the left end point and one for the right end point, respectively. Since the sum of the vertex degrees is twice the number of edges for each graph, taking all boundary values at all vertices corresponds to taking all boundary values for all edges, so this change of perspective just yields a reordering of the boundary values (now sorted by edges and not by vertices).

In order to formulate a variant of the main result of [14], some notation has to be introduced. For a compact metric graph Γ with the total number of edges $E = |\mathcal{E}|$, let $\mathcal{E}_0 \subseteq \mathcal{E}$ be a set of selected edges, which we are going to shrink to zero. For each edge e, denote by $l(e)$ and $r(e)$ the left, respectively, and right end points of the interval with which e is identified, and let

$$\partial\Gamma := \bigcup_{e\in\mathcal{E}} \{l(e), r(e)\};$$

we distinguish all end points, even if they possibly correspond to the same points on the real line according to the parametrizations of the edges. Then, for any function $f \in \widetilde{H}^1(\Gamma)$, we may consider a vector

$$f|_{\partial\Gamma} \in \mathbb{C}^{2E}$$

of evaluations of f at all end points of all edges. Note that $f|_{\partial\Gamma}$ describes the same boundary values as $(F(v))_{v\in\mathcal{V}}$, just sorted differently. Especially, if $f \in \mathrm{dom}\, H$ for some self-adjoint Schrödinger operator H as in Proposition 2.1, then the vertex conditions may be expressed in terms of the vectors $f|_{\partial\Gamma}$ and $f'|_{\partial\Gamma}$ where, in the latter case, the derivatives are taken in inward direction. Note also that $f'|_{\partial\Gamma}$ describes the same boundary values for the derivatives as $(F'(v))_{v\in\mathcal{V}}$, just sorted differently. Furthermore, define

$$D_0 := D_0(\mathcal{E}_0) := \{F \in \mathbb{C}^{2E} : F_{l(e)} = F_{r(e)}, \ e \in \mathcal{E}_0\},$$

$$N_0 := N_0(\mathcal{E}_0) := \{F' \in \mathbb{C}^{2E} : F'_{l(e)} = -F'_{r(e)}, \ e \in \mathcal{E}_0\}.$$

In the following we will consider a decomposition of the edge set of Γ into $\mathcal{E} = \mathcal{E}_+ \cup \mathcal{E}_0$ with $\mathcal{E}_+ \cap \mathcal{E}_0 = \emptyset$ and shrink the lengths of the edges in $\mathcal{E}_0$ to zero, while the edge lengths of the edges in $\mathcal{E}_+$ may also change but will be sent to a positive limit. We let Γ_+ be the subgraph of Γ constituted by the edge set $\mathcal{E}_+$ and indicate the dependence of Γ on the length function by $\Gamma(L)$; if L is chosen formally such that $L(e) = 0$ for all $e \in \mathcal{E}_0$, then $\Gamma(L) = \Gamma_+$.

Let us consider the sequence of self-adjoint operators $H(L)$, where $H(L)$ acts as the Laplacian on $\Gamma(L)$, $L(e) > 0$ for all $e \in \mathcal{E}$, and all operators $H(L)$ are equipped with the same vertex conditions. The natural limit operator $\widetilde{H}$ of $H(L)$ if $L \to \widetilde{L}$, where $\widetilde{L}(e) = 0$ for all $e \in \mathcal{E}_0$ and $\widetilde{L}(e) > 0$ for all $e \in \mathcal{E}_+$, is the Laplacian defined on the restrictions to Γ_+ of all functions f satisfying the chosen vertex conditions, such that, in addition,

$$f|_{\partial\Gamma} \in D_0 \quad \text{and} \quad f'|_{\partial\Gamma} \in N_0. \tag{4.3}$$

The following hypothesis on the vertex conditions of a self-adjoint Laplacian H on a metric graph will be imposed for the next theorem.

Hypothesis 4.14 *Suppose that for all $f \in \operatorname{dom} H$ we have that if $f|_{\partial\Gamma} \in D_0$ and $f'|_{\partial\Gamma} \in N_0$ and if $f|_{\partial\Gamma_+} = f'|_{\partial\Gamma_+} = 0$, then $f|_{\partial\Gamma} = 0$.*

For the special case of the Laplacian, the following theorem collects the results in [14, Theorems 3.1, 3.5, 3.6], where the coupling conditions were formulated in terms of Lagrangian subspaces for the values at the vertices; see also [69].

Theorem 4.15 *Let $\Gamma = \Gamma(L)$ be a compact metric graph, and let $H(L)$ be the Laplacian in $L^2(\Gamma(L))$ subject to arbitrary, L-independent self-adjoint coupling conditions at each vertex. Let Hypothesis 4.14 be satisfied, and let $\widehat{H}$ be the Laplacian defined on the restrictions of functions satisfying the original vertex conditions and (4.3). Then $\widetilde{H}$ is self-adjoint, $H(L)$ converges to $\widetilde{H}$ in generalized norm resolvent sense as $L \to \widetilde{L}$, cf. Remark 4.17, and the spectrum of H converges to the spectrum of $\widetilde{H}$ (in the Hausdorff sense of multisets) as the edge lengths of $\mathcal{E}_0$ shrink to zero.*

Remark 4.16 Hypothesis 4.14 was analyzed in [14] in great detail. Among others, [14, Lemma 3.3] shows that among vertex conditions without Robin part (i.e., $P_{v,\mathrm{R}} = 0$ for all $v \in \mathcal{V}$), Hypothesis 4.14 is satisfied if and only if the zero function is the only function satisfying the prescribed vertex conditions and being both constant on each edge of Γ_0 and identically zero on Γ_+. Moreover, [14, Lemma 3.4] implies that Hypothesis 4.14 is satisfied for all vertex conditions that require continuity at each vertex, i.e., for δ vertex conditions, including the special case of continuity-Kirchhoff conditions.

Remark 4.17 The convergence of $H(L)$ to $\widetilde{H}$ in generalized norm resolvent sense shown in Theorem 4.15 can be found in [74, p. 351], see also [65, Chapter 4], and means that

$$\| J_L(\widetilde{H} + \mathrm{i})^{-1} - (H(L) + \mathrm{i})^{-1} J_L \| \to 0$$

as $L \to \widetilde{L}$, where $\| \cdot \|$ denotes the norm in the space of bounded linear operators from $L^2(\widetilde{\Gamma})$ to $L^2(\Gamma)$ and J_L is the embedding of $L^2(\widetilde{\Gamma})$ to $L^2(\Gamma(L))$ given by

$$(J_L f)_e(x) := \begin{cases} \sqrt{\dfrac{\widetilde{L}(e)}{L(e)}} \, f_e\left(\dfrac{\widetilde{L}(e)}{L(e)} x\right) & x \in (0, L(e)), \ e \in \mathcal{E}_+, \\ 0 & e \in \mathcal{E}_0, \end{cases}$$

i.e., J_L scales on the edges $\mathcal{E}_+$ linearly and extends trivially by zero on the edges $\mathcal{E}_0$.

We illustrate the above result at two examples.

Example 4.18 Consider a "lasso graph" consisting of an interval e_1 and a loop e_2; see Fig. 9. We impose anti-Kirchhoff coupling conditions; this simplifies to a Dirichlet condition at the degree-1 vertex at the end of the interval. We send one or the other edge length to zero:

(a) Let $\mathcal{E}_+ = \{e_1\}$ and $\mathcal{E}_0 = \{e_2\}$. Then Hypothesis 4.14 is satisfied. Indeed, in view of Remark 4.16, if $f = 0$ on e_1 and f is constant on e_2 (which in particular means $f_{e_2}(0) = f_{e_2}(L(e_2))$), then together with the anti-Kirchhoff condition at the vertex, i.e., $f_{e_2}(0) + f_{e_2}(L(e_2)) = 0$, we obtain $f = 0$ on e_2.

   The limiting graph after shrinking the length of e_2 to zero is an interval graph with a Dirichlet condition on the left and a Neumann condition on the right.
(b) Let $\mathcal{E}_+ = \{e_2\}$ and $\mathcal{E}_0 = \{e_1\}$. Then Hypothesis 4.14 is also satisfied. Indeed, again with Remark 4.16, if $f = 0$ on e_2 and f is constant on e_1, the Dirichlet condition on the left end point of e_1 (as well as the anti-Kirchhoff coupling at the right end point of e_1) yields $f_{e_1} = 0$.

   The limiting graph after shrinking the length of e_1 to zero is a loop graph with an anti-Kirchhoff condition at the vertex.

The secular equation for the lasso graph is given by

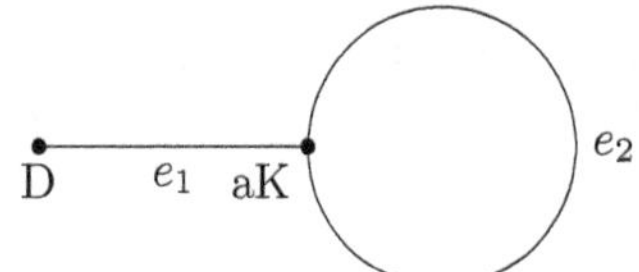

Fig. 9 The lasso graph

Fig. 10 Left: A circle graph with anti-Kirchhoff coupling conditions and the resulting circle graph after shrinking edge e_0 to zero length

$$2\cos(kL(e_1))\cos(kL(e_2)) + 2\cos(kL(e_1)) - \sin(kL(e_1))\sin(kL(e_2)) = 0,$$

so that the eigenvalues are given by $\lambda = k^2$, where k is a solution of the secular equation. The implicit function theorem yields that the solutions k (and therefore the eigenvalues) converge when shrinking one edge. This resembles the spectral convergence in Theorem 4.15.

Example 4.19 Take a circle graph with at least two vertices with anti-Kirchhoff coupling conditions as in Fig. 10. We want to shrink one edge to zero, i.e., for one fixed $e_0 \in \mathcal{E}$, we have $\mathcal{E}_0 = \{e_0\}$ and $\mathcal{E}_+ = \mathcal{E} \setminus \{e_0\}$. By Remark 4.16 it is easy to see that Hypothesis 4.14 is satisfied. Let us determine the coupling for the limiting graph. In order to do that, let us assume that the edges are parametrized counterclockwise, and let $\check{e}$ be the edge preceding e_0 and $\hat{e}$ be the edge following e_0. Then the anti-Kirchhoff condition at the left end point of e_0 reads

$$-f'_{\check{e}}(L(\check{e})) = f'_{e_0}(0), \quad f_{\check{e}}(L(\check{e})) + f_{e_0}(0) = 0,$$

and analogously for the right end point of e_0 we obtain

$$-f'_{e_0}(L(e_0)) = f'_{\hat{e}}(0), \quad f_{e_0}(L(e_0)) + f_{\hat{e}}(0) = 0.$$

Moreover, having boundary values in D_0 yields

$$f_{e_0}(0) = f_{e_0}(L(e_0)),$$

while having boundary values for the (inward) derivatives in N_0 yields

$$f'_{e_0}(0) = f'_{e_0}(L(e_0)).$$

Together, we obtain that shrinking the length of e_0 to zero yields the coupling conditions

$$f_{\check{e}}(L(\check{e})) = f_{\hat{e}}(0), \quad -f'_{\check{e}}(L(\check{e})) = -f'_{\hat{e}}(0);$$

thus we obtain a continuity-Kirchhoff condition relating the edges $\check{e}$ and $\hat{e}$.

Fig. 11 The pumpkin graph
with three edges

We point out that the theorem does not generally apply to graphs with anti-Kirchhoff or δ' vertex conditions, as the following example shows.

Example 4.20 Consider a "pumpkin" graph consisting of two vertices v_1, v_2 and three edges e_1, e_2, e_3 all connecting v_1 and v_2 and having arbitrary positive lengths, see Fig. 11. Impose either anti-Kirchhoff or δ' vertex conditions at v_1 and v_2, and let $\mathcal{E}_0 = \{e_1, e_2\}$ and $\mathcal{E}_+ = \{e_3\}$. Our ambition is, thus, to send $L(e_1)$ and $L(e_2)$ to zero. However, Theorem 4.15 is not applicable. Indeed, the function f that is constantly 1 on e_1, -1 on e_2, and 0 on e_3 belongs to dom (H), i.e., it satisfies the imposed vertex conditions. Moreover, on each edge in $\mathcal{E}_0$, the function has the same value at both end points and zero derivatives, so that $f|_{\partial\Gamma} \in D_0$ and $f'|_{\partial\Gamma} \in N_0$ and $f|_{\partial\Gamma_+} = f'|_{\partial\Gamma_+} = 0$. But, clearly, f has nonzero boundary values on the edges e_1 and e_2; thus Hypothesis 4.14 is violated.

For further illuminating examples, we refer the reader to [14, Section 3].

5 Graph Manipulations Preserving the Total Length

In this section we consider graph manipulations that keep the total length constant and study their influence on the eigenvalues of Schrödinger operators. In some cases also the coupling conditions at the vertices are changed.

5.1 *Changing the Strength of a Coupling*

We start by considering the effect of a change of the strength of a coupling condition. We formulate a general statement for coupling conditions with nontrivial Robin part. The cases of δ and δ' couplings will then be simple corollaries.

Theorem 5.1 *Let Γ be a compact metric graph, $q \in L^\infty(\Gamma)$ real, $v_0 \in \mathcal{V}$, and H the Schrödinger operator on Γ with potential q, arbitrary self-adjoint coupling conditions as in Proposition 2.1 at the vertices $v \in \mathcal{V} \setminus \{v_0\}$, and a self-adjoint coupling condition with nontrivial Robin part, $P_{v_0,\mathrm{R}} \neq 0$, at the vertex v_0. Denote by Λ_{v_0} the self-adjoint, invertible coupling operator in $\mathrm{ran}\, P_{v_0,\mathrm{R}}$, and let $\widetilde{\Lambda}_{v_0}$ be another self-adjoint, invertible operator in $\mathrm{ran}\, P_{v_0,\mathrm{R}}$ such that $\widetilde{\Lambda}_{v_0} \leq \Lambda_{v_0}$. Denote by $\widetilde{H}$ the Schrödinger operator on Γ obtained from H by replacing the coupling operator Λ_{v_0} at v_0 by $\widetilde{\Lambda}_{v_0}$. Then*

$$\lambda_k(\widetilde{H}) \leq \lambda_k(H) \tag{5.1}$$

holds for all $k \in \mathbb{N}$. If the eigenvalue $\lambda_k(H)$ is simple and for the corresponding eigenfunction f and its associated vector $F(v_0)$ of boundary evaluations at the vertex v_0, $P_{v_0,\mathrm{R}}F(v_0) \notin \ker(\Lambda_{v_0} - \widetilde{\Lambda}_{v_0})$, then

$$\lambda_k(\widetilde{H}) < \lambda_k(H).$$

Proof Let h be the quadratic form associated with H and $\widetilde{h}$ be the quadratic form associated with $\widetilde{H}$. Moreover, denote by F the span of the spaces $\ker(H - \lambda_j(H))$ for $1 \le j \le k$. Then $\dim F \ge k$, and for all $f \in F$ we have

$$h(f) - \widetilde{h}(f) = \big\langle \Lambda_{v_0} P_{v_0,\mathrm{R}}F(v), P_{v_0,\mathrm{R}}F(v)\big\rangle - \big\langle \widetilde{\Lambda}_{v_0} P_{v_0,\mathrm{R}}F(v), P_{v_0,\mathrm{R}}F(v)\big\rangle \ge 0.$$
$$(5.2)$$

Thus, the assertion (5.1) follows immediately from the min-max principle.

Assume now that for some k, $\lambda_k(H)$ is simple and equality holds in (5.1). Then

$$\lambda_{k-1}(\widetilde{H}) \le \lambda_{k-1}(H) < \lambda_k(H) = \lambda_k(\widetilde{H});$$

in particular, for all nontrivial f in the space F_{k-1} formed of the span of $\ker(H - \lambda_j(H))$ for $1 \le j \le k - 1$,

$$\widetilde{h}(f) \le \lambda_{k-1}(H) \int_\Gamma |f|^2 < \lambda_k(H) \int_\Gamma |f|^2.$$

Therefore the assumption $\lambda_k(\widetilde{H}) = \lambda_k(H)$ implies equality in (5.2) for all $f \in \ker(H - \lambda_k(H))$. This implies

$$P_{v_0,\mathrm{R}}F(v_0) \in \ker(\Lambda_{v_0} - \widetilde{\Lambda}_{v_0})$$

for all $f \in \ker(H - \lambda_1(H))$ and proves the second assertion. $\qquad\square$

For the cases of δ and δ' coupling conditions at v_0, we obtain the following two corollaries.

Corollary 5.2 ([13, Theorem 3.1.8]) *Let Γ be a compact metric graph, $q \in L^\infty(\Gamma)$ real, and $v_0 \in \mathcal{V}$, and let H be the Schrödinger operator on Γ with potential q, arbitrary self-adjoint coupling conditions at the vertices $v \in \mathcal{V} \setminus \{v_0\}$, and a δ coupling condition with strength $\alpha_{v_0} \in \mathbb{R}$ at v_0. Let $\widetilde{\alpha}_{v_0} < \alpha_{v_0}$, and denote by $\widetilde{H}$ the Schrödinger operator on Γ with the same potential as H, the same coupling conditions at all $v \in \mathcal{V} \setminus \{v_0\}$, and a δ coupling condition with strength $\widetilde{\alpha}_{v_0}$ at v_0. Then*

$$\lambda_k(\widetilde{H}) \le \lambda_k(H)$$

holds for all $k \in \mathbb{N}$. If the eigenvalue $\lambda_k(H)$ is simple and $f(v_0) \neq 0$, then

$$\lambda_k(\widetilde{H}) < \lambda_k(H).$$

In particular, $\lambda_1(\widetilde{H}) < \lambda_1(H)$ if δ coupling conditions are imposed at all vertices.

Proof This follows immediately from Theorem 5.1, since $\Lambda_{v_0} = \frac{\alpha_{v_0}}{\deg(v_0)}$ and functions in dom H are continuous in this case. $\qquad\square$

Thus the eigenvalues are monotone in each strength for δ couplings. However, the situation is more involved for δ' couplings. Note that the following theorem includes the case of a δ' coupling of strength zero, i.e., an anti-Kirchhoff condition.

Corollary 5.3 *Let Γ be a compact metric graph, $q \in L^\infty(\Gamma)$ real, and $v_0 \in V$, and let H be the Schrödinger operator on Γ with potential q, arbitrary self-adjoint coupling conditions at the vertices $v \in V \setminus \{v_0\}$, and a δ' coupling condition with strength $\beta_{v_0} \in \mathbb{R}$ at v_0. Let $\widetilde{\beta}_{v_0} \in \mathbb{R}$. Denote by $\widetilde{H}$ the Schrödinger operator on Γ with potential q with the same coupling conditions as H at all $v \in V \setminus \{v_0\}$ and a δ' coupling condition with strength $\widetilde{\beta}_{v_0}$ at v_0. Suppose that one of the following conditions is satisfied:*

(a) $0 < \beta_{v_0} < \widetilde{\beta}_{v_0}$.
(b) $\beta_{v_0} < \widetilde{\beta}_{v_0} < 0$.
(c) $\widetilde{\beta}_{v_0} < 0 < \beta_{v_0}$.
(d) $\widetilde{\beta}_{v_0} < \beta_{v_0} = 0$.

Then

$$\lambda_k(\widetilde{H}) \leq \lambda_k(H)$$

holds for all $k \in \mathbb{N}$.

Proof If $\beta_{v_0}, \widetilde{\beta}_{v_0} \neq 0$, the statements follow immediately from Theorem 5.1 and $\Lambda_{v_0} = \frac{\deg v_0}{\beta_{v_0}}$. For the case of assumption (d), the quadratic forms associated with H and $\widetilde{H}$ satisfy dom $h \subseteq$ dom $\widetilde{h}$ and $h(f) = \widetilde{h}(f)$ for all $f \in$ dom h. Thus, the min-max principle yields the assertion in this case. $\qquad\square$

5.2 Changing δ Couplings to δ' Couplings

Closely related to the change of the coupling strength at a vertex considered in the previous section is replacing a δ coupling condition by a δ' coupling condition.

Theorem 5.4 *Let Γ be a compact metric graph and $q \in L^\infty(\Gamma)$ real, and let $v_0 \in V$. Denote by H the Schrödinger operator with potential q subject to arbitrary self-adjoint vertex conditions at all $v \in V \setminus \{v_0\}$ and a δ vertex condition with coupling* *strength $\alpha_{v_0} \in \mathbb{R}$ at v_0. Furthermore, denote by $\widetilde{H}$ the Schrödinger operator with the same potential and the same vertex conditions as for H at all vertices except v_0 and a δ' vertex condition with strength $\beta_{v_0} \in \mathbb{R} \setminus \{0\}$ at v_0. Assume that*

$$\frac{\deg(v_0)^2}{\beta_{v_0}} \leq \alpha_{v_0}.$$

Then

$$\lambda_k(\widetilde{H}) \leq \lambda_k(H)$$

holds for all $k \in \mathbb{N}$.

A version of this theorem for Lipschitz partitions of the Euclidean space can be found in [7, Corollary 4.3, Corollary 4.9].

Proof Let h and $\widetilde{h}$ be the quadratic forms associated with H and $\widetilde{H}$, respectively, and let $f \in \mathrm{dom}\, h$. Then $f \in \mathrm{dom}\, \widetilde{h}$ and

$$h(f) - \widetilde{h}(f) = \alpha_{v_0} |f(v_0)|^2 - \frac{1}{\beta_{v_0}} \left| \sum_{j=1}^{\deg(v_0)} F_j(v_0) \right|^2$$

$$= \alpha_{v_0} |f(v_0)|^2 - \frac{\deg(v_0)^2}{\beta_{v_0}} |f(v_0)|^2 \geq 0,$$

which implies the assertion. $\qquad\qquad\qquad\qquad\qquad\qquad\qquad\qquad\square$

The case of an anti-Kirchhoff condition, $\beta_{v_0} = 0$, is more involved; the above proof fails since in this case no inclusion between the domains of the two quadratic forms h and $\widetilde{h}$ prevails. The following example illustrates what can happen.

Example 5.5 Consider a star graph with three edges of arbitrary lengths and the zero potential on them. Impose Neumann vertex conditions at the degree-1 vertices. For the operator H, we impose a continuity-Kirchhoff vertex condition at the central vertex, while $\widetilde{H}$ is equipped with an anti-Kirchhoff condition there. As earlier, any function that is constantly 1 on one edge, -1 on another, and 0 on the third edge belongs to $\ker \widetilde{H}$, and these functions span a two-dimensional vector space. We obtain

$$\lambda_2(\widetilde{H}) = 0 < \lambda_2(H).$$

Consider, on the other hand, the Laplacian on an interval, and fix a Neumann boundary condition at the left end point for both H and $\widetilde{H}$. At the right end point we impose a Kirchhoff (i.e., Neumann) condition for H and an anti-Kirchhoff (i.e.,

Dirichlet) condition for $\widetilde{H}$. Then

$$\lambda_1(H) = 0 < \frac{\pi^2}{4L^2} = \lambda_1(\widetilde{H}),$$

where L denotes the length of the interval.

Complementary to this example, it should be mentioned that in certain situations continuity-Kirchhoff and anti-Kirchhoff vertex conditions lead to the same positive eigenvalues; cf. [55].

5.3 Joining Two Vertices

Given a graph, we may join two vertices by identifying them. More specifically, the graph $\widetilde{\Gamma}$ is obtained from Γ by replacing two vertices v_1, v_2 by a new vertex v_0 with $\deg(v_0) = \deg(v_1) + \deg(v_2)$, such that

$$\mathcal{E}_{v_0} = \mathcal{E}_{v_1} \cup \mathcal{E}_{v_2};$$

see Fig. 12. We review now how the eigenvalues of Schrödinger operators behave under this transformation. We start with the case of δ coupling conditions at the vertices to be joined.

Theorem 5.6 ([32, Theorem 2] and [12, Theorem 5.3]) *Let Γ be a compact metric graph, $q \in L^\infty(\Gamma)$ real, $v_1, v_2 \in \mathcal{V}$ with $v_1 \neq v_2$, and H the Schrödinger operator on Γ with potential q, arbitrary self-adjoint coupling conditions at the vertices $v \in \mathcal{V}\setminus\{v_1, v_2\}$, and δ coupling conditions with strengths $\alpha_{v_1}, \alpha_{v_2} \in \mathbb{R}$ at v_1 and v_2, respectively. Denote by $\widetilde{\Gamma}$ the graph obtained from Γ by joining v_1 and v_2 to form one single vertex v_0. Let $\widetilde{H}$ be the self-adjoint Schrödinger operator in $L^2(\widetilde{\Gamma})$ having the same potential and the same coupling conditions as H at all vertices apart from v_0 and satisfying a δ coupling condition with strength $\alpha_{v_0} := \alpha_{v_1} + \alpha_{v_2}$ at v_0. Then*

$$\lambda_k(H) \leq \lambda_k(\widetilde{H})$$

holds for all $k \in \mathbb{N}$.

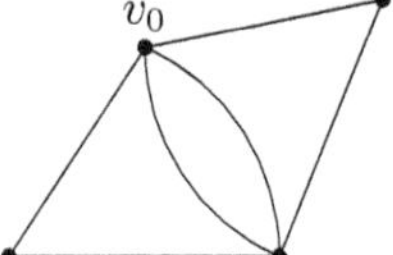

Fig. 12 Joining the vertices v_1 and v_2 to a new vertex v_0

Proof The quadratic forms h and $\widetilde{h}$ associated with H and $\widetilde{H}$, respectively, satisfy $\operatorname{dom}\widetilde{h} \subseteq \operatorname{dom} h$. In particular, each $f \in \operatorname{dom}\widetilde{h}$ satisfies $f(v_0) = f(v_1) = f(v_2)$, and by Lemma 2.3 we get

$$\widetilde{h}(f) - h(f) = \alpha_{v_0}|f(v_0)|^2 - \alpha_{v_1}|f(v_1)|^2 - \alpha_{v_2}|f(v_2)|^2 = 0.$$

By the min-max principle, this leads to the assertion. $\qquad\square$

While joining two δ couplings in the above way may only increase the eigenvalues, the situation is more involved for δ' couplings.

Theorem 5.7 ([68, Theorem 4.2]) *Let Γ be a compact metric graph, $q \in L^\infty(\Gamma)$ real, and $v_1, v_2 \in \mathcal{V}$ with $v_1 \neq v_2$, and let H be the Schrödinger operator in $L^2(\Gamma)$ with potential q, arbitrary self-adjoint coupling conditions at the vertices $v \in \mathcal{V} \setminus \{v_1, v_2\}$, and δ' coupling conditions with strengths $\beta_{v_1}, \beta_{v_2} \in \mathbb{R}$ at v_1 and v_2, respectively. Denote by $\widetilde{\Gamma}$ the graph obtained from Γ by joining v_1 and v_2 to form one single vertex v_0. Furthermore, let $\widetilde{H}$ be the self-adjoint Schrödinger operator in $L^2(\widetilde{\Gamma})$ having the same potential and the same coupling conditions as H at all vertices apart from v_0 and satisfying a δ' coupling condition with strength $\beta_{v_0} := \beta_{v_1} + \beta_{v_2}$ at v_0.*

(i) *If $\beta_{v_1}, \beta_{v_2} \geq 0$, or $\beta_{v_1} \cdot \beta_{v_2} < 0$ and $\beta_{v_0} < 0$, then $\lambda_k(\widetilde{H}) \leq \lambda_k(H)$ for all $k \in \mathbb{N}$.*

(ii) *If $\beta_{v_1}, \beta_{v_2} < 0$, or $\beta_{v_1} \cdot \beta_{v_2} < 0$ and $\beta_{v_0} \geq 0$, then $\lambda_k(H) \leq \lambda_k(\widetilde{H})$ for all $k \in \mathbb{N}$.*

Proof Consider again the quadratic forms h and $\widetilde{h}$ corresponding to H and $\widetilde{H}$, respectively. First, consider the case $\beta_{v_1}, \beta_{v_2}, \beta_{v_0} \neq 0$. Then $\operatorname{dom} h = \operatorname{dom}\widetilde{h}$, and for $f \in \operatorname{dom} h$ we have

$$\widetilde{h}(f) - h(f) = \frac{1}{\beta_{v_0}}\left|\sum_{j=1}^{\deg(v_0)} F_j(v_0)\right|^2 - \frac{1}{\beta_{v_1}}\left|\sum_{j=1}^{\deg(v_1)} F_j(v_1)\right|^2$$

$$- \frac{1}{\beta_{v_2}}\left|\sum_{j=1}^{\deg(v_2)} F_j(v_2)\right|^2 .$$

Now, a case distinction with respect to the signs of $\beta_{v_1}, \beta_{v_2}, \beta_{v_0}$ yields the corresponding statements.

For $\beta_{v_0} = 0$, we have $\beta_{v_1} = -\beta_{v_2}$ and $\operatorname{dom}\widetilde{h} \subseteq \operatorname{dom} h$ and $\sum_{j=1}^{\deg(v_0)} F_j(v_0) = 0$ for $f \in \operatorname{dom}\widetilde{h}$, and therefore

$$\sum_{j=1}^{\deg(v_1)} F_j(v_1) = - \sum_{j=1}^{\deg(v_2)} F_j(v_2).$$

Thus, $\widetilde{h}(f) - h(f) = 0$ for all $f \in \mathrm{dom}\,\widetilde{h}$, and the assertion follows.

In case $\beta_{v_1} = 0$ or $\beta_{v_2} = 0$, we have $\mathrm{dom}\,h \subseteq \mathrm{dom}\,\widetilde{h}$ and again $\widetilde{h}(f) - h(f) = 0$ for all $f \in \mathrm{dom}\,h$. $\qquad\qquad\square$

5.4　Unfolding Parallel and Pendant Edges

In this section we consider a transformation, called unfolding, which replaces several parallel edges or several pendant edges by one single edge. We say that edges $e_1, \ldots, e_r \in \mathcal{E}$ are *parallel* if all of them are incident to the same two vertices. That is, either there exist $v_1, v_2 \in \mathcal{V}$ such that $v_1 \neq v_2$ and $\{e_1, \ldots, e_r\} \subseteq \mathcal{E}_{v_1} \cap \mathcal{E}_{v_2}$, or each of the edges $e_1, \ldots, e_r$ is a loop attached to the same vertex $v \in \mathcal{V}$.

Definition 5.8 Let Γ be a compact metric graph in which the edges $e_1, \ldots, e_r$ are parallel. We say that the graph $\widetilde{\Gamma}$ is obtained from Γ by *unfolding the parallel edges* $e_1, \ldots, e_r$ if $\widetilde{\Gamma}$ has the same set of vertices as Γ, the same set of edges and edge lengths as Γ except for $e_1, \ldots, e_r$, and one edge $\widehat{e}$ of length $L(e_1) + \cdots + L(e_r)$ incident to the same vertices as each of the edges $e_1, \ldots, e_r$; cf. Fig. 13.

The following theorem will be formulated for δ vertex conditions. It was shown more generally, including the lowest nontrivial eigenvalue for continuity-Kirchhoff vertex conditions, in [11, Theorem 3.18]. In contrast to [11], we admit potentials on the edges and demonstrate that the proof from [11] carries over to this case.

Theorem 5.9 *Let Γ be a compact metric graph in which the edges $e_1, \ldots, e_r$ are parallel. Let H be a Schrödinger operator on Γ with real potential $q \in L^\infty(\Gamma)$ on the edges such that*

$$q|_{e_j} \geq 0, \quad j = 1, \ldots, r,$$

and with δ vertex conditions at all vertices. Furthermore, let $\widetilde{\Gamma}$ be the graph obtained by unfolding the parallel edges $e_1, \ldots, e_r$, and let $\widetilde{H}$ be the Schrödinger operator on $\widetilde{\Gamma}$ with the same potential as for H on all edges except for the new edge $\widehat{e}$ and the constant zero potential on $\widehat{e}$. Moreover, for $\widetilde{H}$ we assume that on all vertices δ

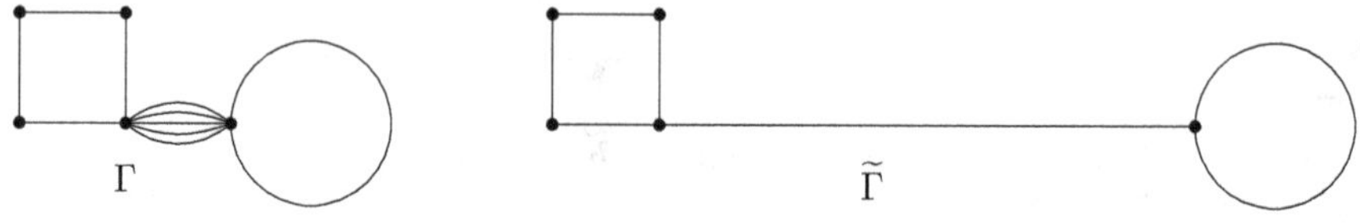

Fig. 13 Unfolding parallel edges in Γ

coupling conditions with the same strengths as for H are imposed. Then

$$\lambda_1(\widetilde{H}) \leq \lambda_1(H)$$

holds. If $\lambda_1(H) \neq 0$, then even $\lambda_1(\widetilde{H}) < \lambda_1(H)$.

Proof We will prove the theorem in the case $r = 2$, i.e., for unfolding two parallel edges e_1, e_2. The general case can be obtained by iterating the following procedure. Let v_1 and v_2 be the vertices to which e_1 and e_2 are incident. We consider only the case that v_1 and v_2 are distinct vertices; the case in which $v_1 = v_2$, i.e., e_1 and e_2 are loops, is completely analogous.

By Theorem 3.1 we can choose $f \in \ker(H - \lambda_1(H))$ strictly positive on Γ. Consider the restrictions f_1 and f_2 of f to e_1 and e_2, respectively. Let, without loss of generality,

$$a := f_1(0) = f_2(0) > 0, \qquad b := f_1(L(e_1)) = f_2(L(e_2)) > 0,$$

and assume without loss of generality $a < b$. Moreover, choose $x_0 \in [0, L(e_1)]$ such that

$$f_1(x_0) = a \quad \text{and} \quad f_1([x_0, L(e_1)]) = [a, b].$$

We distinguish two cases.

Case 1: $\lambda_1(H) \geq 0$. In this case, the quadratic form h corresponding to H satisfies $h(f) \geq 0$. Define a function $\widetilde{f}$ on the graph $\widetilde{H}$ by letting it be equal to f on all edges except $\widehat{e}$ (after identification of these edges with edges of Γ in the natural way). On $\widehat{e}$, identified as $[0, L(e_1) + L(e_2)]$, we define $\widetilde{f}$ by inserting the edge e_2 and the function f_2 on it into e_1 at the point x_0 and defining $\widetilde{f}$ to be equal to f_1 to the left of x_0 (if not $x_0 = 0$) and constantly equal to b to the right. That is,

$$\widetilde{f_{\widehat{e}}}(x) = \begin{cases} f_1(x), & x \in [0, x_0], \\ f_2(x - x_0), & x \in [x_0, x_0 + L(e_2)], \\ b, & x \in [x_0 + L(e_2), L(e_1) + L(e_2)]. \end{cases}$$

This construction is made such that the function $\widetilde{f}$ belongs to $\widetilde{H}^1(\widetilde{\Gamma})$ and satisfies

$$\widetilde{f}(v_1) = a = f(v_1) \quad \text{and} \quad \widetilde{f}(v_2) = b = f(v_2);$$

in particular, it belongs to the domain of the quadratic form $\widetilde{h}$ corresponding to $\widetilde{H}$. Furthermore,

$$\widetilde{h}(\widetilde{f}) = h(f) - \int_{e_1 \cup e_2} (|f'|^2 + q|f|^2) + \int_{\widehat{e}} |\widetilde{f}'|^2 \leq h(f),$$

where we have used that q is nonnegative on $e_1 \cup e_2$ and

$$\int_{e_1 \cup e_2} |f'|^2 \geq \int_{\widehat{e}} |\widetilde{f}'|^2. \tag{5.3}$$

Moreover,

$$\int_{\widetilde{\Gamma}} |\widetilde{f}|^2 = \int_{\Gamma} |f|^2 - \int_{e_1 \cup e_2} |f|^2 + \int_{\widehat{e}} |\widetilde{f}|^2 \geq \int_{\Gamma} |f|^2 \tag{5.4}$$

since

$$\int_{\widehat{e}} |\widetilde{f}|^2 = \int_0^{x_0} |f_1|^2 + \int_0^{L(e_2)} |f_2|^2 + b^2 (L(e_1) - x_0) \geq \int_{e_1} |f_1|^2 + \int_{e_2} |f_2|^2.$$

From (5.3), (5.4), and $h(f) \geq 0$, we obtain

$$\lambda_1(\widetilde{H}) \leq \frac{\widetilde{h}(\widetilde{f})}{\int_{\widetilde{\Gamma}} |\widetilde{f}|^2} \leq \frac{h(f)}{\int_{\Gamma} |f|^2} = \lambda_1(H). \tag{5.5}$$

Case 2: $\lambda_1(H) < 0$. In this case we do a little modification of the construction in Case 1 by defining

$$\widetilde{f_{\widehat{e}}}(x) = \begin{cases} f_1(x), & x \in [0, x_0], \\ a, & x \in [x_0, L(e_1)], \\ f_2(x - L(e_1)), & x \in [L(e_1), L(e_1) + L(e_2)]. \end{cases}$$

Then computations analogous to those in Case 1 yield

$$\widetilde{h}(\widetilde{f}) \leq h(f) < 0$$

and

$$\int_{\widetilde{\Gamma}} |\widetilde{f}|^2 \leq \int_{\Gamma} |f|^2.$$

The last two relations lead again to (5.5).

If $\lambda_1(H) \neq 0$, then f_1 is not constant and, hence, the inequality (5.5), respectively, its counterpart in Case 2, is strict. This completes the proof. $\qquad \square$

The argument carried out in the above proof relies crucially on the continuity of the eigenfunction at the vertices. The latter allows to construct a test function that is continuous on the whole unfolded edge $\widehat{e}$. For vertex conditions that do not require continuity at the vertices, the assertion of Theorem 5.9 need not be true, as the next example shows.

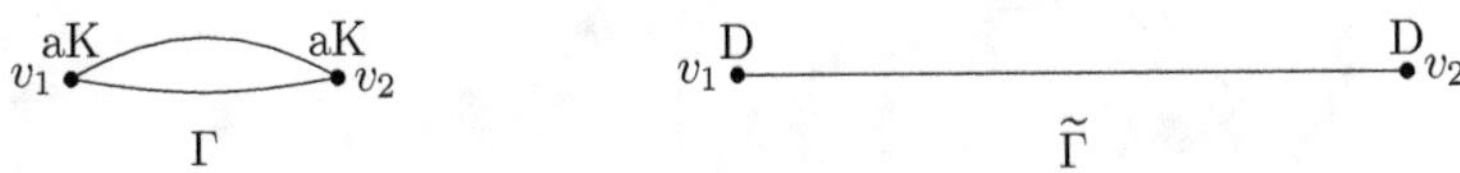

Fig. 14 Unfolding parallel edges in the graph of Example 5.10

Example 5.10 Consider a graph Γ consisting of two vertices v_1, v_2 and two parallel edges of arbitrary finite lengths ℓ_1, ℓ_2 connecting them. We let H be the Schrödinger operator on Γ corresponding to $q = 0$ identically on the edges and anti-Kirchhoff vertex conditions at both v_1 and v_2. Then $\lambda_1(H) = 0$, the corresponding eigenspace being spanned by the function that is constantly 1 on the first edge and -1 on the second one. On the other hand, after unfolding Γ, we obtain the graph $\widetilde{\Gamma}$ isomorphic to an interval of length $\ell_1 + \ell_2$, see Fig. 14, and the Laplacian $\widetilde{H}$ on $\widetilde{\Gamma}$ subject to anti-Kirchhoff vertex conditions equals the Laplacian on the interval with Dirichlet boundary conditions at both end points. In particular,

$$\lambda_1(\widetilde{H}) > 0 = \lambda_1(H).$$

We would like to point out that the exact same construction works if the anti-Kirchhoff conditions in the above example are replaced by δ' vertex conditions with positive strengths.

Open Problem 5.11 *Example 5.10 shows that for δ' vertex conditions in some cases unfolding parallel edges leads to an increase of the principal eigenvalue. Can one prove a general statement of this type for δ' vertex conditions, possibly depending on the signs of the coupling strengths?*

Next we consider a related surgical graph manipulation introduced in [11], called unfolding pendant edges; we call an edge e *pendant* if it is incident to a vertex of degree 1.

Definition 5.12 Let Γ be a compact metric graph in which the edges $e_1, \ldots, e_r$ are pendant and all incident to the same vertex v_0; that is, e_j is incident to v_0 and a vertex v_j of degree 1. We say that the graph $\widetilde{\Gamma}$ is obtained from Γ by *unfolding the pendant edges* $e_1, \ldots, e_r$ if $\widetilde{\Gamma}$ has the same set of vertices as Γ except for $v_1, \ldots, v_r$ and the same set of edges except for $e_1, \ldots, e_r$ and, in addition, an edge $\widehat{e}$ of length $L(e_1) + \cdots + L(e_r)$ connecting v_0 to a new vertex $\widehat{v}$ of degree 1; cf. Fig. 15.

The following theorem is a variant of Theorem 3.18 (4) in [11], where δ or Dirichlet conditions imposed at all vertices were considered and the case of the second eigenvalue for continuity-Kirchhoff conditions was included; on the other hand, the theorem was stated in the potential-free case in [11].

Theorem 5.13 *Let Γ be a compact metric graph in which the edges $e_1, \ldots, e_r$ are pendant and incident to the same vertex v_0, and let v_j denote the vertex distinct from v_0 to which e_j is incident, $j = 1, \ldots, r$. Let H be a Schrödinger operator on Γ with*

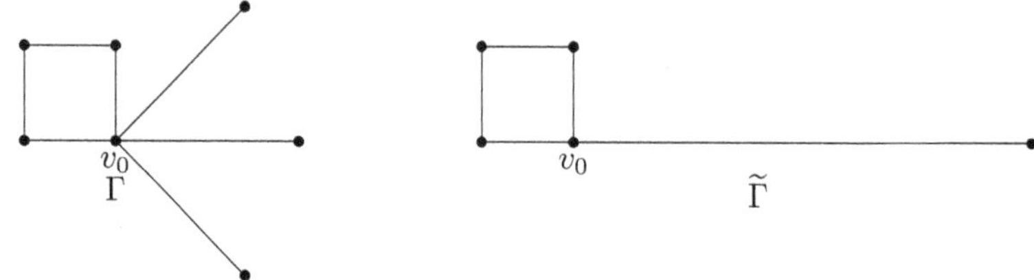

Fig. 15 Unfolding pendant edges in Γ

real potential $q \in L^\infty(\Gamma)$ on the edges such that

$$q|_{e_j} \geq 0, \quad j = 1, \ldots, r,$$

and a δ vertex condition with strength α_v at each vertex $v \in \mathcal{V}$. We assume

$$\alpha_{v_j} \geq 0, \quad j = 1, \ldots, r.$$

Furthermore, let $\widetilde{\Gamma}$ be the graph obtained by unfolding the pendant edges $e_1, \ldots, e_r$ into one edge $\widehat{e}$ incident to a new vertex $\widehat{v}$ of degree 1, and let $\widetilde{H}$ be the Schrödinger operator on $\widetilde{\Gamma}$ with the same potentials as for H on all edges except the new edge $\widehat{e}$ and the constant zero potential on $\widehat{e}$. Moreover, for $\widetilde{H}$ we assume that on all vertices except $\widehat{v}$ (but including v_0) δ vertex conditions with the same strengths as for H are imposed, and $\widehat{v}$ is equipped with a Neumann boundary condition. Then

$$\lambda_1(\widetilde{H}) \leq \lambda_1(H)$$

holds. If $\lambda_1(H) \neq 0$, then even $\lambda_1(\widetilde{H}) < \lambda_1(H)$ holds.

Proof The proof is very similar to the proof of Theorem 5.9 and is only sketched here. We may assume here that $r = 2$, i.e., only two pendant edges are unfolded. The general case can be obtained through multiple applications of this case. Starting with a uniformly positive function $f \in \ker(H - \lambda_1(H))$, we consider a test function $\widetilde{f}$ on $\widetilde{\Gamma}$ which is equal to f on all edges except $\widehat{e}$, employing the natural identification of those edges with edges in the original graph Γ. On the new edge $\widehat{e}$, we define $\widetilde{f}$ depending on whether $\lambda_1(H)$ is nonnegative or negative. If $\lambda_1(H) \geq 0$, choose $x_0 \in e_1 \cup e_2$ such that $f|_{e_1 \cup e_2}$ takes its maximum at x_0; without loss of generality we assume $x_0 \in e_1$. Then, on $\widehat{e}$, identified with the interval $[0, L(e_1) + L(e_2)]$, we set

$$\widetilde{f}_{\widehat{e}}(x) = \begin{cases} f_1(x), & x \in [0, x_0], \\ f_1(x_0), & x \in [x_0, x_0 + L(e_2)], \\ f_1(x - L(e_2)), & x \in [x_0 + L(e_2), L(e_1) + L(e_2)]; \end{cases}$$

in other words, we transplant the edge e_2 into e_1 at the position x_0 and replace the original eigenfunction on e_2 by a suitable constant. This construction yields a function $\widetilde{f}$ on $\widetilde{\Gamma}$ which is continuous inside each edge and at the joint vertex v_0 and belongs to the domain of the quadratic form $\widetilde{h}$ associated with $\widetilde{H}$. Now a computation analogous to the one in the proof of Theorem 5.9 leads to

$$\lambda_1(\widetilde{H}) \leq \frac{\widetilde{h}(\widetilde{f})}{\int_{\widetilde{\Gamma}} |\widetilde{f}|^2} \leq \frac{h(f)}{\int_{\Gamma} |f|^2} = \lambda_1(H).$$

The case where $\lambda_1(H) \leq 0$ can be treated analogously; the only difference is that x_0 is chosen such that $f|_{e_1 \cup e_2}$ takes its minimum—instead of maximum—there.

If $\lambda_1(H) \neq 0$, then f_2 is not constant, and thus the above computations even yield strict inequality for the Rayleigh quotients. This implies the strict eigenvalue inequality. $\qquad\qquad\square$

The construction in the previous proof makes use of the fact that removing pendant edges (for transplantation to a different position) does neither change the vertex condition required in the domain of the quadratic form at the vertex v_0 nor the vertex term in the quadratic form. This is fundamentally different for, e.g., δ' or anti-Kirchhoff vertex conditions. In fact, the following example shows that for such conditions the statement of the previous theorem may fail.

Example 5.14 Consider a star graph Γ consisting of two pendant edges with lengths ℓ_1, ℓ_2 attached to the same vertex v_0 and the Laplacian on Γ with an anti-Kirchhoff vertex condition at v_0 and Neumann vertex conditions at the degree-1 vertices. Then $\lambda_1(H) = 0$, and the eigenspace is spanned by the function which is constantly 1 on one edge and -1 on the other. However, unfolding the pendant edges and keeping the anti-Kirchhoff condition at v_0 result, effectively, in an interval with a Dirichlet boundary condition at one end and a Neumann boundary condition at the other end, see Fig. 16. However, the Laplacian $\widetilde{H}$ subject to these conditions satisfies

$$\lambda_1(\widetilde{H}) = \frac{\pi^2}{4(\ell_1 + \ell_2)^2} > 0 = \lambda_1(H).$$

The exact same construction works if the anti-Kirchhoff condition is replaced by a δ' condition with positive strength. In this case the same spectral effect occurs.

Open Problem 5.15 *In Example 5.14 we have seen a situation where unfolding pendant edges for δ' vertex conditions increases the first eigenvalue. Can one show a general statement of this type, possibly depending on the signs of the strengths in the δ' couplings?*

Fig. 16 Unfolding pendant edges in Example 5.14

6 Hadamard-Type Formulas

In this section we study Hadamard-type formulas, i.e., the variation of eigenvalues and eigenvectors with respect to perturbations of the vertex conditions.

Remark 6.1 In general, the vertex conditions are parametrized by certain matrices, i.e., by elements of a Banach space, and differentiating has to be understood in the sense of, e.g., the Fréchet derivative: Let X be a Banach space, $U \subseteq X$ open, and $F: U \to \mathbb{C}$ Fréchet differentiable; cf. [75, Section 2.1]. Then F' is a mapping from U to $\mathcal{L}(X, \mathbb{C})$, where $\mathcal{L}(X, \mathbb{C})$ denotes the space of bounded linear functionals on X. Thus, for $x \in U$, $F'(x): X \to \mathbb{C}$ is linear and bounded. In our case, X will be finite-dimensional, $X = \mathbb{C}^{d \times d}$ for some d, and then for $x \in U$ we can represent $F'(x)$ by its matrix of partial derivatives.

Theorem 6.2 *Let Γ be a compact metric graph, $q \in L^{\infty}(\Gamma)$ real, $v_0 \in \mathcal{V}$, and H the Schrödinger operator on Γ with potential q, arbitrary self-adjoint coupling conditions as in Proposition 2.1 at the vertices $v \in \mathcal{V} \setminus \{v_0\}$, and a self-adjoint coupling condition with nontrivial Robin part, $P_{v_0,R} \neq 0$, at the vertex v_0. Denote by Λ_{v_0} the self-adjoint, invertible coupling operator in $\mathrm{ran}\, P_{v_0,R}$. Let λ be a simple eigenvalue of H and f be a corresponding normalized eigenfunction. Then λ and f are differentiable w.r.t. Λ_{v_0}, and we have*

$$\lambda'(\Lambda_{v_0}): \widetilde{\Lambda_{v_0}} \mapsto \left\langle \widetilde{\Lambda_{v_0}} P_{v_0,R} F(v_0), P_{v_0,R} F(v_0) \right\rangle,$$

that is,

$$\lambda'(\Lambda_{v_0})(\widetilde{\Lambda_{v_0}}) = \left\langle \widetilde{\Lambda_{v_0}} P_{v_0,R} F(v_0), P_{v_0,R} F(v_0) \right\rangle$$

for all operators $\widetilde{\Lambda_{v_0}}$ in $\mathrm{ran}\, P_{v_0,R}$.

Remark 6.3 We would like to point out that $\mathrm{ran}\, P_{v_0,R}$ is a finite-dimensional vector space, implying that Λ_{v_0} can be represented as a Hermitian matrix. Hence, differentiating with respect to Λ_{v_0} in fact is just differentiating on a finite-dimensional space.

Proof of Theorem 6.2 By [12, Theorem 3.8 and Theorem 3.10] and the different equivalent ways to describe vertex conditions (see, e.g., [48, Theorem 5]), the eigenvalues and eigenfunctions are differentiable with respect to Λ_{v_0}. Let $d := \dim \mathrm{ran}\, P_{v_0,R} > 0$ and $j, k \in \{1, \ldots, d\}$. We abbreviate $\partial_{jk} := \frac{\partial}{\partial (\Lambda_{v_0})_{j,k}}$ for the partial derivatives.

We check that partial derivatives of f w.r.t. components of the parameter again belong to $\mathrm{dom}\, h$. It suffices to check that the derivative is in $\widetilde{H}^1(\Gamma)$, as the condition $P_{v,D} F(v) = 0$ for all $v \in \mathcal{V}$ is a closed condition and thus stable under taking partial derivatives w.r.t. parameters. Note that on each edge $e \in \mathcal{E}$, f is a linear combination of two basis functions solving the eigenvalue equation $-f'' + qf = \lambda f$; the dependence of f on Λ_{v_0} is only present in the coefficients of these basis functions. Thus, on each edge, $\partial_{jk} f$ is again in H^1 on each edge (as edges have finite length). Thus, $\partial_{jk} f \in \widetilde{H}^1(\Gamma)$, and hence $\partial_{jk} f \in \operatorname{dom} h$.

Since f is normalized, we obtain

$$2 \operatorname{Re}\langle f, \partial_{jk} f \rangle = 0, \quad j, k = 1, \ldots, d.$$

Moreover, as $h(f) = \lambda$, we get

$$\partial_{jk}\lambda(\Lambda_{v_0}) = \left(P_{v_0,\mathrm{R}} F(v_0)\right)_k \overline{\left(P_{v_0,\mathrm{R}} F(v_0)\right)}_j + 2 \operatorname{Re} h(f, \partial_{jk} f).$$

Now, since $f \in \operatorname{dom} H$ with $Hf = \lambda f$ and $\partial_{jk} f \in \operatorname{dom} h$, we observe

$$\operatorname{Re} h(f, \partial_{jk} f) = \operatorname{Re}\langle Hf, \partial_{jk} f \rangle = \lambda \operatorname{Re}\langle f, \partial_{jk} f \rangle = 0,$$

and therefore

$$\partial_{jk}\lambda(\Lambda_{v_0}) = \left(P_{v_0,\mathrm{R}} F(v_0)\right)_k \overline{\left(P_{v_0,\mathrm{R}} F(v_0)\right)}_j.$$

We note that this partial derivative is actually constant.

With respect to the canonical basis $\left((\delta_{jk,lm})_{l,m\in\{1,\ldots,d\}} : j, k \in \{1, \ldots, d\}\right)$ of $\mathbb{C}^{d \times d}$, we observe

$$\lambda'(\Lambda_{v_0})((\delta_{jk,lm})) = \partial_{jk}\lambda(\Lambda_{v_0}) = \left(P_{v_0,\mathrm{R}} F(v_0)\right)_k \overline{\left(P_{v_0,\mathrm{R}} F(v_0)\right)}_j$$

$$= \langle (\delta_{jk,lm}) P_{v_0,\mathrm{R}} F(v_0), P_{v_0,\mathrm{R}} F(v_0) \rangle.$$

This yields the assertion. $\qquad\qquad\qquad\qquad\qquad\qquad\qquad\qquad\qquad\qquad\qquad\square$

We will now specialize to δ and δ' coupling conditions; we will formulate this as two corollaries.

Corollary 6.4 ([12, Proposition 4.2]) *Let Γ be a compact metric graph, $q \in L^\infty(\Gamma)$ real, $v_0 \in V$, and H the Schrödinger operator on Γ with potential q, arbitrary self-adjoint coupling conditions as in Proposition 2.1 at the vertices $v \in V \setminus \{v_0\}$, and a δ coupling condition of strength α at the vertex v_0. Let λ be a simple eigenvalue and f be a corresponding normalized eigenfunction. Then λ and f are differentiable w.r.t. α, and we have*

$$\frac{d\lambda}{d\alpha}(\alpha) = |f(v_0)|^2.$$

Proof The proof follows from Theorem 6.2 and the fact that for the δ coupling condition we have that Λ_{v_0} is multiplication by $\frac{\alpha}{\deg(v_0)}$. $\qquad\qquad\square$

Corollary 6.5 *Let Γ be a compact metric graph, $q \in L^\infty(\Gamma)$ real, $v_0 \in V$, and H the Schrödinger operator on Γ with potential q, arbitrary self-adjoint coupling* *conditions as in Proposition 2.1 at the vertices $v \in \mathcal{V} \setminus \{v_0\}$, and a δ' coupling condition of strength β at the vertex v_0. Let λ be a simple eigenvalue and f be a corresponding normalized eigenfunction. Then λ and f are differentiable w.r.t. $\frac{1}{\beta}$ and w.r.t. β, and we have*

$$\frac{d\lambda}{d\frac{1}{\beta}}(\beta) = \left| \sum_{j=1}^{\deg(v_0)} F_j(v_0) \right|^2 \quad and \quad \frac{d\lambda}{d\beta}(\beta) = -\frac{1}{\beta^2} \left| \sum_{j=1}^{\deg(v_0)} F_j(v_0) \right|^2.$$

Proof The proof of the first equality follows from Theorem 6.2 and the fact that for the δ' coupling condition we have that Λ_{v_0} is multiplication by $\frac{\deg v_0}{\beta}$. The derivative of λ w.r.t. β can then be obtained by the chain rule. $\qquad\square$

7   Bounds for the Lowest Eigenvalue

In this final section we review some known bounds on the ground state eigenvalue of a Schrödinger operator on a metric graph, for special choices of vertex conditions. We consciously exclude the case of continuity-Kirchhoff vertex conditions, where $\lambda_1(H) = 0$, but would like to mention that a large body of literature dealing with estimates for the lowest positive eigenvalue (also called spectral gap) and its higher eigenvalues exists. We refer the reader to [3, 11, 17, 18, 22, 35, 36, 49, 50, 52, 62, 63, 66, 67] and the references therein.

For δ coupling conditions, one has the following lower bound, where the total positive and, respectively, negative interaction strengths of H are defined as

$$I_+ := \int_\Gamma q_+ + \sum_{v:\alpha_v > 0} \alpha_v \quad and \quad I_- := \int_\Gamma q_- - \sum_{v:\alpha_v < 0} \alpha_v, \tag{7.1}$$

denoting by $q_+, q_- \geq 0$ the positive and negative parts of the potential q.

Theorem 7.1 ([32, Theorem 1]) *Let Γ be a compact, connected metric graph with total length L, and let H be the Schrödinger operator on Γ with real-valued potential $q \in L^\infty(\Gamma)$ and a δ coupling condition of strength α_v at each vertex $v \in \mathcal{V}$. Moreover, let the total positive and negative interaction strengths of H be given in (7.1). Then the first eigenvalue $\lambda_1(H)$ satisfies*

$$\lambda_1(H) \geq \lambda_1(\widehat{H}),$$

where $\widehat{H}$ is the Schrödinger operator with potential zero (i.e., the Laplacian) on the interval of length L with a δ vertex condition (i.e., Robin boundary condition) of strength I_+ at one end point and a δ vertex condition of strength I_- at the other end point. In other words, $\lambda_1(H)$ is bounded from below by k^2, where k is the smallest

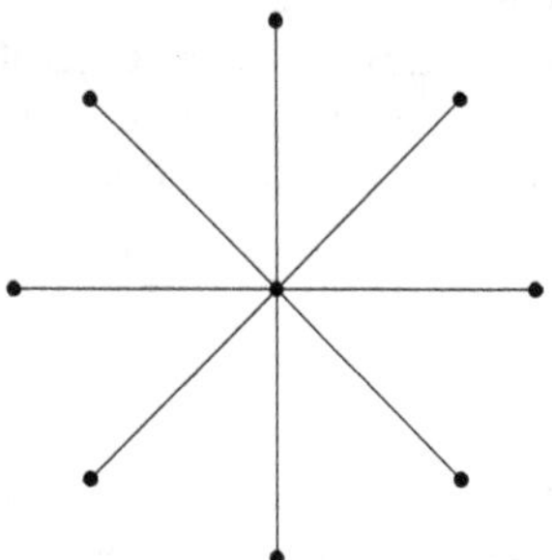

$$I_+ \qquad\qquad I_-$$

Fig. 17 The graph and δ strengths for $\widehat{H}$

solution to the secular equation

$$\left(k + \frac{I_- I_+}{k}\right)\tan(kL) = I_+ - I_-;$$

cf. Fig. 17.

The latter theorem may be viewed as an isoperimetric result: Among the Schrödinger operators with δ coupling conditions on all graphs of fixed length and fixed total positive and negative interaction strengths, the first eigenvalue gets minimal on an interval without edge potential.

The following example indicates that an upper bound for $\lambda_1(H)$, in the case of δ vertex conditions, depending only on the strengths of the δ couplings and the total length of the graph might exist. For the first nontrivial eigenvalue of the Laplacian with continuity-Kirchhoff vertex conditions, such a bound cannot exist, as simple counterexamples show; see, e.g., [36]. However, the lowest eigenvalue is always zero in this case, hence bounded.

Example 7.2 Consider a star graph of total length $L = 1$ with E edges of length $l = 1/E$, see Fig. 18. We impose Neumann boundary conditions at all vertices of degree 1 and a δ coupling condition of strength $\alpha = 1$ at the star vertex. If we parametrize the edges by E copies of the interval $(0, l)$, with the end point l corresponding to the star vertex, then the restriction f_e of f to any edge $e \in \mathcal{E}$ is given by

$$f_e(x) = A_e \cos(kx), \quad x \in (0, l),$$

Fig. 18 The equilateral E-star graph for $E = 8$

where $\lambda = k^2$ is the eigenvalue. Moreover, the vertex conditions at the star vertex can be phrased

$$f_e(l) \text{ is independent of } e \quad \text{and} \quad -\sum_{e \in \mathcal{E}} f_e'(l) = f_e(l), \quad e \in \mathcal{E}.$$

By a simple computation, this gives rise to two cases, either

$$k \in \frac{\pi}{2l}\mathbb{N}$$

or k solves

$$\frac{1}{k} = E \tan\left(\frac{k}{E}\right).$$

As the right-hand side converges to k as $E \to \infty$, there is a sequence of solutions k_E of the above problem converging to 1 as $E \to \infty$. Hence the lowest eigenvalue remains bounded.

A trivial upper bound for the lowest eigenvalue of the Laplacian with δ coupling conditions may be obtained by inserting any constant function into the Rayleigh quotient:

Proposition 7.3 *Let H be a Schrödinger operator on a compact metric graph Γ with real-valued potential q and a δ coupling condition with strength α_v at each vertex $v \in \mathcal{V}$. Then*

$$\lambda_1(H) \leq \frac{1}{L(\Gamma)}\left(\int_\Gamma q + \sum_{v \in \mathcal{V}} \alpha_v\right),$$

where $L(\Gamma)$ denotes the total length of Γ.

Equality in the above estimate holds only if $q = 0$ and $\alpha_v = 0$ for all $v \in \mathcal{V}$, the case in which $\lambda_1(H) = 0$. However, in general the estimate may be rather rough, especially in the presence of large coupling coefficients: If $\alpha_v \to +\infty$ for all v, then $\lambda_1(H)$ converges to the lowest eigenvalue of the Schrödinger operator with Dirichlet boundary conditions at each vertex—in the potential-free case this is

$$\frac{\pi^2}{L_{\max}^2} \leq \frac{E^2\pi^2}{L(\Gamma)^2},$$

where $L_{\max}$ is the largest edge length in Γ. However, the bound in Proposition 7.3 tends to $+\infty$ in this case.

The following proposition indicates that a careful study of the eigenvalues of flower graphs may give rise to better bounds.

6. J. Behrndt, M. Langer, V. Lotoreichik, J. Rohleder, Spectral enclosures for non-self-adjoint extensions of symmetric operators. J. Funct. Anal. **275**, 1808–1888 (2018)
7. J. Behrndt, P. Exner, V. Lotoreichik, Schrödinger operators with δ and δ'-interactions on Lipschitz surfaces and chromatic numbers of associated partitions. Rev. Math. Phys. **26**(8), 1450015 (2014), 43 pp.
8. J. Behrndt, A. Luger, On the number of negative eigenvalues of the Laplacian on a metric graph. J. Phys. A **43**, 474006 (2010), 11 pp.
9. G. Berkolaiko, M. Ettehad, Three-dimensional elastic beam frames: rigid joint conditions in variational and differential formulation. Stud. Appl. Math. **148**(4), 1586–1623 (2022)
10. G. Berkolaiko, J.B. Kennedy, P. Kurasov, D. Mugnolo, Edge connectivity and the spectral gap of combinatorial and quantum graphs. J. Phys. A **50**(36), 365201 (2017), 29 pp.
11. G. Berkolaiko, J.B. Kennedy, P. Kurasov, D. Mugnolo, Surgery principles for the spectral analysis of quantum graphs. Trans. Am. Math. Soc. **372**(7), 5153–5197 (2019)
12. G. Berkolaiko, P. Kuchment, *Dependence of the Spectrum of a Quantum Graph on Vertex Conditions and Edge Lengths*. Spectral Geometry, 117–137, Proceedings of Symposia in Pure Mathematics, vol. 84 (American Mathematical Society, Providence, 2012)
13. G. Berkolaiko, P. Kuchment, *Introduction to Quantum Graphs*. Mathematical Surveys and Monographs, 186 (American Mathematical Society, Providence, 2013)
14. G. Berkolaiko, Y. Latushkin, S. Sukhtaiev, Limits of quantum graph operators with shrinking edges. Adv. Math. **352**, 632–669 (2019)
15. P. Bifulco, J. Kerner, Comparing the spectrum of Schrödinger operators on quantum graphs. arXiv preprint 2212.13954 (2022), 10 pp.
16. J. Bolte, J. Harrison, Spectral statistics for the dirac operator on graphs. J. Phys. A: Math. Gen. **36**, 2747–2769 (2003)
17. D. Borthwick, L. Corsi, K. Jones, Sharp diameter bound on the spectral gap for quantum graphs. Proc. Am. Math. Soc. **149**(7), 2879–2890 (2021)
18. D. Borthwick, E.M. Harrell II, H. Yu, Gaps between consecutive eigenvalues for compact metric graphs. preprint, arXiv:2301.07149
19. C. Cattaneo, The spectrum of the continuous Laplacian on a graph. Monatsh. Math. **124**(3), 215–235 (1997)
20. P. Exner, M. Jex, On the ground state of quantum graphs with attractive δ coupling. Phys. Lett. A **376**, 713–717 (2012)
21. P. Exner, A. Kostenko, M. Malamud, H. Neidhardt, Spectral theory of infinite quantum graphs. Ann. Henri Poincaré **19**(11), 3457–3510 (2018)
22. L. Friedlander, Extremal properties of eigenvalues for a metric graph. Ann. Inst. Fourier **55**, 199–211 (2005)
23. F. Gregorio, D. Mugnolo, Bi-Laplacians on graphs and networks. J. Evol. Equ. **20**(1), 191–232 (2020)
24. F. Gregorio, D. Mugnolo, Higher-order operators on networks: hyperbolic and parabolic theory. Integral Equ. Oper. Theory **92**(6), 50 (2020), 22 pp.
25. E.M. Harrell II, A.V. Maltsev, On Agmon metrics and exponential localization for quantum graphs. Commun. Math. Phys. **359**(2), 429–448 (2018)
26. E.M. Harrell II, A.V. Maltsev, Localization and landscape functions on quantum graphs. Trans. Am. Math. Soc. **373**(3), 1701–1729 (2020)
27. M. Hofmann, J.B. Kennedy, D. Mugnolo, M. Plümer, On Pleijel's nodal domain theorem for quantum graphs. Ann. Henri Poincaré **22**(11), 3841–3870 (2021)
28. A. Hussein, Bounds on the negative eigenvalues of Laplacians on finite metric graphs. Integral Equ. Oper. Theory **76**(3), 381–401 (2013)
29. A. Hussein, D. Krejčiřík, P. Siegl, Non-self-adjoint graphs. Trans. Am. Math. Soc. **367**(4), 2921–2957 (2015)
30. A. Hussein, D. Mugnolo, Quantum graphs with mixed dynamics: the transport/diffusion case. J. Phys. A **46**(23), 235202 (2013), 19 pp.
31. U. Kant, T. Klauß, J. Voigt, M. Weber, Dirichlet forms for singular one-dimensional operators and on graphs. J. Evol. Equ. **9**, 637–659 (2009)

32. G. Karreskog, P. Kurasov, I. Trygg Kupersmidt, Schrödinger operators on graphs: symmetrization and Eulerian cycles. Proc. Am. Math. Soc. **144**, 1197–1207 (2016)

33. T. Kato, *Perturbation Theory for Linear Operators* (Springer, Berlin, 1995)

34. M. Keller, D. Lenz, R. Wojciechowski, *Graphs and Discrete Dirichlet Spaces*. Grundlehren der mathematischen Wissenschaften [Fundamental Principles of Mathematical Sciences] 358 (Springer, Cham, 2021)

35. J.B. Kennedy, *A Family of Diameter-Based Eigenvalue Bounds for Quantum Graphs*. Discrete and continuous models in the theory of networks, 213–239, Operator Theory: Advances and Applications, 281 (Birkhäuser/Springer, Cham, 2020)

36. J.B. Kennedy, P. Kurasov, G. Malenová, D. Mugnolo, On the spectral gap of a quantum graph. Ann. Henri Poincaré **17**(9), 2439–2473 (2016)

37. J.B. Kennedy, R. Lang, On the eigenvalues of quantum graph Laplacians with large complex δ couplings. Port. Math. **77**(2), 133–161 (2020)

38. J.B. Kennedy, J. Rohleder, On the hot spots of quantum graphs. Commun. Pure Appl. Anal. **20**(9), 3029–3063 (2021)

39. A. Kostenko, D. Mugnolo, N. Nicolussi, Self-adjoint and Markovian extensions of infinite quantum graphs. J. Lond. Math. Soc. (2) **105**(2), 1262–1313 (2022)

40. A. Kostenko, N. Nicolussi, Laplacians on infinite graphs: discrete vs. continuous. Proc. 8ECM (2021), in press

41. A. Kostenko, N. Nicolussi, Laplacians on infinite graphs. Mem. Eur. Math. Soc., in press

42. A. Kostenko, N. Nicolussi, Spectral estimates for infinite quantum graphs. Calc. Var. Partial Differ. Equ. **58**(1), 15 (2019), 40 pp.

43. A. Kostenko, N. Nicolussi, Quantum graphs on radially symmetric antitrees. J. Spectr. Theory **11**(2), 411–460 (2021)

44. V. Kostrykin, R. Schrader, Kirchhoff's rule for quantum wires. J. Phys. A: Math. Gen. **32**, 595–630 (1999)

45. V. Kostrykin, J. Potthoff, R. Schrader, *Laplacians on Metric Graphs: Eigenvalues, Resolvents and Semigroups*. Quantum Graphs and Their Applications, Contemporary Mathematics, 415 (American Mathematical Society, Providence, 2006), pp. 201–225.

46. V. Kostrykin, J. Potthoff, R. Schrader, Contraction semigroups on metric graphs, in *Analysis on Graphs and its Applications*. Proceedings of Symposia in Pure Mathematics, 77, ed. by P. Exner et al. (American Mathematical Society, Providence, 2008), pp. 423–458

47. P. Kuchment, Quantum graphs: I. Some basic structures. Waves Random Media **14**(1), S107—S128 (2004)

48. P. Kuchment, Quantum graphs: an introduction and a brief survey, in *Analysis on Graphs and Its Applications*. Proceedings of Symposia in Pure Mathematics, 77, ed. by P. Exner et al. (American Mathematical Society, Providence, 2008), pp. 291–314

49. P. Kurasov, On the spectral gap for Laplacians on metric graphs. Acta Phys. Polonica A **124**, 1060–1062 (2013)

50. P. Kurasov, *Spectral Gap for Complete Graphs: Upper and Lower Estimates*. Mathematical Technology of Networks. Springer Proceedings in Mathematics and Statistics, 128 (Springer, Cham, 2015), pp. 121–132

51. P. Kurasov, On the ground state for quantum graphs. Lett. Math. Phys. **109**(11), 2491–2512 (2019)

52. P. Kurasov, G. Malenová, S. Naboko, Spectral gap for quantum graphs and their connectivity. J. Phys. A **46**, 275309 (2013)

53. P. Kurasov, J. Muller, *On the Spectral Gap for Networks of Beams*. Schrödinger Operators, Spectral Analysis and Number Theory. Springer Proceedings in Mathematics and Statistics, 348 (Springer, Cham, 2021), pp. 169–179

54. P. Kurasov, S. Naboko, Rayleigh estimates for differential operators on graphs. J. Spectral Theory **4**, 211–219 (2014)

55. P. Kurasov, J. Rohleder, Laplacians on bipartite metric graphs. Oper. Matrices **14**, 535–553 (2020)

56. P. Kurasov, A. Serio, Optimal potentials for quantum graphs. Ann. Henri Poincaré **20**(5), 1517–1542 (2019)
57. D. Mugnolo, *Semigroup Methods for Evolution Equations on Networks* (Springer, Berlin, 2014)
58. D. Mugnolo, D. Noja, C. Seifert, Airy-type evolution equations on star graphs. Anal. PDE **11**(7), 1625–1652 (2018)
59. D. Mugnolo, M. Plümer, Lower estimates on eigenvalues of quantum graphs. Oper. Matrices **14**(3), 743–765 (2020)
60. D. Mugnolo, M. Plümer, On torsional rigidity and ground-state energy of compact quantum graphs. Calc. Var. Partial Differ. Equ. **62**(1), 27 (2023), 37 pp.
61. J. Muller, J. Rohleder, The Krein–von Neumann extension for Schrödinger operators on metric graphs. Complex Anal. Oper. Theory **15**(2), 27 (2021), 41 pp.
62. S. Nicaise, Spectre des réseaux topologiques finis. Bull. Sci. Math. **111**, 401–413 (1987)
63. M. Plümer, Upper eigenvalue bounds for the Kirchhoff Laplacian on embedded metric graphs. J. Spectr. Theory **11**(4), 1857–1894 (2021)
64. M. Plümer, M. Täufer, On fully supported eigenfunctions of quantum graphs. Lett. Math. Phys. **111**(6), 153 (2021), 23 pp.
65. O. Post, *Spectral Analysis of Graph-like Spaces*. Springer Lecture Notes 2039, 2012
66. J. Rohleder, Eigenvalue estimates for the Laplacian on a metric tree. Proc. Am. Math. Soc. **145**, 2119–2129 (2017)
67. J. Rohleder, Quantum trees which maximize higher eigenvalues are unbalanced. Proc. Am. Math. Soc. Ser. B **9**, 50–59 (2022)
68. J. Rohleder, C. Seifert, Spectral monotonicity for Schrödinger operators on metric graphs. Oper. Theory Adv. Appl. **281**, 291–310 (2020)
69. C. Schubert, C. Seifert, J. Voigt, M. Waurick, Boundary systems and (skew-)self-adjoint operators on infinite metric graphs. Math. Nachr. **288**(14–15), 1776–1785 (2015)
70. C. Seifert, J. Voigt, Dirichlet forms for singular diffusion on graphs. Oper. Matrices **5**(4), 723–734 (2011)
71. C. Seifert, *The Linearised Korteweg–de Vries Equation on General Metric Graphs*. The Diversity and Beauty of Applied Operator Theory. Operator Theory: Advances and Applications, 268 (Birkhäuser/Springer, Cham, 2018), pp. 449–458
72. J. von Below, A characteristic equation associated to an eigenvalue problem on c^2-networks. Linear Algebra Appl. **71**, 309–325 (1985)
73. M. Waurick, M. Kaliske, *On the Well-Posedness of Evolutionary Equations on Infinite Graphs*. Spectral Theory, Mathematical System Theory, Evolution Equations, Differential and Difference Equations. Operator Theory: Advances and Applications, 221 (Birkhäuser/Springer Basel AG, Basel, 2012), pp. 653–666
74. J. Weidmann, Lineare Operatoren in Hilberträumen. Teil 1, Mathematische Leitfäden. [Mathematical Textbooks] (B. G. Teubner, Stuttgart, 2000)
75. E. Zeidler, *Applied Functional Analysis*. Applied Mathematical Sciences, vol. 109 (Springer, New York, 1995)

Asymptotic Stability of Port-Hamiltonian Systems

Marcus Waurick (iD) and Hans Zwart

1 Introduction

In this chapter we discuss the stability of port-Hamiltonian partial differential equations (p.d.e.s) of the form

$$\frac{\partial x}{\partial t}(\zeta, t) = \sum_{k=0}^{N} P_k \frac{\partial^k}{\partial \zeta^k} \left(\mathcal{H}(\zeta) x(\zeta, t) \right) \tag{1}$$

on the spatial interval $[a, b]$. Here, $P_k \in \mathbb{C}^{n \times n}$ satisfy $P_k^* = (-1)^{k+1} P_k$ ($k \in \{0, \cdots, N\}$), and P_N is invertible. The *Hamiltonian density* $\mathcal{H} \colon (a, b) \to \mathbb{C}^{n \times n}$ is uniformly bounded, $\mathcal{H}(\zeta)^* = \mathcal{H}(\zeta)$, and there exists an $m > 0$ such that $\mathcal{H}(\zeta) \geq mI$ for almost all $\zeta \in [a, b]$. The above p.d.e. is completed with boundary

M. Waurick (✉)
Institut für Angewandte Analysis, TU Bergakademie Freiberg, Freiberg, Germany
e-mail: marcus.waurick@math.tu-freiberg.de

H. Zwart
Department of Applied Mathematics, University of Twente, Twente, The Netherlands

Department of Mechanical Engineering, Technische Universiteit Eindhoven, Eindhoven, The Netherlands

© The Author(s), under exclusive license to Springer Nature Switzerland AG 2024
F. L. Schwenninger, M. Waurick (eds.), *Systems Theory and PDEs*,
Trends in Mathematics, https://doi.org/10.1007/978-3-031-64991-2_4

conditions given as

$$W_B \begin{bmatrix} (\mathcal{H}x)(b,t) \\ \vdots \\ \frac{\partial^{N-1}(\mathcal{H}x)}{\partial \zeta^{N-1}}(b,t) \\ (\mathcal{H}x)(a,t) \\ \vdots \\ \frac{\partial^{N-1}(\mathcal{H}x)}{\partial \zeta^{N-1}}(a,t) \end{bmatrix} = 0, \tag{2}$$

where W_B is an $nN \times 2nN$ matrix which we assume to be of full rank.

One possible approach to address properties of the above system is to invoke the theory of strongly continuous semigroups of bounded linear operators, in that one reformulates the above as an abstract ordinary differential equation in a (infinite-dimensional) state space X. For this, a suitable functional analytic setting is conveniently formulated in the weighted Hilbert space $X = L_{2,\mathcal{H}}((a,b); \mathbb{C}^n)$, which coincides with $L_2((a,b); \mathbb{C}^n)$ as set and is endowed with the weighted inner product given by

$$\langle f, g \rangle_{\mathcal{H}} = \frac{1}{2} \int_a^b g(\zeta)^* \mathcal{H}(\zeta) f(\zeta) d\zeta. \tag{3}$$

X is known as the *energy space*. In [7] necessary and sufficient conditions are given when the operator associated with the above p.d.e. plus boundary conditions generates a contraction semigroup. Since the notation will be used in our stability result, we repeat this theorem. We define the $nN \times nN$ matrix Q and the $2nN \times 2nN$ matrix R_{ext} as

$$Q = [Q_{ij}], i, j \in \{1, \cdots, N\} \text{ with } Q_{ij} = \begin{cases} 0 & i+j > N \\ (-1)^{i-1} P_k & i+j-1 = k \end{cases} \tag{4}$$

and

$$R_{ext} = \frac{1}{\sqrt{2}} \begin{bmatrix} Q & -Q \\ I_{nN} & I_{nN} \end{bmatrix}.$$

Note that R_{ext} is invertible since P_N is invertible.

Theorem 1.1 ([7–9]) *Consider the operator*

$$A = \left(P_N \frac{d^N}{d\zeta^N} + \cdots + P_1 \frac{d}{d\zeta} + P_0 \right) \mathcal{H} \tag{5}$$

with domain

$$\mathrm{dom}(A)$$

$$= \{x \in L_{2,\mathcal{H}}((a,b); \mathbb{C}^n) \mid \mathcal{H}x \in H^N((a,b); \mathbb{C}^n), W_B \begin{bmatrix} (\mathcal{H}x)(b) \\ \vdots \\ \frac{d^{N-1}\mathcal{H}x}{d\zeta^{N-1}}(b) \\ (\mathcal{H}x)(a) \\ \vdots \\ \frac{d^{N-1}\mathcal{H}x}{d\zeta^{N-1}}(a) \end{bmatrix} = 0\}, \tag{6}$$

where W_B is a full rank $nN \times 2nN$ matrix. Then the following conditions are equivalent:

(i) A generates a contraction semigroup X.
(ii) W_B satisfies

$$W_B R_{ext}^{-1} \begin{bmatrix} 0 & I_{nN} \\ I_{nN} & 0 \end{bmatrix} \left(W_B R_{ext}^{-1} \right)^* \geq 0. \tag{7}$$

(iii) For all $x \in \mathrm{dom}(A)$, there holds

$$\langle Ax, x \rangle_{\mathcal{H}} + \langle x, Ax \rangle_{\mathcal{H}} \leq 0.$$

Although the characterisation of contraction semigroups is given for the operator associated with the p.d.e. (1) with boundary conditions (2), other properties have only been studied for the case $N = 1$, see, e.g., [9]. Recently, inspired by the findings in [10], for that class the characterisation of exponential stability was found. In order to state this result, we need to consider the following parametrised ordinary differential equation:

$$\frac{dv}{d\zeta}(\zeta) = i\omega P_1^{-1}(\mathcal{H}(\zeta)^{-1} + P_0)v(\zeta) \text{ for } \zeta \in (a,b) \text{ and } \omega \in \mathbb{R}. \tag{8}$$

Let Φ_ω denote its fundamental solution,[1] i.e.,

$$\Phi_\omega \colon [a,b] \to \mathbb{C}^{n \times n} \text{ with } \Phi_\omega(a) = I_n, \tag{9}$$

---

[1] In Appendix A we show that this exists and is uniquely determined, even when the right-hand side of (8) is not continuous.

and satisfy (8). The characterisation of exponential stability now reads as follows.

Theorem 1.2 ([12]) *Let $(A, \operatorname{dom}(A))$ be given as in Theorem 1.1 for $N = 1$ and be such that A generates a contraction semigroup on X. Furthermore, assume that*

$$\sup_{\omega \in \mathbb{R}} \|\Phi_\omega\|_\infty = \sup_{\omega \in \mathbb{R}} \sup_{\zeta \in (a,b)} \|\Phi_\omega(\zeta)\| < \infty. \tag{10}$$

Then the following conditions are equivalent:

(i) $(T(t))_{t \geq 0}$ *is exponentially stable.*

(ii) $Q_\omega := W_B \begin{bmatrix} \Phi_\omega(b) \\ I_n \end{bmatrix}$ *is invertible and* $\sup_{\omega \in \mathbb{R}} \|Q_\omega^{-1}\| < \infty.$

Inspired by this theorem, we characterise the semi-uniform/asymptotic stability of the contraction semigroup. For this, we invoke the celebrated Batty–Duyckaerts theorem, [1]. This result can be formulated for the general class, and so we do not assume that $N = 1$. We introduce the following extension of the differential equation, see also (8):

$$\frac{dv}{d\zeta}(\zeta) = \mathcal{P}_\lambda(\zeta)v(\zeta) \text{ for } \zeta \in (a, b) \text{ and } \lambda \in \mathbb{C}, \tag{11}$$

where v takes its values in $\mathbb{C}^{nN}$ and

$$\mathcal{P}_\lambda(\zeta) = \begin{bmatrix} 0 & I & 0 \cdots & 0 \\ 0 & 0 & I \ 0 & \vdots \\ \vdots & & \ddots \ \ddots & 0 \\ 0 & \cdots & \cdots \ 0 & I \\ \lambda P_N^{-1} \mathcal{H}(\zeta)^{-1} - P_N^{-1} P_0 & -P_N^{-1} P_1 & \cdots \cdots & -P_N^{-1} P_{N-1} \end{bmatrix}. \tag{12}$$

Let Ψ_λ denote its fundamental solution,[2] i.e.,

$$\Psi_\lambda : [a, b] \to \mathbb{C}^{nN \times nN} \text{ with } \Psi_\lambda(a) = I_{nN}, \tag{13}$$

and satisfy (11). Note that for $N = 1$ we have that $\Psi_{i\omega} = \Phi_\omega$, see (8) and (9).

Theorem 1.3 *Assume that $(A, \operatorname{dom}(A))$ as given in (5) and (6) generates a contraction semigroup $(T(t))_{t \geq 0}$ on X. Then the following conditions are equivalent:*

(i) $(T(t))_{t \geq 0}$ *is semi-uniformly stable (with decay rate $f : [0, \infty) \to [0, \infty)$), i.e.,* $f(t) \to 0$ *as $t \to \infty$ and for all $x \in \operatorname{dom}(A)$*

---

[2] In Appendix A we show that this exists, even when the right-hand side of (11) is not continuous.

$$\|T(t)x\|_{\mathcal{H}} \le f(t)\|x\|_{\mathrm{dom}(A)} \quad t \ge 0.$$

(ii) $(T(t))_{t\ge 0}$ *is asymptotically stable, i.e., for all* $x \in X$, *there holds* $T(t)x \to 0$
   as $t \to \infty$;

(iii) *The operator* A *with domain* $\mathrm{dom}(A)$ *has no eigenvalues on the imaginary
   axis.*

(iv) *The square matrix* $Q(i\omega) := W_B \begin{bmatrix} \Psi_{i\omega}(b) \\ I_{nN} \end{bmatrix}$ *is invertible for all* $\omega \in \mathbb{R}$.

(v) *For all* $\omega \in \mathbb{R}$, *the set*

$$\{v_a \in \mathbb{C}^{nN} \mid Q(i\omega)v_a = 0 \text{ and } v_a^* \left[\, \Psi_{i\omega}(b)^* \; I_{nN} \, \right] \begin{bmatrix} Q & 0 \\ 0 & -Q \end{bmatrix} \begin{bmatrix} \Psi_{i\omega}(b) \\ I_{nN} \end{bmatrix} v_a = 0\}$$

contains only the zero element.

The proof of this theorem will be given in the following section. Furthermore,
we also adopt the insights drawn from this equivalence to shed some more
light on the characterisation of exponential stability from [12]. We exemplify our
results in Sects. 4 and 5. We provide a conclusion afterwards. This manuscript
is supplemented with an appendix gathering known material for non-autonomous
o.d.e.s with bounded and measurable coefficients.

2 Characterisation of Asymptotic/Semi-uniform Stability

We recall the celebrated Batty–Duyckaerts result on semi-uniform stability.

Theorem 2.1 (Batty–Duyckaerts, [1]; see, [2, Theorem 3.4]) *Let* $(T(t))_{t\ge 0}$ *be a
bounded* C_0-*semigroup on a Banach space* X, *with generator* A, $\mu \in \rho(A)$. *Then
the following conditions are equivalent:*

(i) $\sigma(A) \cap i\mathbb{R} = \emptyset$.

(ii) $\lim_{t\to\infty} \|T(t)(\mu - A)^{-1}\| = 0$.

(iii) *There exists a function* $f : [0,\infty) \to [0,\infty)$ *with* $f(t) \to 0$ *as* $t \to \infty$ *and
   such that*

$$\forall x_0 \in \mathrm{dom}(A) : \|T(t)x\| \le f(t)\|x_0\|_{\mathrm{dom}(A)},$$

i.e., T *is* semi-uniformly stable (with decay rate f).

In order to apply the Batty–Duyckaerts result, we provide some (standard)
observations.

Theorem 2.2 *The operator* A *with domain* $\mathrm{dom}(A)$ *as given by (5) and (6) has
compact resolvent on the energy space* X.

Proof The assertion for $\mathcal{H} = I_n$ follows upon using the Arzelà–Ascoli Theorem since $\mathrm{dom}(A)$ embeds continuously into $H^N((a, b); \mathbb{C}^n)$. The general case then follows since X is isomorphic to $L^2((a, b); \mathbb{C}^n)$. $\qquad\square$

Lemma 2.3 *Let $\mathcal{B}$ be a closed, densely defined, linear operator on a Hilbert space X, with $\rho(\mathcal{B}) \neq \emptyset$. If $\mathrm{dom}(\mathcal{B}) \hookrightarrow X$ is compact, then*

$$\sigma(\mathcal{B}) = \{\lambda \in \mathbb{C} \mid 0 < \dim(\ker(\lambda I - \mathcal{B})) < \infty\} =: \sigma_p(\mathcal{B}).$$

Proof For the proof we invoke standard theory of (unbounded) Fredholm operators, see, e.g., [5, Theorem 3.6] for an overview and suitable references. Let $\mu \in \rho(\mathcal{B})$ and $\lambda \in \sigma(\mathcal{B})$. Since $\mu I - \mathcal{B}$ is continuously invertible, $\mu I - \mathcal{B}$ is a Fredholm operator. The compactness of $\mathrm{dom}(\mathcal{B}) \hookrightarrow X$ yields that $(\mu I - \mathcal{B})^{-1}$ is compact and, hence, $(\lambda - \mu)I$ is a relatively compact perturbation of $\mu I - \mathcal{B}$ and so $\lambda I - \mu I + (\mu I - \mathcal{B}) = \lambda I - \mathcal{B}$ is Fredholm and $\mathrm{ind}(\mu I - \mathcal{B}) = \mathrm{ind}(\lambda I - \mathcal{B})$. In particular, $\dim(\ker(\lambda I - \mathcal{B})) < \infty$. Since $\lambda \in \sigma(\mathcal{B})$, $\lambda I - \mathcal{B}$ fails to be continuously invertible, which can, thus, only happen if $\ker(\lambda I - \mathcal{B}) \neq \{0\}$, whence $\dim(\ker(\lambda I - \mathcal{B})) > 0$. $\qquad\square$

The application of Theorem 2.1 to port-Hamiltonian systems reads as follows.

Theorem 2.4 *Assume $(A, \mathrm{dom}(A))$ as in Theorem 1.1 generates a contraction semigroup $(T(t))_{t\geq 0}$. Then the following conditions are equivalent:*

(i) $(T(t))_{t\geq 0}$ is semi-uniformly stable.
(ii) $\sigma_p(A) \cap i\mathbb{R} = \emptyset$.

Proof The result is a straightforward consequence of Theorem 2.1 in conjunction with the observations in Theorem 2.2 and Lemma 2.3 to determine the type of spectrum A can possibly have. $\qquad\square$

From this result we see that the eigenvalues of A will play an essential role. Therefore, we characterise these in the following lemma.

Lemma 2.5 *Let $(A, \mathrm{dom}(A))$ be given by (5) and (6), then λ is an eigenvalue if and only if $\mathcal{Q}(\lambda) := W_B \begin{bmatrix} \Psi_\lambda(b) \\ I_{nN} \end{bmatrix}$ is not invertible, see (13).*

Proof Let $\lambda \in \mathbb{C}$ be an eigenvalue of $(A, \mathrm{dom}(A))$, i.e., $Ax = \lambda x$ for $0 \neq x \in \mathrm{dom}(A)$. Using (5), we have that

$$\lambda x = Ax \Rightarrow \lambda x(\zeta) = \left(P_N \frac{d^N}{d\zeta^N} + \cdots + P_1 \frac{d}{d\zeta} + P_0 \right) (\mathcal{H}x)(\zeta).$$

Introducing $v_1(\zeta) := (\mathcal{H}x)(\zeta)$, we can rewrite the above o.d.e. as

$$\left(P_N \frac{d^N}{d\zeta^N} + \cdots + P_1 \frac{d}{d\zeta} + P_0 \right) v_1(\zeta) - \lambda \mathcal{H}^{-1}(\zeta)v_1(\zeta) = 0. \tag{14}$$

Next we introduce

$$v(\zeta) = \begin{bmatrix} v_1(\zeta) \\ \vdots \\ \frac{d^{N-1}v_1}{d\zeta^{N-1}}(\zeta) \end{bmatrix}$$

and so with $\mathcal{P}_\lambda$ as defined in (12) we can formulate the Nth order o.d.e. (14) as the first-order o.d.e.

$$\frac{dv}{d\zeta}(\zeta) = \mathcal{P}_\lambda(\zeta)v(\zeta),$$

and thus $v(\zeta) = \Psi_\lambda(\zeta)v(a)$.

Since x was an eigenvector, it is an element of $\mathrm{dom}(A)$, and thus $v_1 = \mathcal{H}x$ satisfies the boundary conditions as given in (6). We can equivalently formulate these conditions by using v. This gives

$$W_B \begin{bmatrix} v(b) \\ v(a) \end{bmatrix} = 0 \Leftrightarrow W_B \begin{bmatrix} \Psi_\lambda(b) \\ I_{nN} \end{bmatrix} v(a) = 0.$$

If $v(a)$ would be zero, then the function v would be zero and so x. This is not possible since x is an eigenvector. So for the above to hold, the square matrix $W_B \begin{bmatrix} \Psi_\lambda(b) \\ I_{nN} \end{bmatrix}$ must be singular.

Similarly, when $W_B \begin{bmatrix} \Psi_\lambda(b) \\ I_{nN} \end{bmatrix}$ is singular, then we choose $v(a)$ in its kernel, and we define $v(\zeta) := \Psi_\lambda(\zeta)v(a)$. It is straightforward to see that $x := \mathcal{H}^{-1}v_1$ is in the domain of A and satisfies $Ax = \lambda x$. $\qquad\square$

Proof of Theorem 1.3

In this section we prove Theorem 1.3.

- **(i)** $\Rightarrow$ **(ii)** Let $x_0 \in X$, and choose $\varepsilon > 0$. Since the domain of A lies dense in X, we can find an $x_1 \in \mathrm{dom}(A)$ such that $\|x_1 - x_0\| \leq \varepsilon$. Next choose $t_1 > 0$ such that $\|T(t_1)x_1\| \leq \varepsilon$. Now

$$\|T(t_1)x_0\| \leq \|T(t_1)(x_0 - x_1)\| + \|T(t_1)x_1\| \leq \|x_0 - x_1\| + \|T(t_1)x_1\| \leq 2\varepsilon,$$

where we have used that $(T(t))_{t\geq0}$ is a contraction semigroup. Using this once more, we find for $t \geq t_1$

$$\|T(t)x_0\| = \|T(t - t_1)T(t_1)x_0\| \leq \|T(t_1)x_0\| \leq 2\varepsilon,$$

which shows that the semigroup is asymptotically stable.

- **(ii)** $\Rightarrow$ **(iii)** If λ is an eigenvalue on the imaginary axis, then for the corresponding eigenvector x_0 there holds $T(t)x_0 = e^{\lambda t}x_0$. This will not converge to zero as $t \to \infty$. Hence since not (iii) implies not (ii), we have shown the implication form (ii) to (iii).
- **(iii)** $\Rightarrow$ **(i)** See Theorem 2.4.
- **(iii)** $\Leftrightarrow$ **(iv)** It follows from Lemma 2.5.
- **(iv)** $\Leftrightarrow$ **(v)** It is clear that part (iv) implies part (v). So we concentrate on the other direction. We do this by showing that if $v_a \in \mathbb{C}^{nN}$ is such that $\mathcal{Q}(i\omega)v_a = 0$, then it also satisfies the second equality in item (v).

Let $0 \neq v_a \in \mathbb{C}^{nN}$ be such that $\mathcal{Q}(i\omega)v_a = 0$. We define $v(\zeta)$ as the solution of (11) with initial condition $v(a) = v_a$. Then it is easy to see that v can be written as

$$v(\zeta) = \begin{bmatrix} v_1(\zeta) \\ \vdots \\ \frac{d^{N-1}v_1}{d\zeta^{N-1}}(\zeta) \end{bmatrix}$$

with v_1 satisfying (14) for $\lambda = i\omega$. Since $\mathcal{H}$ is symmetric, we find that

$$0 = v_1(\zeta)^* \left(i\omega\mathcal{H}(\zeta)^{-1}v_1(\zeta)\right) + \left(i\omega\mathcal{H}(\zeta)^{-1}v_1(\zeta)\right)^* v_1(\zeta)$$

$$= v_1(\zeta)^* \left(\left(P_N\frac{d^N}{d\zeta^N} + \cdots + P_1\frac{d}{d\zeta} + P_0\right)v_1(\zeta)\right) +$$

$$\left(\left(P_N\frac{d^N}{d\zeta^N} + \cdots + P_1\frac{d}{d\zeta} + P_0\right)v_1(\zeta)\right)^* v_1(\zeta).$$

Integrating this expression from $\zeta = a$ till b and using [7, Lemma 3.1], we find that

$$0 = v(b)^* Qv(b) - v(a)^* Qv(a).$$

Since $v(a) = v_a$ and $v(b) = \Psi_{i\omega}(b)v_a$, this is the second equality in part (v).

3 Exponential Stability Revisited

Before we turn to our example, we shortly revisit exponential stability and reformulate the conditions from [12]. We start off with an elementary observation on families of invertible matrices.

Proposition 3.1 *Let $Q: \mathbb{R} \to \mathbb{C}^{n \times n}$ be bounded and continuous. Assume that for all $t \in \mathbb{R}$, $Q(t)$ is invertible. Then the following conditions are equivalent:*

(i) $\sup_{t \in \mathbb{R}} \| Q(t)^{-1} \| < \infty$.
(ii) *For all $v \in \mathbb{C}^n$, $v \neq 0$,* $\liminf_{t \to \pm\infty} \| Q(t)v \| > 0$.

Proof (i)$\Rightarrow$(ii): Let $M := \sup_{t \in \mathbb{R}} \| Q(t)^{-1} \|$ and $v \in \mathbb{C}^n$, $v \neq 0$. Then, for all $t \in \mathbb{R}$,

$$\| v \| = \| Q(t)^{-1} Q(t)v \| \leq M \| Q(t)v \|.$$

Taking the lim inf on both sides of the inequality yields the assertion.
(ii)$\Rightarrow$(i): Consider

$$B := \overline{\{ Q(t); t \in \mathbb{R} \}} \subseteq \mathbb{C}^{n \times n}.$$

By the boundedness of Q, B is bounded and closed, hence compact. Next, we show that for all $Q_0 \in B$, $|\det Q_0| > 0$. For this let $Q_0 \in B$. Then there exists a sequence $(t_k)_{k \in \mathbb{N}}$ such that $Q(t_k) \to Q_0$ as $k \to \infty$. If $(t_k)_{k \in \mathbb{N}}$ accumulates at some $s \in \mathbb{R}$, we find a subsequence (not relabelled) such that $\lim_{k \to \infty} t_k = s$. By continuity, we infer $Q_0 = Q(s)$, which is invertible by assumption. Hence, $\det Q_0 \neq 0$. If $(t_k)_{k \in \mathbb{N}}$ accumulates at $-\infty$ or ∞, we may assume without restriction (the other case is similar) that we find a subsequence (again not relabelled) such that $\lim_{k \to \infty} t_k = \infty$. Let $v \in \mathbb{C}^n$, $v \neq 0$. We infer, using (ii),

$$\| Q_0 v \| = \lim_{k \to \infty} \| Q(t_k)v \| \geq \liminf_{t \to \pm\infty} \| Q(t)v \| > 0.$$

Hence, 0 cannot be an eigenvalue of Q_0, that is, $\det Q_0 \neq 0$. By compactness of B and continuity of $\det$, we thus obtain that $|\det R| \geq \varepsilon > 0$ for all $R \in B$ and some $\varepsilon > 0$. Using the boundedness of Q and Cramer's rule, the condition in (i) follows. $\square$

In the light of the characterisation of semi-uniform stability in Theorem 1.3, we shall revisit Theorem 1.2. For this, recall $\Psi_{i\omega}$ from (13) and $Q(i\omega)$ from Theorem 1.3. Note that by Proposition A.3, the mapping $\omega \mapsto \Psi_{i\omega}$ and thus $\omega \mapsto Q(i\omega)$ are continuous.

Theorem 3.2 *Assume that $(A, \mathrm{dom}(A))$ as given in (5) and (6) generates a contraction semigroup $(T(t))_{t \geq 0}$ on X with $N = 1$. Assume that $\sup_{\omega \in \mathbb{R}} \| \Psi_{i\omega} \| < \infty$.*

Then the following conditions are equivalent:

(i) $(T(t))_{t\geq 0}$ *is exponentially stable.*

(ii) For all sequences $(\omega_k)_{k\in\mathbb{N}}$ in $\mathbb{R}$, the set

$$\{v_a \in \mathbb{C}^n \mid \liminf_{k\to\infty}\left(\|\mathcal{Q}(i\omega_k)v_a\| + \|v_a^*\,[\,\Psi_{i\omega_k}(b)^*\ I_n\,]\begin{bmatrix} \mathcal{Q} & 0 \\ 0 & -\mathcal{Q}\end{bmatrix}\begin{bmatrix}\Psi_{i\omega_k}(b) \\ I_n\end{bmatrix}v_a\|\right)$$

$$= 0\}$$

contains only the zero element.

(iii) For all sequences $(\omega_k)_{k\in\mathbb{N}}$ in $\mathbb{R}$, the set

$$\{v_a \in \mathbb{C}^n \mid \liminf_{k\to\infty}\|\mathcal{Q}(i\omega_k)v_a\| = 0\}$$

contains only the zero element.

Proof Note that, by Theorem 1.2, we see that $(T(t))_{t\geq 0}$ is exponentially stable if and only if for all $\omega \in \mathbb{R}$, $\mathcal{Q}(i\omega)$ is invertible and $\sup_{\omega\in\mathbb{R}}\|\mathcal{Q}(i\omega)^{-1}\| < \infty$. Since $\Psi_i. : \mathbb{R} \to \mathbb{C}^{n\times n}$ is bounded by assumption, it follows from its definition that the function $\mathcal{Q}$ is bounded as well, see Theorem 1.3(iv). Next we turn to the proof of the actual theorem.

(i)$\Rightarrow$(ii). Note that by our observation from the beginning $\mathcal{Q}(i\omega)$ is invertible for all $\omega \in \mathbb{R}$. Hence, by Theorem 1.3,

$$\{v_a \in \mathbb{C}^n \mid \|\mathcal{Q}(i\omega)v_a\| + \|v_a^*\,[\,\Psi_{i\omega}(b)^*\ I_n\,]\begin{bmatrix} \mathcal{Q} & 0 \\ 0 & -\mathcal{Q}\end{bmatrix}\begin{bmatrix}\Psi_{i\omega}(b) \\ I_n\end{bmatrix}v_a\| = 0\} = \{0\}$$

for all $\omega \in \mathbb{R}$.

Now, let $(\omega_k)_{k\in\mathbb{N}}$ be a bounded sequence in $\mathbb{R}$ and $v_a \in \mathbb{R}^n$. Then

$$\liminf_{k\to\infty}\left(\|\mathcal{Q}(i\omega_k)v_a\| + \|v_a^*\,[\,\Psi_{i\omega_k}(b)^*\ I_n\,]\begin{bmatrix} \mathcal{Q} & 0 \\ 0 & -\mathcal{Q}\end{bmatrix}\begin{bmatrix}\Psi_{i\omega_k}(b) \\ I_n\end{bmatrix}v_a\|\right) = 0$$

is equivalent to

$$\|\mathcal{Q}(i\omega)v_a\| + \|v_a^*\,[\,\Psi_{i\omega}(b)^*\ I_n\,]\begin{bmatrix} \mathcal{Q} & 0 \\ 0 & -\mathcal{Q}\end{bmatrix}\begin{bmatrix}\Psi_{i\omega}(b) \\ I_n\end{bmatrix}v_a\| = 0$$

for one accumulation value ω of $(\omega_k)_{k\in\mathbb{N}}$. Since the latter implies $v_a = 0$, we obtain condition (ii) for all bounded sequences. If $(\omega_k)_{k\in\mathbb{N}}$ is unbounded, we may assume without loss of generality that $(\omega_k)_{k\in\mathbb{N}} \to \infty$ (any bounded subsequence has been dealt with already, and the case $(\omega_k)_{k\in\mathbb{N}} \to -\infty$ is similar). Since by exponential stability, we get $\sup_{\omega\in\mathbb{R}}\|\mathcal{Q}(i\omega)^{-1}\| < \infty$, Proposition 3.1 yields for every non-zero v_a that $\liminf_{k\to\infty}\|\mathcal{Q}(i\omega)v_a\| > 0$ and so condition (ii) holds for all sequences $(\omega_k)_{k\in\mathbb{N}}$.

(ii)⇒(iii). Let v_a be such that $\liminf_{k\to\infty}\|\mathcal{Q}(i\omega)v_a\| = 0$ for some sequence $(\omega_k)_{k\in\mathbb{N}}$. Then by the reformulation as in the proof of Theorem 1.3 (iv)⇔(v), we deduce that

$$\liminf_{k\to\infty}\left(\|\mathcal{Q}(i\omega_k)v_a\| + \|v_a^*\,[\,\Psi_{i\omega_k}(b)^*\ I_n\,]\begin{bmatrix} Q & 0 \\ 0 & -Q \end{bmatrix}\begin{bmatrix} \Psi_{i\omega_k}(b) \\ I_n \end{bmatrix}v_a\|\right) = 0.$$

For this particularly note that the reformulation in the proof of Theorem 1.3 (iv)⇔(v) yields that given a sequence $(\omega_k)_{k\in\mathbb{N}}$ with $\lim_{k\to\infty}\|\mathcal{Q}(i\omega_k)v_a\| = 0$ we obtain $\lim_{k\to\infty}\|v_a^*\,[\,\Psi_{i\omega_k}(b)^*\ I_n\,]\begin{bmatrix} Q & 0 \\ 0 & -Q \end{bmatrix}\begin{bmatrix} \Psi_{i\omega_k}(b) \\ I_n \end{bmatrix}v_a\| = 0$ for the *same* sequence. Hence, $v_a = 0$, by (ii).

(iii)⇒(i). Considering constant sequences, we deduce that (ii) implies Theorem 1.3 (iv). Thus, $\mathcal{Q}(i\omega)$ is invertible for all $\omega \in \mathbb{R}$. Finally, by Proposition 3.1, we deduce that $\sup_{\omega\in\mathbb{R}}\|\mathcal{Q}(i\omega)^{-1}\| < \infty$. $\qquad\square$

4 Example: Network of Vibrating Strings

In this section we study in detail the stability of a (small) network of vibrating strings. The network consists of three interconnected (undamped) vibrating strings, which are connected to a damper, see also Fig. 1.

The model of every string is described using the state variables $\rho_k\frac{\partial w_k}{\partial t}$ (the momentum) and $\frac{\partial w_k}{\partial\zeta}$ (the strain) and is given by

$$\frac{\partial x_k}{\partial t}(\zeta, t) = \frac{\partial}{\partial t}\begin{bmatrix} \rho_k\frac{\partial w_k}{\partial t} \\ \frac{\partial w_k}{\partial\zeta} \end{bmatrix}(\zeta, t) = \begin{bmatrix} 0 & 1 \\ 1 & 0 \end{bmatrix}\frac{\partial}{\partial\zeta}\left(\begin{bmatrix} \frac{1}{\rho_k(\zeta)} & 0 \\ 0 & T_k(\zeta) \end{bmatrix}x_k(\zeta, t)\right) \qquad (15)$$

$$= P_{1,k}(\mathcal{H}_k x_k)(\zeta, t), \qquad k \in \{\mathrm{I}, \mathrm{II}, \mathrm{III}\}. \qquad (16)$$

In string I, we let the ζ run from left to right, but in strings II and III, we let it go in the opposite direction. We have the following six boundary conditions, see also Fig. 1:

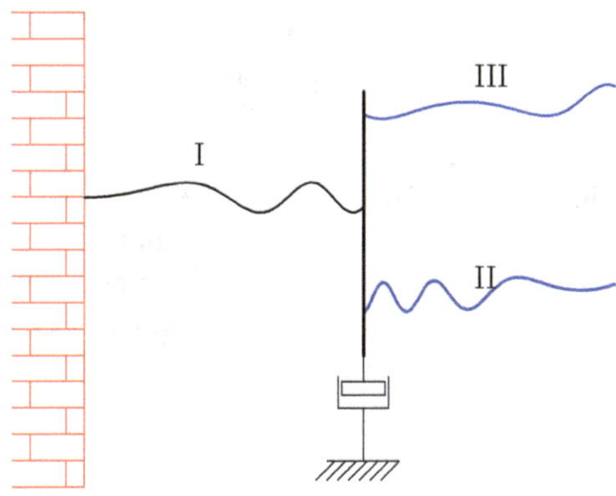

Fig. 1 Coupled vibrating strings with damping

$$\frac{\partial w_{\mathrm{I}}}{\partial t}(a, t) = 0 \tag{17}$$

$$T_{\mathrm{II}}(a)\frac{\partial w_{\mathrm{II}}}{\partial \zeta}(a, t) = T_{\mathrm{III}}(a)\frac{\partial w_{\mathrm{III}}}{\partial \zeta}(a, t) = 0 \tag{18}$$

$$\frac{\partial w_{\mathrm{I}}}{\partial t}(b, t) = \frac{\partial w_{\mathrm{II}}}{\partial t}(b, t) = \frac{\partial w_{\mathrm{III}}}{\partial t}(b, t) \tag{19}$$

$$-\beta\frac{\partial w_{\mathrm{I}}}{\partial t}(b, t) = T_{\mathrm{I}}(b)\frac{\partial w_{\mathrm{I}}}{\partial \zeta}(b, t) + T_{\mathrm{II}}(b)\frac{\partial w_{\mathrm{II}}}{\partial \zeta}(b, t) + T_{\mathrm{III}}(b)\frac{\partial w_{\mathrm{III}}}{\partial \zeta}(b, t), \tag{20}$$

where $\beta > 0$ denotes the damping coefficient of the damper.

Formulating the system as (1) is done by putting the three states into one six-dimensional vector. For this system the P_1 and $\mathcal{H}$ are given as the block diagonal matrices with the three $P_{1,k}$'s and $\mathcal{H}_k$'s on the diagonal, respectively, that is,

$$P_1 = \begin{bmatrix} P_{1,\mathrm{I}} & & \\ & P_{1,\mathrm{II}} & \\ & & P_{1,\mathrm{III}} \end{bmatrix} \text{ and } \mathcal{H} = \begin{bmatrix} \mathcal{H}_{\mathrm{I}} & & \\ & \mathcal{H}_{\mathrm{II}} & \\ & & \mathcal{H}_{\mathrm{III}} \end{bmatrix}. \tag{21}$$

We investigate the asymptotic stability of the above model by employing part (v) of Theorem 1.3. For this we need the solution of the differential equation (11)–(12). Using that $N = 1$ and the special form of $\mathcal{H}$ and P_1, this can be written as three differential equations on $[a, b]$ as

$$\frac{dv_k}{d\zeta}(\zeta) = i\omega \begin{bmatrix} 0 & T_k(\zeta)^{-1} \\ \rho_k(\zeta) & 0 \end{bmatrix} v_k(\zeta), \text{ with } v_k(a) = v_{a,k}, \quad k \in \{\mathrm{I, II, III}\}. \tag{22}$$

The boundary conditions of (17) and (18) give that

$$v_{a,\mathrm{I}} = \begin{bmatrix} 0 \\ T_{\mathrm{I}}(a)\frac{\partial w_{\mathrm{I}}}{\partial \zeta}(a) \end{bmatrix} := \begin{bmatrix} 0 \\ F_{\mathrm{I}}(a) \end{bmatrix}, \quad v_{a,k} = \begin{bmatrix} \frac{\partial w_k}{\partial t}(a) \\ 0 \end{bmatrix} := \begin{bmatrix} v_{a,k} \\ 0 \end{bmatrix}, \quad k \in \{\mathrm{II, III}\}. \tag{23}$$

The other three boundary conditions imply that

$$v_k(b) = \begin{bmatrix} \frac{\partial w_k}{\partial t}(b) \\ T_k(b)\frac{\partial w_k}{\partial \zeta}(b) \end{bmatrix} := \begin{bmatrix} v_b \\ F_k(b) \end{bmatrix}, \quad k \in \{\mathrm{I, II, III}\}, \tag{24}$$

with $-\beta v_b = F_{\mathrm{I}}(b) + F_{\mathrm{II}}(b) + F_{\mathrm{III}}(b)$.

Having provided the reformulation of (11)–(12), for convenience of the reader, we provide the consequence of our main theorem in this particular situation. For this, we quickly check the generation property of the port-Hamiltonian operator.

Theorem 4.1 *For $k \in \{\mathrm{I}, \mathrm{II}, \mathrm{III}\}$, let $T_k, \rho_k \in L_\infty(a, b)$, and assume there exists $\varepsilon_0 > 0$ such that for a.e. $\zeta \in (a, b)$, $T_k(\zeta), \rho_k(\zeta) \geq \varepsilon_0$. With (21) in $L_{2,\mathcal{H}}((a, b); \mathbb{C}^6)$, consider the operator*

$$A = P_1 \frac{d}{d\zeta} \mathcal{H}$$

with domain

$$\mathrm{dom}(A) = \{x \in L_{2,\mathcal{H}}((a, b); \mathbb{C}^6) \mid \mathcal{H}x \in H^1((a, b); \mathbb{C}^6), \; W_B \begin{pmatrix} (\mathcal{H}x)(b) \\ (\mathcal{H}x)(a) \end{pmatrix} = 0\},$$

where W_B is given by (17)–(20), that is,

$$W_B = \left(\begin{array}{cccccc|cccccc} 1 & 0 & -1 & 0 & 0 & 0 & 0 & 0 & 0 & 0 & 0 & 0 \\ 1 & 0 & 0 & 0 & -1 & 0 & 0 & 0 & 0 & 0 & 0 & 0 \\ \beta & 1 & 0 & 1 & 0 & 1 & 0 & 0 & 0 & 0 & 0 & 0 \\ \hline 0 & 0 & 0 & 0 & 0 & 0 & 1 & 0 & 0 & 0 & 0 & 0 \\ 0 & 0 & 0 & 0 & 0 & 0 & 0 & 0 & 0 & 1 & 0 & 0 \\ 0 & 0 & 0 & 0 & 0 & 0 & 0 & 0 & 0 & 0 & 0 & 1 \end{array}\right).$$

Then $(A, \mathrm{dom}(A))$ is the generator of contraction semigroup.

Proof We apply Theorem 1.1. For this, it is easy to see that W_B has full rank. Hence, using $Q = P_1$ and $R_{ext} = \frac{1}{\sqrt{2}} \begin{bmatrix} P_1 & -P_1 \\ I_6 & I_6 \end{bmatrix}$, we need to establish inequality (7). For this, using $P_1^2 = I_6$ and

$$R_{ext}^{-1} = \frac{1}{\sqrt{2}} \begin{bmatrix} P_1 & I_6 \\ -P_1 & I_6 \end{bmatrix},$$

we compute

$$R_{ext}^{-1} \begin{bmatrix} 0 & I_6 \\ I_6 & 0 \end{bmatrix} R_{ext}^{-*} = \begin{bmatrix} P_1 & 0 \\ 0 & -P_1 \end{bmatrix};$$

thus

$$W_B R_{ext}^{-1} \begin{bmatrix} 0 & I_6 \\ I_6 & 0 \end{bmatrix} (W_B R_{ext}^{-1})^* = W_B \begin{bmatrix} P_1 & 0 \\ 0 & -P_1 \end{bmatrix} W_B^*$$

$$
= \left[\left[\begin{bmatrix} 1 & 0 \\ 1 & 0 \end{bmatrix} \begin{bmatrix} -1 & 0 \\ 0 & 0 \end{bmatrix} \begin{bmatrix} 0 & 0 \\ -1 & 0 \end{bmatrix} \right] \left[\begin{bmatrix} 0 & 0 \\ 0 & 0 \end{bmatrix} \begin{bmatrix} 0 & 0 \\ 0 & 0 \end{bmatrix} \begin{bmatrix} 0 & 0 \\ 0 & 0 \end{bmatrix} \right] \right.
$$

$$
\left. \begin{bmatrix} \beta & 1 \\ 0 & 0 \end{bmatrix} \begin{bmatrix} 0 & 1 \\ 0 & 0 \end{bmatrix} \begin{bmatrix} 0 & 1 \\ 0 & 0 \end{bmatrix} \quad \begin{bmatrix} 0 & 0 \\ 1 & 0 \end{bmatrix} \begin{bmatrix} 0 & 0 \\ 0 & 0 \end{bmatrix} \begin{bmatrix} 0 & 0 \\ 0 & 0 \end{bmatrix} \right.
$$

$$
\left. \begin{bmatrix} 0 & 0 \\ 0 & 0 \end{bmatrix} \begin{bmatrix} 0 & 0 \\ 0 & 0 \end{bmatrix} \begin{bmatrix} 0 & 0 \\ 0 & 0 \end{bmatrix} \quad \begin{bmatrix} 0 & 0 \\ 0 & 0 \end{bmatrix} \begin{bmatrix} 0 & 1 \\ 0 & 0 \end{bmatrix} \begin{bmatrix} 0 & 0 \\ 0 & 1 \end{bmatrix} \right] \times
$$

$$
\times \begin{bmatrix} P_1 & 0 \\ 0 & -P_1 \end{bmatrix} W_B^*
$$

$$
= \begin{bmatrix} \begin{bmatrix} 1 & 0 \\ 1 & 0 \end{bmatrix} \begin{bmatrix} -1 & 0 \\ 0 & 0 \end{bmatrix} \begin{bmatrix} 0 & 0 \\ -1 & 0 \end{bmatrix} \\ \begin{bmatrix} \beta & 1 \\ 0 & 0 \end{bmatrix} \begin{bmatrix} 0 & 1 \\ 0 & 0 \end{bmatrix} \begin{bmatrix} 0 & 1 \\ 0 & 0 \end{bmatrix} \\ 0 \qquad 0 \qquad 0 \end{bmatrix} \begin{bmatrix} P_{1,\mathrm{I}} & & \\ & P_{1,\mathrm{II}} & \\ & & P_{1,\mathrm{III}} \end{bmatrix} \begin{bmatrix} \begin{bmatrix} 1 & 1 \\ 0 & 0 \end{bmatrix} \begin{bmatrix} \beta & 0 \\ 1 & 0 \end{bmatrix} & 0 \\ \begin{bmatrix} -1 & 0 \\ 0 & 0 \end{bmatrix} \begin{bmatrix} 0 & 0 \\ 1 & 0 \end{bmatrix} & 0 \\ \begin{bmatrix} 0 & -1 \\ 0 & 0 \end{bmatrix} \begin{bmatrix} 0 & 0 \\ 1 & 0 \end{bmatrix} & 0 \end{bmatrix}
$$

$$
- \begin{bmatrix} 0 & 0 & 0 \\ \begin{bmatrix} 0 & 0 \\ 1 & 0 \end{bmatrix} & 0 & 0 \\ 0 & \begin{bmatrix} 0 & 1 \\ 0 & 0 \end{bmatrix} \begin{bmatrix} 1 & 0 \\ 0 & 0 \end{bmatrix} \end{bmatrix} \begin{bmatrix} P_{1,\mathrm{I}} & & \\ & P_{1,\mathrm{II}} & \\ & & P_{1,\mathrm{III}} \end{bmatrix} \begin{bmatrix} 0 & \begin{bmatrix} 0 & 1 \\ 0 & 0 \end{bmatrix} & 0 \\ 0 & 0 & \begin{bmatrix} 0 & 0 \\ 1 & 0 \end{bmatrix} \\ 0 & 0 & \begin{bmatrix} 1 & 0 \\ 0 & 0 \end{bmatrix} \end{bmatrix}
$$

$$
= \begin{bmatrix} 0 & 0 & 0 & 0 & 0 & 0 \\ 0 & 0 & 0 & 0 & 0 & 0 \\ 0 & 0 & 2\beta & 0 & 0 & 0 \\ 0 & 0 & 0 & 0 & 0 & 0 \\ 0 & 0 & 0 & 0 & 0 & 0 \\ 0 & 0 & 0 & 0 & 0 & 0 \end{bmatrix},
$$

which is non-negative definite as $\beta \geq 0$. Note that we could also have shown this by checking the ineqality in part (iii) of Theorem 1.1. $\qquad \square$

Thus, we are in the position to apply Theorem 1.3. For this, let $\Psi_\lambda : [a, b) \to \mathbb{C}^{6 \times 6}$ with $\Psi_\lambda(a) = I_6$ be the fundamental solution of

$$
\frac{dv}{d\zeta}(\zeta) = \lambda P_1^{-1} \mathcal{H}(\zeta)^{-1} v(\zeta) \quad \text{for } \zeta \in (a, b) \text{ and } \lambda \in \mathbb{C}. \tag{25}
$$

As a result, we deduce the following characterisation of stability.

Theorem 4.2 *Consider $(A, \mathrm{dom}(A))$ as provided in Theorem 4.1, and let $(T(t))_{t \geq 0}$ be the generated contraction semigroup. Then the following conditions are equivalent:*

(i) $(T(t))_{t\geq 0}$ *is semi-uniformly stable.*
(ii) *For all* $\omega \in \mathbb{R}$, *the set*

$$\{v_a \in \mathbb{C}^6$$

$$\left| W_B \begin{bmatrix} \Psi_{i\omega}(b) \\ I_6 \end{bmatrix} v_a = 0 \text{ and } v_a^* [\, \Psi_{i\omega}(b)^* \ I_6 \,] \begin{bmatrix} P_1 & 0 \\ 0 & -P_1 \end{bmatrix} \begin{bmatrix} \Psi_{i\omega}(b) \\ I_6 \end{bmatrix} v_a = 0 \}$$

contains only the zero element.
(iii) *For all* $\omega \in \mathbb{R}$, *the following holds: Let* $v : [a, b] \to \mathbb{C}^6 = (\mathbb{C}^2)^3$ *be a solution*
of (25) with $\lambda = i\omega$. *Denote* $v(a) = \left(\begin{bmatrix} v_{a,k} \\ F_k(a) \end{bmatrix} \right)_{k \in \{I,II,III\}}$ *and similarly for*
$v(b)$. *If*

$$v_{a,I} = v_{b,I}$$

$$= v_{b,II} = v_{b,III} = 0, \ F_{II}(a) = F_{III}(a) = 0, \ F_I(b) + F_{II}(b) + F_{III}(b) = 0,$$

then $v = 0$.

Proof By Theorem 1.3 it suffices to show that the latter two items are equivalent. For this note that any solution of (25) for $\lambda = i\omega$ can be represented by $v(\cdot) = \Psi_{i\omega}(\cdot)v_a$ for some $v_a \in \mathbb{C}^6$ (and vice versa). In this case, $v(a) = v_a$ and $v(b) = \Psi_{i\omega}(b)v_a$. Hence, it follows that the equations in the set of condition (ii) can be equivalently expressed as

$$W_B \begin{bmatrix} v(b) \\ v(a) \end{bmatrix} = 0 \text{ and } v(b)^* P_1 v(b) - v(a)^* P_1 v(a) = 0. \tag{26}$$

To obtain the claim, it thus suffices to show that the equalities in (26) are equivalent to the ones claimed in condition (iii). For this note that the conditions encoded in $W_B \begin{bmatrix} v(b) \\ v(a) \end{bmatrix} = 0$ read

$$v_{a,I} = 0, \quad F_{II}(a) = F_{III}(a) = 0$$

$$v_b := v_{b,I} = v_{b,II} = v_{b,III}$$

$$-\beta v_b = F_I(b) + F_{II}(b) + F_{III}(b).$$

Next, using the form of P_1 and v and the boundary conditions, we reformulate $v(b)^* P_1 v(b) - v(a)^* P_1 v(a) = 0$:

$$v(b)^* P_1 v(b) - v(a)^* P_1 v(a)$$

$$= 2 \sum_{k \in \{I,II,III\}} \left(\begin{bmatrix} v_{b,k} \\ F_k(b) \end{bmatrix}^* \begin{bmatrix} 0 & 1 \\ 1 & 0 \end{bmatrix} \begin{bmatrix} v_{b,k} \\ F_k(b) \end{bmatrix} - \begin{bmatrix} v_{a,k} \\ F_k(a) \end{bmatrix}^* \begin{bmatrix} 0 & 1 \\ 1 & 0 \end{bmatrix} \begin{bmatrix} v_{a,k} \\ F_k(a) \end{bmatrix} \right)$$

$$= 2 \sum_{k\in\{I,II,III\}} \operatorname{Re} v_{b,k}^* F_k(b) - \operatorname{Re} v_{a,k}^* F_k(a) = 2 \sum_{k\in\{I,II,III\}} \operatorname{Re} v_{b,k}^* F_k(b)$$

$$= 2 \operatorname{Re} v_b^* (F_I(b) + F_{II}(b) + F_{III}(b)).$$

Hence, $v(b)^* P_1 v(b) - v(a)^* P_1 v(a) = 0$ is the same as saying

$$\operatorname{Re} v_b^* (F_I(b) + F_{II}(b) + F_{III}(b)) = 0.$$

Combining this with the equality $-\beta v_b = F_I(b) + F_{II}(b) + F_{III}(b)$, we see that this is equivalent to

$$v_b = 0 \text{ and } F_I(b) + F_{II}(b) + F_{III}(b) = 0.$$

Indeed, sufficiency of the latter system for the former being obvious, the necessity of the latter follows upon substituting the first equation of the former into its second equation to obtain $-\beta \operatorname{Re} v_b^* v_b = 0$ leading to $v_b = 0$ as $\beta > 0$ and, hence, $F_I(b) + F_{II}(b) + F_{III}(b) = 0$ by the first equation. $\square$

Remark 4.3 In the situation of Theorem 4.2, assume that there exists $\omega \in \mathbb{R}$ such that in condition (iii), we find $v \neq 0$ satisfying the equations for the boundary values of v. Then $i\omega$ is an eigenvalue of A. Indeed, this follows from Lemma 2.5 together with the equivalence of (iv) and (v) in Theorem 1.3 and the respective proof.

Next, we provide necessary conditions for the validity of condition (iii) in Theorem 4.2. In any case, challenges only arise for $\omega \neq 0$, as the next result confirms.

Lemma 4.4 *Let* $v \colon [a, b] \to \mathbb{C}^6 = (\mathbb{C}^2)^3$ *be a solution of* (25) *with* $\lambda = 0$. *Denote* $v(a) = \left(\begin{bmatrix} v_{a,k} \\ F_k(a) \end{bmatrix} \right)_{k\in\{I,II,III\}}$ *and similarly for* $v(b)$. *If*

$$v_{a,I} = v_{b,I}$$

$$= v_{b,II} = v_{b,III} = 0, \ F_{II}(a) = F_{III}(a) = 0, \ F_I(b) + F_{II}(b) + F_{III}(b) = 0,$$

then $v = 0$.

Proof It is easy to see that for $\omega = 0$ the solution of (22) is constant, i.e.,

$$v_k(\zeta) = \begin{bmatrix} v_k \\ F_k \end{bmatrix}, \quad \zeta \in [a, b], \quad k \in \{I, II, III\}.$$

Since $v_k = v_{a,k} = v_{b,k}$ for all $k \in \{I, II, III\}$, by assumption it follows that $v_k = 0$ for all $k \in \{I, II, III\}$. Similarly, we infer $F_k = 0$ for $k \in \{II, III\}$, and from $F_I(b) + F_{II}(b) + F_{III}(b) = 0$, it thus follows that $F_{III} = 0$, which leads to $v = 0$. $\square$

For the next observation, we recall that, by (21), the matrices P_1 and $\mathcal{H}(\zeta)$ are block diagonal with 2×2-blocks sitting in its diagonal. Thus, the fundamental solution of (25) is, too, block diagonal. Hence, for any $\omega \in \mathbb{R}$, $\Psi_{i\omega}$ can be represented by

$$\Psi_{i\omega,k}(\zeta) = \begin{bmatrix} \Psi^{11}_{i\omega,k}(\zeta) & \Psi^{12}_{i\omega,k}(\zeta) \\ \Psi^{21}_{i\omega,k}(\zeta) & \Psi^{22}_{i\omega,k}(\zeta) \end{bmatrix}, \qquad k \in \{\mathrm{I}, \mathrm{II}, \mathrm{III}\}, \tag{27}$$

each $\Psi_{i\omega,k}$ being the fundamental solution of (22), that is,

$$\frac{dv_k}{d\zeta}(\zeta) = i\omega \begin{bmatrix} 0 & T_k(\zeta)^{-1} \\ \rho_k(\zeta) & 0 \end{bmatrix} v_k(\zeta), \qquad k \in \{\mathrm{I}, \mathrm{II}, \mathrm{III}\}. \tag{28}$$

With the latter in mind, we find a criterion eventually ensuring the validity of condition (iii) in Theorem 4.2.

Proposition 4.5 *Let $\omega \in \mathbb{R}$, and let $v \colon [a, b] \to \mathbb{C}^6$ be a solution of (25) with $\lambda = i\omega$. Denote $v(a) = \left(\begin{bmatrix} v_{a,k} \\ F_k(a) \end{bmatrix} \right)_{k\in\{\mathrm{I},\mathrm{II},\mathrm{III}\}}$ and similarly for $v(b)$, and assume*

$$v_{a,\mathrm{I}} = v_{b,\mathrm{I}}$$

$$= v_{b,\mathrm{II}} = v_{b,\mathrm{III}} = 0, \quad F_{\mathrm{II}}(a) = F_{\mathrm{III}}(a) = 0, \quad F_{\mathrm{I}}(b) + F_{\mathrm{II}}(b) + F_{\mathrm{III}}(b) = 0.$$

If

$$|\Psi^{12}_{i\omega,\mathrm{I}}(b)\Psi^{11}_{i\omega,\mathrm{II}}(b)| + |\Psi^{12}_{i\omega,\mathrm{I}}(b)\Psi^{11}_{i\omega,\mathrm{III}}(b)| + |\Psi^{11}_{i\omega,\mathrm{II}}(b)\Psi^{11}_{i\omega,\mathrm{III}}(b)| > 0, \tag{29}$$

then $v = 0$.

Proof By the mentioned block structure, it follows that we can write

$$v_k(\zeta) = \Psi_{i\omega,k}(\zeta)v_{a,k} = \begin{bmatrix} \Psi^{11}_{i\omega,k}(\zeta) & \Psi^{12}_{i\omega,k}(\zeta) \\ \Psi^{21}_{i\omega,k}(\zeta) & \Psi^{22}_{i\omega,k}(\zeta) \end{bmatrix} \begin{bmatrix} v_{a,k} \\ F_k(a) \end{bmatrix}, \qquad k \in \{\mathrm{I}, \mathrm{II}, \mathrm{III}\}.$$

In particular,

$$\begin{bmatrix} v_{b,k} \\ F_k(b) \end{bmatrix} = \begin{bmatrix} \Psi^{11}_{i\omega,k}(b) & \Psi^{12}_{i\omega,k}(b) \\ \Psi^{21}_{i\omega,k}(b) & \Psi^{22}_{i\omega,k}(b) \end{bmatrix} \begin{bmatrix} v_{a,k} \\ F_k(a) \end{bmatrix}.$$

By assumption, we obtain

$$0 = v_{b,\mathrm{I}} = \Psi^{11}_{i\omega,\mathrm{I}}(b)v_{a,\mathrm{I}} + \Psi^{12}_{i\omega,\mathrm{I}}(b)F_{\mathrm{I}}(a) = \Psi^{12}_{i\omega,\mathrm{I}}(b)F_{\mathrm{I}}(a),$$

$$0 = v_{b,\mathrm{II}} = \Psi^{11}_{i\omega,\mathrm{II}}(b)v_{a,\mathrm{II}} + \Psi^{12}_{i\omega,\mathrm{II}}(b)F_{\mathrm{II}}(a) = \Psi^{11}_{i\omega,\mathrm{II}}(b)v_{a,\mathrm{II}},$$

$$0 = v_{b,\text{III}} = \Psi_{i\omega,\text{III}}^{11}(b)v_{a,\text{III}} + \Psi_{i\omega,\text{II}}^{12}(b)F_{\text{III}}(a) = \Psi_{i\omega,\text{III}}^{11}(b)v_{a,\text{III}}.$$

Using the condition in (29), we obtain one of the following alternatives:

$$v_{a,\text{II}} = v_{a,\text{III}} = 0 \text{ or } F_{\text{I}}(a) = v_{b,\text{II}} = 0 \text{ or } F_{\text{I}}(a) = v_{b,\text{III}} = 0.$$

We consider $v_{a,\text{II}} = v_{a,\text{III}} = 0$ first. For this note that, by assumption, $F_{\text{II}}(a) = F_{\text{III}}(a) = 0$ and, so, $v_k = 0$ for $k \in \{\text{II, III}\}$. In particular, we find $F_{\text{II}}(b) = F_{\text{III}}(b) = 0$. Thus, the assumed equality $F_{\text{I}}(b) + F_{\text{II}}(b) + F_{\text{III}}(b) = 0$ leads to $F_{\text{I}}(b) = 0$ and, in conjunction with $v_{b,\text{I}} = 0$, leads to $v_{\text{I}} = 0$ by uniqueness of the solution of (28) for given data.

Next, we consider $F_{\text{I}}(a) = 0$ and $v_{a,\text{II}} = 0$ or $v_{a,\text{III}} = 0$. It suffices to consider the first alternative as the other case follows a similar rationale. In any case, $F_{\text{I}}(a) = 0$ together with the assumed $v_{a,\text{I}} = 0$ implies $v_{\text{I}} = 0$. The condition $v_{a,\text{II}} = 0$ together with $F_{\text{II}}(a) = 0$ leads to $v_{\text{II}} = 0$. As a consequence, we find $F_{\text{I}}(b) = F_{\text{II}}(b) = 0$. Hence, by assumption, $F_{\text{III}}(b) = 0$, which together with $v_{b,\text{III}} = 0$ leads to $v_{\text{III}} = 0$ again by a uniqueness argument. $\qquad\square$

We now have a criterion at hand that characterises semi-uniform (or, equivalently, asymptotic) stability for the semigroup $(T(t))_{t\geq0}$ in question:

Theorem 4.6 *For $\omega \in \mathbb{R}$, let $\Psi_{i\omega,k}$ be as in (27), $k \in \{\text{I, II, III}\}$. If for all $\omega \in \mathbb{R} \setminus \{0\}$, we have*

$$|\Psi_{i\omega,\text{I}}^{12}(b)\Psi_{i\omega,\text{II}}^{11}(b)| + |\Psi_{i\omega,\text{I}}^{12}(b)\Psi_{i\omega,\text{III}}^{11}(b)| + |\Psi_{i\omega,\text{II}}^{11}(b)\Psi_{i\omega,\text{III}}^{11}(b)| > 0,$$

then the semigroup $(T(t))_{t\geq0}$ given in Theorem 4.1 is semi-uniformly stable.
If, on the other hand,

$$|\Psi_{i\omega,\text{I}}^{12}(b)\Psi_{i\omega,\text{II}}^{11}(b)| + |\Psi_{i\omega,\text{I}}^{12}(b)\Psi_{i\omega,\text{III}}^{11}(b)| + |\Psi_{i\omega,\text{II}}^{11}(b)\Psi_{i\omega,\text{III}}^{11}(b)| = 0$$

for one $\omega \in \mathbb{R}$, then $i\omega$ is an eigenvalue of $(A, \text{dom}(A))$ given in Theorem 4.1 and $(T(t))_{t\geq0}$ is not semi-uniformly stable.

Proof The result follows upon application of Theorem 4.2. In the first case, by Lemma 4.4 (for $\omega = 0$) and Proposition 4.5 (for $\omega \neq 0$), condition (iii) of Theorem 4.2 is satisfied and, therefore, $(T(t))_{t\geq0}$ is semi-uniformly stable.
Conversely, assume

$$|\Psi_{i\omega,\text{I}}^{12}(b)\Psi_{i\omega,\text{II}}^{11}(b)| + |\Psi_{i\omega,\text{I}}^{12}(b)\Psi_{i\omega,\text{III}}^{11}(b)| + |\Psi_{i\omega,\text{II}}^{11}(b)\Psi_{i\omega,\text{III}}^{11}(b)| = 0$$

for one $\omega \in \mathbb{R}$. By Remark 4.3, it suffices to construct a non-zero solution v satisfying the boundary equations in condition (iii) of Theorem 4.2. The assumption implies that two of the three numbers $\Psi_{i\omega,\text{I}}^{12}(b)$, $\Psi_{i\omega,\text{II}}^{11}(b)$, $\Psi_{i\omega,\text{III}}^{11}(b)$ are zero. We discuss the cases $\Psi_{i\omega,\text{I}}^{12}(b) = \Psi_{i\omega,\text{II}}^{11}(b) = 0$ and $\Psi_{i\omega,\text{III}}^{11}(b) = \Psi_{i\omega,\text{II}}^{11}(b) = 0$, and the remaining one can be dealt with similarly to the first.

Consider the first case, i.e., $\Psi^{12}_{i\omega,\mathrm{I}}(b) = \Psi^{11}_{i\omega,\mathrm{II}}(b) = 0$. Since $\Psi_{i\omega,\mathrm{I}}$ is a fundamental solution, $\Psi_{i\omega,\mathrm{I}}(b)$ is invertible; thus, by $\Psi^{12}_{i\omega,\mathrm{I}}(b) = 0$, it follows that $\Psi^{22}_{i\omega,\mathrm{I}}(b) \neq 0$. For the construction of a solution, it suffices to provide a non-trivial initial value $v(a) = \left(\begin{bmatrix} v_{a,k} \\ F_k(a) \end{bmatrix} \right)_{k\in\{\mathrm{I,II,III}\}}$, which leads to the satisfaction of the boundary conditions in condition (iii) of Theorem 4.2. For this, we put $v_{a,\mathrm{I}} = v_{a,\mathrm{III}} = 0$ and $F_{\mathrm{II}}(a) = F_{\mathrm{III}}(a) = 0$. Moreover, we set $v_{a,\mathrm{II}} = 1$ and $F_{\mathrm{I}}(a) = (\Psi^{22}_{i\omega,\mathrm{I}}(b))^{-1}(-\Psi^{21}_{i\omega,\mathrm{II}}(b))$. Then, $v(a) \neq 0$ and, hence, $v \neq 0$. Moreover, evidently, $v_{b,\mathrm{III}} = F_{\mathrm{III}}(b) = 0$ since $v_{\mathrm{III}} = 0$. Next,

$$v_{b,\mathrm{I}} = \Psi^{11}_{i\omega,\mathrm{I}}(b)v_{a,\mathrm{I}} + \Psi^{12}_{i\omega,\mathrm{I}}(b)F_{\mathrm{I}}(a) = 0$$

$$v_{b,\mathrm{II}} = \Psi^{11}_{i\omega,\mathrm{II}}(b)v_{a,\mathrm{II}} + \Psi^{12}_{i\omega,\mathrm{II}}(b)F_{\mathrm{II}}(a) = 0.$$

It remains to check whether $F_{\mathrm{I}}(b) + F_{\mathrm{II}}(b) + F_{\mathrm{III}}(b) = 0$. The last term of the left-hand side being zero, we compute

$$F_{\mathrm{I}}(b) + F_{\mathrm{II}}(b) = \Psi^{21}_{i\omega,\mathrm{I}}(b)v_{a,\mathrm{I}} + \Psi^{22}_{i\omega,\mathrm{I}}(b)F_{\mathrm{I}}(a) + \Psi^{21}_{i\omega,\mathrm{II}}(b)v_{a,\mathrm{II}} + \Psi^{22}_{i\omega,\mathrm{II}}(b)F_{\mathrm{II}}(a)$$

$$= 0 + \Psi^{22}_{i\omega,\mathrm{I}}(b)((\Psi^{22}_{i\omega,\mathrm{I}}(b))^{-1}(-\Psi^{21}_{i\omega,\mathrm{II}}(b))) + \Psi^{21}_{i\omega,\mathrm{II}}(b) + 0 = 0,$$

which establishes all boundary conditions satisfied by $v \neq 0$.

Finally, consider the case $\Psi^{11}_{i\omega,\mathrm{III}}(b) = \Psi^{11}_{i\omega,\mathrm{II}}(b) = 0$. Since $\Psi_{i\omega,\mathrm{III}}$ is a fundamental solution, $\Psi^{11}_{i\omega,\mathrm{III}}(b) = 0$ implies $\Psi^{21}_{i\omega,\mathrm{III}}(b) \neq 0$. With this observation, we may choose $v_{a,\mathrm{I}} = F_{\mathrm{I}}(a) = F_{\mathrm{II}}(a) = F_{\mathrm{III}}(a) = 0$ and $v_{a,\mathrm{II}} = 1$ and $v_{a,\mathrm{III}} = -(\Psi^{21}_{i\omega,\mathrm{III}}(b))^{-1}\Psi^{21}_{i\omega,\mathrm{II}}(b)$. Then $v(a) \neq 0$ and, thus, $v \neq 0$. Also, it is not difficult to see that $v_{b,k} = 0$ for all $k \in \{\mathrm{I, II, III}\}$. Moreover, $F_{\mathrm{I}}(b) = 0$. It remains to check $F_{\mathrm{II}}(b) + F_{\mathrm{III}}(b) = 0$. For this, we compute

$$F_{\mathrm{II}}(b) + F_{\mathrm{III}}(b)$$

$$= \Psi^{21}_{i\omega,\mathrm{II}}(b)v_{a,\mathrm{II}} + \Psi^{22}_{i\omega,\mathrm{II}}(b)F_{\mathrm{II}}(a) + \Psi^{21}_{i\omega,\mathrm{III}}(b)v_{a,\mathrm{III}} + \Psi^{22}_{i\omega,\mathrm{III}}(b)F_{\mathrm{III}}(a)$$

$$= \Psi^{21}_{i\omega,\mathrm{II}}(b) + 0 - \Psi^{21}_{i\omega,\mathrm{III}}(b)(\Psi^{21}_{i\omega,\mathrm{III}}(b))^{-1}\Psi^{21}_{i\omega,\mathrm{II}}(b) + 0 = 0,$$

which proves the assertion. $\square$

Theorem 4.6 is formulated in terms of the fundamental solution of (25). Since this fundamental solution is a rather complicated object for variable coefficients $\mathcal{H}(\zeta)$, a more detailed characterisation can only be provided if we assume additional properties of $\mathcal{H}$. As an exemplary case, we consider the most elementary one next, the case of constant coefficients. We shall see that already in this case the conditions characterising asymptotic stability are quite subtle.

For the rest of this section, we assume that $\mathcal{H}$ is constant, that is,

$$\mathcal{H}_k(\zeta) = \begin{bmatrix} \frac{1}{\rho_k} & 0 \\ 0 & T_k \end{bmatrix} \tag{30}$$

for some $\rho_k, T_k > 0$ for all $k \in \{\mathrm{I}, \mathrm{II}, \mathrm{III}\}$. The criterion in Theorem 4.6 can be expressed in terms of certain ratios of the coefficients. Therefore, for $k \in \{\mathrm{I}, \mathrm{II}, \mathrm{III}\}$, we introduce $c_k > 0$ satisfying

$$c_k^2 = \frac{\rho_k}{T_k}. \tag{31}$$

Note that c_k is the inverse of the characteristic speed in the strings. By differentiation and uniqueness of fundamental solutions, it is not difficult to see that for $k \in \{\mathrm{I}, \mathrm{II}, \mathrm{III}\}$ we have

$$\Psi_{i\omega,k}(\zeta) = \begin{bmatrix} \cos(c_k\omega(\zeta - a)) & \frac{ic_k}{\rho_k}\sin(c_k\omega(\zeta - a)) \\ ic_k T_k \sin(c_k\omega(\zeta - a)) & \cos(c_k\omega(\zeta - a)) \end{bmatrix}. \tag{32}$$

Thus

$$\Psi_{i\omega,\mathrm{I}}^{12}(b) = (ic_{\mathrm{I}}/\rho_{\mathrm{I}})\sin(c_{\mathrm{I}}\,\omega(b - a)) \tag{33}$$

$$\Psi_{i\omega,\mathrm{II}}^{11}(b) = \cos(c_{\mathrm{II}}\omega(b - a)) \tag{34}$$

$$\Psi_{i\omega,\mathrm{III}}^{11}(b) = \cos(c_{\mathrm{III}}\omega(b - a)). \tag{35}$$

For the next result we introduce

$$\mathbb{O}/\mathbb{O} := \{\frac{1 + 2k}{1 + 2\ell}; k, \ell \in \mathbb{Z}\}$$

$$\mathbb{E}/\mathbb{O} := \{\frac{2k}{1 + 2\ell}; k, \ell \in \mathbb{Z}\}.$$

Proposition 4.7 *For $\omega \in \mathbb{R}$, consider (32), and let $c_k > 0$ be given by (31) for $k \in \{\mathrm{I}, \mathrm{II}, \mathrm{III}\}$. Then the following conditions are equivalent:*

(i) There is $\omega \in \mathbb{R}$ with

$$|\Psi_{i\omega,\mathrm{I}}^{12}(b)\Psi_{i\omega,\mathrm{II}}^{11}(b)| + |\Psi_{i\omega,\mathrm{I}}^{12}(b)\Psi_{i\omega,\mathrm{III}}^{11}(b)| + |\Psi_{i\omega,\mathrm{II}}^{11}(b)\Psi_{i\omega,\mathrm{III}}^{11}(b)| = 0.$$

(ii) At least one of the following statements is true:

$$\frac{c_{\mathrm{I}}}{c_{\mathrm{II}}} \in \mathbb{E}/\mathbb{O}, \qquad \frac{c_{\mathrm{I}}}{c_{\mathrm{III}}} \in \mathbb{E}/\mathbb{O}, \qquad \frac{c_{\mathrm{II}}}{c_{\mathrm{III}}} \in \mathbb{O}/\mathbb{O}.$$

In either case, there exists infinitely many $\omega \in \mathbb{R}$ with the property in (i).

Proof Before we dive into the actual proof, we deduce the following equivalences from the form of the fundamental solutions given in (33)–(35):

$$\Psi^{12}_{i\omega,\mathrm{I}}(b) = 0 \iff \sin(c_{\mathrm{I}}\,\omega(b-a)) = 0 \iff \omega \in \frac{\pi}{c_{\mathrm{I}}(b-a)}\mathbb{Z}$$

$$\Psi^{11}_{i\omega,\mathrm{II}}(b) = 0 \iff \cos(c_{\mathrm{II}}\omega(b-a)) = 0 \iff \omega \in \frac{\pi}{2c_{\mathrm{II}}(b-a)} + \frac{\pi}{c_{\mathrm{II}}(b-a)}\mathbb{Z}$$

$$\Psi^{11}_{i\omega,\mathrm{III}}(b) = 0 \iff \cos(c_{\mathrm{III}}\omega(b-a)) = 0$$

$$\iff \omega \in \frac{\pi}{2c_{\mathrm{III}}(b-a)} + \frac{\pi}{c_{\mathrm{III}}(b-a)}\mathbb{Z}.$$

With these observations, we find $\omega \in \mathbb{R}$ with $\Psi^{12}_{i\omega,\mathrm{I}}(b) = \Psi^{11}_{i\omega,\mathrm{II}}(b) = 0$ if and only if there are integers $k, \ell \in \mathbb{Z}$ with

$$\frac{\pi}{c_{\mathrm{I}}(b-a)}k = \frac{\pi}{2c_{\mathrm{II}}(b-a)} + \frac{\pi}{c_{\mathrm{II}}(b-a)}\ell, \text{ that is, } \frac{c_{\mathrm{I}}}{c_{\mathrm{II}}} = \frac{2k}{1+2\ell}.$$

Similarly, we find $\omega \in \mathbb{R}$ with $\Psi^{12}_{i\omega,\mathrm{I}}(b) = \Psi^{11}_{i\omega,\mathrm{III}}(b) = 0$ if and only if there are integers $k, \ell \in \mathbb{Z}$ with

$$\frac{c_{\mathrm{I}}}{c_{\mathrm{III}}} = \frac{2k}{1+2\ell}.$$

Finally, there exists $\omega \in \mathbb{R}$ with $\Psi^{11}_{i\omega,\mathrm{II}}(b) = \Psi^{11}_{i\omega,\mathrm{III}}(b) = 0$ if and only if there are integers $k, \ell \in \mathbb{Z}$ such that

$$\frac{\pi}{2c_{\mathrm{II}}(b-a)} + \frac{\pi}{c_{\mathrm{II}}(b-a)}k = \frac{\pi}{2c_{\mathrm{III}}(b-a)} + \frac{\pi}{c_{\mathrm{III}}(b-a)}\ell, \text{ that is, } \frac{c_{\mathrm{II}}}{c_{\mathrm{III}}} = \frac{1+2k}{1+2\ell}.$$

Finally, assume that either (hence both) statements (i), and (ii), are true. We carry out the argument only for the case $\frac{c_{\mathrm{I}}}{c_{\mathrm{II}}} \in \mathbb{E}/\mathbb{O}$ as the other cases can be dealt with similarly. Note that the above proof shows that any pair $(k, \ell) \in \mathbb{N} \times \mathbb{N}$ such that

$$\frac{c_{\mathrm{I}}}{c_{\mathrm{II}}} = \frac{2k}{1+2\ell} \tag{36}$$

yields an ω with the desired properties. Note that two distinct pairs lead to different ω's. Thus, if (k, ℓ) satisfies (36), then for every odd q, we get that $(kq, \frac{q-1}{2} + \ell q)$ yields

$$\frac{2kq}{1+2(\frac{q-1}{2}+\ell q)} = \frac{2kq}{(1+2\ell)q} = \frac{2k}{1+2\ell} = \frac{c_{\mathrm{I}}}{c_{\mathrm{II}}},$$

which eventually shows the assertion. $\qquad\square$

Finally, we can summarise our characterisation of semi-uniform stability/asymptotic stability for the case of constant coefficients.

Theorem 4.8 *Let $(A, \mathrm{dom}(A))$ and $(T(t))_{t\geq 0}$ be given by Theorem 4.1. Assume that $\mathcal{H}$ is constant, that is, given by (30), and let Ψ be given by (27). Then the following conditions are equivalent:*

(i) $(T(t))_{t\geq 0}$ *is not semi-uniformly stable.*
(ii) There exists $\omega \in \mathbb{R}$ with

$$|\Psi_{i\omega,\mathrm{I}}^{12}(b)\Psi_{i\omega,\mathrm{II}}^{11}(b)| + |\Psi_{i\omega,\mathrm{I}}^{12}(b)\Psi_{i\omega,\mathrm{III}}^{11}(b)| + |\Psi_{i\omega,\mathrm{II}}^{11}(b)\Psi_{i\omega,\mathrm{III}}^{11}(b)| = 0.$$

(iii) At least one of the following statements is true

$$\frac{c_{\mathrm{I}}}{c_{\mathrm{II}}} \in \mathbb{E}/\mathbb{O}, \qquad \frac{c_{\mathrm{I}}}{c_{\mathrm{III}}} \in \mathbb{E}/\mathbb{O}, \qquad \frac{c_{\mathrm{II}}}{c_{\mathrm{III}}} \in \mathbb{O}/\mathbb{O},$$

where $c_k = \sqrt{\rho_k/T_k}$, $k \in \{\mathrm{I}, \mathrm{II}, \mathrm{III}\}$.
(iv) A has infinitely many eigenvalues on the imaginary axis.

Proof The equivalence of (i) and (ii) follows from Theorem 4.6.

The equivalence of (ii) and (iii) follows from Proposition 4.7.

The condition in (iv) implies (i) by Theorem 1.3. Consequently, by what we have already shown, (iv) implies (ii), which, by Proposition 4.7, leads to the existence of infinitely many $\omega \in \mathbb{R}$ satisfying (ii). By the concluding statement in Theorem 4.6, it follows that A has infinitely many eigenvalues on the imaginary axis, which concludes the proof. $\qquad\Box$

As an immediate corollary we obtain the following:

Corollary 4.9 *In the situation of Theorem 4.8, the following conditions are equivalent:*

(i) $(T(t))_{t\geq 0}$ *is semi-uniformly stable.*
(ii) $(T(t))_{t\geq 0}$ *is asymptotically stable.*
(iii) $\frac{c_{\mathrm{I}}}{c_{\mathrm{II}}}, \frac{c_{\mathrm{I}}}{c_{\mathrm{III}}} \notin \mathbb{E}/\mathbb{O}$ *and* $\frac{c_{\mathrm{II}}}{c_{\mathrm{III}}} \notin \mathbb{O}/\mathbb{O}$.

Proof Conditions (i) and (ii) are equivalent by Theorem 1.3; the remaining equivalence follows from Theorem 4.8. $\qquad\Box$

5   Vibrating Strings and Exponential Stability

We complement the characterisation of exponential stability for our system of vibrating strings in the following. For this, we use the reformulation of exponential stability from Sect. 3. Recall from [12, Section 6] that condition $\sup_{\omega\in\mathbb{R}} \|\Psi_{i\omega}\| < \infty$

is satisfied if both ρ_k and T_k are of bounded variation for all $k \in \{I, II, III\}$. This is particularly the case, if $\mathcal{H}$ is constant.

We start off with a result similar to that of Theorem 4.2. For this we denote for
$$v_a = \left(\begin{bmatrix} v_{a,k} \\ F_k(a) \end{bmatrix} \right)_{k \in \{I,II,III\}} \in \mathbb{C}^6 \text{ and } \omega \in \mathbb{R}, \, v_{v_a}^{(\omega)} : [a, b] \to \mathbb{C}^6 = (\mathbb{C}^2)^3 \text{ to be a}$$
solution of (25) with $\lambda = i\omega$ and $v_{v_a}^{(\omega)}(a) = v_a$.

Proposition 5.1 *Let $(A, \mathrm{dom}(A))$ and $(T(t))_{t \geq 0}$ be given by Theorem 4.1. Assume that $\sup_{\omega \in \mathbb{R}} \|\Psi_{i\omega}\| < \infty$, where $\Psi_{i\omega}$ is given by (25). Furthermore, let $(\omega_n)_{n \in \mathbb{N}}$ be a sequence in $\mathbb{R}$. Then the following conditions are equivalent:*

(i) The set

$$\{ v_a \in \mathbb{C}^6 \mid \liminf_{n \to \infty} \left(\left\| W_B \begin{bmatrix} \Psi_{i\omega_n}(b) \\ I_6 \end{bmatrix} v_a \right\| + \right.$$

$$\left. |v_a^* [\, \Psi_{i\omega_n}(b)^* \ I_6 \,] \begin{bmatrix} P_1 & 0 \\ 0 & -P_1 \end{bmatrix} \begin{bmatrix} \Psi_{i\omega_n}(b) \\ I_6 \end{bmatrix} v_a | \right) = 0 \}$$

contains only the zero element.

(ii) Let $v_a \in \mathbb{C}^6$ and $v^{(n)} := v_{v_a}^{(\omega_n)}; \, v^{(n)}(b) := \left(\begin{bmatrix} v_{b,k}^{(n)} \\ F_k^{(n)}(b) \end{bmatrix} \right)_{k \in \{I,II,III\}}$. If

$$v_{a,I} = F_{II}(a) = F_{III}(a) = 0 \text{ and}$$

$$\liminf_{n \to \infty} \left(|v_{b,I}^{(n)} - v_{b,II}^{(n)}| + |v_{b,I}^{(n)} - v_{b,III}^{(n)}| + |\beta v_{b,I}^{(n)} + F_I^{(n)}(b) + F_{II}^{(n)}(b) + F_{III}^{(n)}(b)| \right.$$

$$\left. + |v^{(n)}(b)^* P_1 v^{(n)}(b) - v_a^* P_1 v_a| \right) = 0,$$

then $v_a = 0$.

(iii) Let $v_a \in \mathbb{C}^6$ and $v^{(n)} := v_{v_a}^{(\omega_n)}; \, v^{(n)}(b) := \left(\begin{bmatrix} v_{a,k}^{(b)} \\ F_k^{(n)}(b) \end{bmatrix} \right)_{k \in \{I,II,III\}}$. If

$$v_{a,I} = F_{II}(a) = F_{III}(a) = 0 \quad \text{and}$$

$$\liminf_{n \to \infty} \left(|v_{b,I}^{(n)}| + |v_{b,II}^{(n)}| + |v_{b,III}^{(n)}| + |F_I^{(n)}(b) + F_{II}^{(n)}(b) + F_{III}^{(n)}(b)| \right) = 0,$$

then $v_a = 0$.

Proof Recall that the solution of (25) for $\lambda = i\omega_n$ is represented by $v^{(n)}(\cdot) = \Psi_{i\omega_n}(\cdot)v_a$. In this case, $v^{(n)}(b) = \Psi_{i\omega_n}(b)v_a$ for all $n \in \mathbb{N}$.

(i)$\Leftrightarrow$(ii). We reformulate the expression in the lim inf in the set given in (i). Let v_a belong to this set. By passing to a subsequence, which is not relabelled, we may assume that the lim inf in (ii) can be replaced by a limit. Since we considered non-negative terms in the lim inf-expression only, we obtain that the individual terms are null sequences. Thus, the terms in the norms read

$$\lim_{n\to\infty} W_B \begin{bmatrix} v^{(n)}(b) \\ v_a \end{bmatrix} = 0 \text{ and} \tag{37}$$

$$\lim_{n\to\infty} v^{(n)}(b)^* P_1 v^{(n)}(b) - v_a^* P_1 v_a = 0, \tag{38}$$

respectively.

Next, since $v_a = \left(\begin{bmatrix} v_{a,k} \\ F_k(a) \end{bmatrix} \right)_{k\in\{I,II,III\}}$ and $v^{(n)}(b) := \left(\begin{bmatrix} v_{b,k}^{(n)} \\ F_k^{(n)}(b) \end{bmatrix} \right)_{k\in\{I,II,III\}}$, the

conditions encoded in $\lim_{n\to\infty} \| W_B \begin{bmatrix} v^{(n)}(b) \\ v_a \end{bmatrix} \| = 0$ read

$$v_{a,I} = 0, \quad F_{II}(a) = F_{III}(a) = 0$$

$$\lim_{n\to\infty} v_{b,I}^{(n)} - v_{b,II}^{(n)} = \lim_{n\to\infty} v_{b,I}^{(n)} - v_{b,III}^{(n)} = 0$$

$$\lim_{n\to\infty} \beta v_{b,I}^{(n)} + F_I^{(n)}(b) + F_{II}^{(n)}(b) + F_{III}^{(n)}(b) = 0.$$

Similarly, we may reformulate the conditions in (ii). Indeed, by passing to subsequence, where one write limits instead of lim inf, we obtain the limit expression in (i) for this subsequence.

(iii)$\Leftrightarrow$(ii) We begin by deriving some equalities. For $n \in \mathbb{N}$,

$$v^{(n)}(b)^* P_1 v^{(n)}(b) - v_a^* P_1 v_a$$

$$= 2 \sum_{k\in\{I,II,III\}} \left(\begin{bmatrix} v_{b,k}^{(n)} \\ F_k^{(n)}(b) \end{bmatrix}^* \begin{bmatrix} 0 & 1 \\ 1 & 0 \end{bmatrix} \begin{bmatrix} v_{b,k}^{(n)} \\ F_k^{(n)}(b) \end{bmatrix} - \begin{bmatrix} v_{a,k} \\ F_k(a) \end{bmatrix}^* \begin{bmatrix} 0 & 1 \\ 1 & 0 \end{bmatrix} \begin{bmatrix} v_{a,k} \\ F_k(a) \end{bmatrix} \right)$$

$$= 2 \sum_{k\in\{I,II,III\}} \mathrm{Re} \left[(v_{b,k}^{(n)})^* F_k^{(n)}(b) \right] - \mathrm{Re}[v_{a,k}^* F_k(a)]$$

$$= 2 \sum_{k\in\{I,II,III\}} \mathrm{Re} \left[(v_{b,k}^{(n)})^* F_k^{(n)}(b) \right],$$

where we have used the (common) conditions on v_a. Hence

$$v^{(n)}(b)^* P_1 v^{(n)}(b) - v_a^* P_1 v_a$$

$$= v^{(n)}(b)^* P_1 v^{(n)}(b) = 2 \sum_{k\in\{I,II,III\}} \mathrm{Re} \left[(v_{b,k}^{(n)})^* F_k^{(n)}(b) \right]. \tag{39}$$

Furthermore, since $v^{(n)}(b) = \Psi_{i\omega_n}(b)v_a$ and $\Psi_{i\omega}$ is assumed to be bounded, we have that the sequences $(v_{b,k}^{(n)}, F_k^{(n)}(b))_{n\in\mathbb{N}}$, $k \in \{I, II, III\}$, are bounded.

Next we prove that (iii)$\Rightarrow$(ii). By possibly passing to a subsequence, we may assume that the lim inf can be replaced by lim.

From (39), we deduce that

$$|v^{(n)}(b)^* P_1 v^{(n)}(b) - v_a^* P_1 v_a| \leq 2 \sum_{k \in \{I,II,III\}} |(v_{b,k}^{(n)})||F_k^{(n)}(b)|.$$

Combining this with the property that $(F_k^{(n)}(b))_{n \in \mathbb{N}}$ is a bounded sequence for all $k \in \{I, II, III\}$ and that $\lim_{n \to \infty} v_{b,k}^{(n)} = 0$, we find that $\lim_{n \to \infty} |v^{(n)}(b)^* P_1 v^{(n)}(b) - v_a^* P_1 v_a| = 0$.

The inequalities

$$|v_{b,I}^{(n)} - v_{b,II}^{(n)}| + |v_{b,I}^{(n)} - v_{b,III}^{(n)}| \leq 2|v_{b,I}^{(n)}| + |v_{b,II}^{(n)}| + |v_{b,III}^{(n)}| \text{ and}$$

$$|\beta v_{b,I}^{(n)} + F_I^{(n)}(b) + F_{II}^{(n)}(b) + F_{III}^{(n)}(b)| \leq |\beta v_{b,I}^{(n)}| + |F_I^{(n)}(b) + F_{II}^{(n)}(b) + F_{III}^{(n)}(b)|$$

yield the remaining convergences. Thus, (iii) is sufficient for (ii).

Next we show that (ii)$\Rightarrow$(iii). Again, we may assume, by possibly passing to a subsequence, not restricting the generality, that lim inf can be replaced by lim. Next, starting with the term on the right-hand side of (39), we compute for $n \in \mathbb{N}$

$$\sum_{k \in \{I,II,III\}} (v_{b,k}^{(n)})^* F_k^{(n)}(b)$$

$$= (v_{b,I}^{(n)})^* \left[F_I^{(n)}(b) + F_{II}^{(n)}(b) + F_{III}^{(n)}(b) \right] +$$

$$[v_{b,II}^{(n)} - v_{b,I}^{(n)}]^* F_{II}^{(n)}(b) + [v_{b,III}^{(n)} - v_{b,I}^{(n)}]^* F_{III}^{(n)}(b)$$

$$= (v_{b,I}^{(n)})^* \left[\beta v_{b,I}^{(n)} + F_I^{(n)}(b) + F_{II}^{(n)}(b) + F_{III}^{(n)}(b) \right] - \beta |v_{b,I}^{(n)}|^2 +$$

$$[v_{b,II}^{(n)} - v_{b,I}^{(n)}]^* F_{II}^{(n)}(b) + [v_{b,III}^{(n)} - v_{b,I}^{(n)}]^* F_{III}^{(n)}(b). \tag{40}$$

Since $(v_{b,I}^{(n)})_{n \in \mathbb{N}}$, $(F_{II}^{(n)}(b))_{n \in \mathbb{N}}$, and $(F_{III}^{(n)}(b))_{n \in \mathbb{N}}$ are bounded sequences, the conditions in item (ii) combined with (39) and (40) imply that $\lim_{n \to \infty} \beta |v_{b,I}^{(n)}|^2 = 0$. Using that $\beta \neq 0$, this gives that $\lim_{n \to \infty} |v_{b,I}^{(n)}|^2 = 0$. Substituting this property into the equalities in item (ii) gives the equalities in item (iii). $\qquad\square$

Theorem 5.2 *Let $(A, \text{dom}(A))$ and $(T(t))_{t \geq 0}$ be given by Theorem 4.1. Furthermore, assume that $\sup_{\omega \in \mathbb{R}} \|\Psi_{i\omega}\| < \infty$, where $\Psi_{i\omega}$ is given by (25). Then the following conditions are equivalent:*

(i) $(T(t))_{t \geq 0}$ is exponentially stable.

(ii) For all sequences in $(\omega_k)_{k\in\mathbb{N}}$ in $\mathbb{R}$,

$$\liminf_{k\to\infty}\left(|\Psi^{12}_{i\omega_k,\mathrm{I}}(b)\Psi^{11}_{i\omega_k,\mathrm{II}}(b)| + |\Psi^{12}_{i\omega_k,\mathrm{I}}(b)\Psi^{11}_{i\omega_k,\mathrm{III}}(b)|\right.$$
$$\left. + |\Psi^{11}_{i\omega_k,\mathrm{II}}(b)\Psi^{11}_{i\omega_k,\mathrm{III}}(b)|\right) > 0.$$

Proof The set-up is now similar to the proof of Theorem 4.6. We use Theorem 3.2 for the characterisation of exponential stability and Proposition 5.1 for the reformulation of (ii) in Theorem 3.2.

(i)$\Rightarrow$(ii) We prove that not (ii) implies not (i). For this, we distinguish two cases. If $(\omega_k)_{k\in\mathbb{N}}$ is a bounded sequence in $\mathbb{R}$ with

$$\liminf_{k\to\infty}\left(|\Psi^{12}_{i\omega_k,\mathrm{I}}(b)\Psi^{11}_{i\omega_k,\mathrm{II}}(b)|+|\Psi^{12}_{i\omega_k,\mathrm{I}}(b)\Psi^{11}_{i\omega_k,\mathrm{III}}(b)|+|\Psi^{11}_{i\omega_k,\mathrm{II}}(b)\Psi^{11}_{i\omega_k,\mathrm{III}}(b)|\right) = 0,$$

then, by the continuity of $\omega \mapsto \Psi_{i\omega}$,

$$\left(|\Psi^{12}_{i\omega,\mathrm{I}}(b)\Psi^{11}_{i\omega,\mathrm{II}}(b)| + |\Psi^{12}_{i\omega,\mathrm{I}}(b)\Psi^{11}_{i\omega,\mathrm{III}}(b)| + |\Psi^{11}_{i\omega_k,\mathrm{II}}(b)\Psi^{11}_{i\omega,\mathrm{III}}(b)|\right) = 0$$

for one accumulation value ω of $(\omega_k)_{k\in\mathbb{N}}$. Hence, by Theorem 4.6, the semigroup is not semi-uniformly stable and, thus, not exponentially stable.

It remains to consider the case $(\omega_k)_{k\in\mathbb{N}}$ is an unbounded sequence in $\mathbb{R}$ with no bounded subsequence such that

$$\liminf_{k\to\infty}\left(|\Psi^{12}_{i\omega_k,\mathrm{I}}(b)\Psi^{11}_{i\omega_k,\mathrm{II}}(b)|+|\Psi^{12}_{i\omega_k,\mathrm{I}}(b)\Psi^{11}_{i\omega_k,\mathrm{III}}(b)|+|\Psi^{11}_{i\omega_k,\mathrm{II}}(b)\Psi^{11}_{i\omega_k,\mathrm{III}}(b)|\right) = 0.$$

Without restriction we may assume that $\omega_k \to \infty$. Moreover, using the boundedness of $\omega \mapsto \Psi_{i\omega}$, we may choose a subsequence (not relabelled) such that $(\Psi_{i\omega_k}(b))_{k\in\mathbb{N}}$ converges as $k \to \infty$; we let

$$P^{ms}_l := \lim_{k\to\infty} \Psi^{ms}_{i\omega_k,l}(b) \quad (l \in \{\mathrm{I}, \mathrm{II}, \mathrm{III}\},\ m, s \in \{1, 2\}).$$

The assumption implies that two of the three numbers P^{12}_{I}, P^{11}_{II}, P^{11}_{III} are zero.

Consider the case $P^{12}_{\mathrm{I}} = P^{11}_{\mathrm{II}} = 0$. By [12, Lemma 3.2], $\omega \mapsto (x \mapsto \Psi_{i\omega}(x)^{-1})$ is uniformly bounded. Thus, $P^{22}_{\mathrm{I}} \neq 0$, otherwise $\|\Psi_{i\omega}(b)^{-1}\| \to \infty$. Next, we define $v_{a,\mathrm{I}} = v_{a,\mathrm{III}} = 0$ and $F_{\mathrm{II}}(a) = F_{\mathrm{III}}(a) = 0$. Moreover, $v_{a,\mathrm{II}} = 1$ and $F_{\mathrm{I}}(a) = (P^{22}_{\mathrm{I}})^{-1}(-P^{21}_{\mathrm{II}})$. Then, $v_a \neq 0$ and, hence, $v^{(n)} := \Psi^{ms}_{i\omega_n}(\cdot)v_a \neq 0$. Then, $v^{(n)}_{b,\mathrm{III}} = F^{(n)}_{\mathrm{III}}(b) = 0$ since $v^{(n)}_{\mathrm{III}} = 0$.

Next,

$$v^{(n)}_{b,\mathrm{I}} = \Psi^{11}_{i\omega_n,\mathrm{I}}(b)v_{a,\mathrm{I}} + \Psi^{12}_{i\omega_n,\mathrm{I}}(b)F_{\mathrm{I}}(a) \to 0 \quad (n \to \infty)$$

$$v^{(n)}_{b,\mathrm{II}} = \Psi^{11}_{i\omega_n,\mathrm{II}}(b)v_{a,\mathrm{II}} + \Psi^{12}_{i\omega_n,\mathrm{II}}(b)F_{\mathrm{II}}(a) \to 0 \quad (n \to \infty).$$

Next, we show $F_{\mathrm{I}}^{(n)}(b) + F_{\mathrm{II}}^{(n)}(b) + F_{\mathrm{III}}^{(n)}(b) \to 0$ as $n \to \infty$. For this, it suffices to consider

$$
\begin{aligned}
F_{\mathrm{I}}^{(n)}(b) &+ F_{\mathrm{II}}^{(n)}(b) \\
&= \Psi_{i\omega_n,\mathrm{I}}^{21}(b)v_{a,\mathrm{I}} + \Psi_{i\omega_n,\mathrm{I}}^{22}(b)F_{\mathrm{I}}(a) + \Psi_{i\omega_n,\mathrm{II}}^{21}(b)v_{a,\mathrm{II}} + \Psi_{i\omega_n,\mathrm{II}}^{22}(b)F_{\mathrm{II}}(a) \\
&= 0 + \Psi_{i\omega_n,\mathrm{I}}^{22}(b)(P_{\mathrm{I}}^{22})^{-1}(-P_{\mathrm{II}}^{21})) + \Psi_{i\omega_n,\mathrm{II}}^{21}(b) + 0 \to 0 \quad (n \to \infty),
\end{aligned}
$$

which establishes all conditions in (iii) of Proposition 5.1 albeit $v_a \neq 0$. Hence, Theorem 3.2 shows that $(T(t))_{t\geq 0}$ is not exponentially stable.

The remaining two cases are either a variant of the one just developed or can be dealt with analogously to the second case considered in Theorem 4.6 subject to the modifications along the lines of the case just carried out here.

(ii)$\Rightarrow$(i) The statement can be shown along the same lines as Proposition 4.5 in order to establishing condition (iii) in Proposition 5.1. Thus, $(T(t))_{t\geq 0}$ is exponentially stable by Theorem 3.2 $\qquad\square$

Remark 5.3 Note that in the case of constant $\mathcal{H}$ it suffices to consider sequences tending to ∞ (or $-\infty$) in (ii) in Theorem 5.2, by Proposition 4.7.

In the remaining part of this section, we restrict ourselves to constant $\mathcal{H}$; that is, we use (30) and recall $c_{\mathrm{I}}, c_{\mathrm{II}}, c_{\mathrm{III}}$ from (31). We emphasise that in this case, $\omega \mapsto \Psi_{i\omega}$ is bounded, as $\mathcal{H}$ is of bounded variation, see [12]. The next example contains a set-up yielding exponential stability.

Example 5.4 Let $c_{\mathrm{I}} = c_{\mathrm{II}} = 1$ and $c_{\mathrm{III}} = 2$, $a = 0$, $b = 1$. Then the corresponding set-up of vibrating strings is exponentially stable. Appealing to (33), (34), and (35) together with Theorem 5.2, we may only consider for $\omega \in \mathbb{R}$ the expression

$$
\eta(\omega) := |\sin(\omega)\cos(\omega)|^2 + |\sin(\omega)\cos(2\omega)|^2 + |\cos(\omega)\cos(2\omega)|^2
$$

and analyse whether this is uniformly bounded away from zero as ω ranges over $\mathbb{R}$.

Using elementary formulas for sin and cos and putting $s := \sin(\omega)$ and $c := \cos(\omega)$, we deduce for all ω

$$
\eta(\omega) = s^2 c^2 + (c^2 - s^2)^2 = s^2(1 - s^2) + (1 - 2s^2)^2 =: f(s),
$$

which as a sum of non-negative terms can only be zero, if either terms vanish. It is not hard to show that the minima of the polynomial f are for $s^2 = 1/2$. So $f(s) \geq 1/4$ for all $s^2 \in [0, 1]$ and thus

$$
\inf_{\omega \in \mathbb{R}} \eta(\omega) \geq \frac{1}{4}.
$$

Using Theorem 5.2, we conclude that this system of vibrating strings yields an exponentially stable semigroup. $\qquad\square$

We conclude this section by providing a setting, where the semigroup corresponding to the port-Hamiltonian system of Fig. 1 is semi-uniformly stable but not exponentially stable. The core is Kronecker's theorem from number theory (see, e.g., [6, Theorem 2.39]). For this we define $\mathbb{T} := \{z \in \mathbb{C} \mid |z| = 1\}$. For $m \in \mathbb{N}$ and $a = (a_1, \ldots, a_m) \in \mathbb{T}^m$, we define

$$\phi_a : \mathbb{T}^m \to \mathbb{T}^m, (z_1, \ldots, z_m) \mapsto (a_1 z_1, \ldots, a_m z_m).$$

Theorem 5.5 (Kronecker) *Let $(a_1, \ldots, a_m) \in \mathbb{T}^m$. Then the following conditions are equivalent:*

(i) For all $z \in \mathbb{T}^m$, the set $\{\phi_a^k(z) \mid k \in \mathbb{N}\}$ is dense in $\mathbb{T}^m$.
(ii) For all $k_1, \ldots, k_m \in \mathbb{Z}$ with $a_1^{k_1} a_2^{k_2} \cdots a_m^{k_m} = 1$, we have $k_1 = k_2 = \ldots = k_m = 0$.

Proof The proof is a (straightforward) combination of [6, Example 2.18 and Theorem 2.21]. $\qquad\square$

We turn back to our example of vibrating strings and assume that $\mathcal{H}$ is constant, that is, given by (30) and recall $c_k = \sqrt{\rho_k / T_k}$, $k \in \{\text{I, II, III}\}$. In this case, we may recall (33)–(35).

We define for $x, y \in \mathbb{R}$ the equivalence relation $x \sim y$ via $x - y \in \mathbb{Z}$ and $\widetilde{\mathbb{T}} := \mathbb{R}/\sim$. We say that $x, y \in \mathbb{R}$ are *linearly dependent in the $\mathbb{Z}$-module $\widetilde{\mathbb{T}}$,* if there exist integers $k, n \in \mathbb{Z}$ such that $kx \sim ny$. In case they are not linearly dependent, they are called *linearly independent in the $\mathbb{Z}$-module $\widetilde{\mathbb{T}}$.*

Theorem 5.6 *Let $(A, \mathrm{dom}(A))$ and $(T(t))_{t \geq 0}$ be given by Theorem 4.1 with $\mathcal{H}$ constant.*

Assume that all of the pairs $(c_{\mathrm{I}}, c_{\mathrm{II}})$, $(c_{\mathrm{I}}, c_{\mathrm{III}})$, and $(c_{\mathrm{II}}, c_{\mathrm{III}})$ are linearly independent in the $\mathbb{Z}$-module $\widetilde{\mathbb{T}}$. Then $(T(t))_{t \geq 0}$ is semi-uniformly stable, but not exponentially stable.

Proof As all the pairs are independent in the $\mathbb{Z}$-module, condition (iii) in Corollary 4.9 is satisfied yielding that $(T(t))_{t \geq 0}$ is semi-uniformly stable.

By Theorem 5.5 the set

$$\{(e^{i c_{\mathrm{I}} 2\pi n}, e^{i c_{\mathrm{II}} 2\pi n}); n \in \mathbb{N}\} = \{\phi^n_{(e^{i c_{\mathrm{I}} 2\pi}, e^{i c_{\mathrm{II}} 2\pi})}(1, 1); n \in \mathbb{N}\} \subseteq \mathbb{T}^2$$

is dense $\mathbb{T}^2$. In particular, $(1, i)$ can be approximated arbitrarily well, which yields that

$$\{(\sin(c_{\mathrm{I}} 2\pi n), \cos(c_{\mathrm{II}} 2\pi n)); n \in \mathbb{N}\}$$

accumulates at 0. Hence, $\omega_n := 2\pi n/(b - a)$ defines a sequence such that

$$\liminf_{n \to \infty} \left(|\Psi^{12}_{i\omega_n, \mathrm{I}}(b) \Psi^{11}_{i\omega_n, \mathrm{II}}(b)| + |\Psi^{12}_{i\omega_n, \mathrm{I}}(b) \Psi^{11}_{i\omega_n, \mathrm{III}}(b)| + |\Psi^{11}_{i\omega_n, \mathrm{II}}(b) \Psi^{11}_{i\omega_n, \mathrm{III}}(b)| \right) = 0,$$

which implies that the semigroup is not exponentially stable by Theorem 5.2. $\qquad\square$

6 Conclusion

For port-Hamiltonian systems on a one-dimensional spatial domain, we have presented equivalent characterisations of asymptotic stability. One of the characterisations gives that the system is asymptotically stable when there are no eigenvalues on the imaginary axis. However, in general these eigenvalues are not easy to calculate, and so there is still a need for (easy) sufficient conditions, like there is for exponential stability, see [13]. Furthermore, our example shows that stability is very sensitive with respect to its (constant) parameters, and thus asymptotic stability lacks robustness. We supported this observation also by looking at exponential stability. A logical follow-up question would be if this is also the case for spatial varying parameters.

There is a strong link between observability/controllability of systems and the stabilisation of it, see, e.g., [3] and the references therein. For wave equations on a network, like we have in Sect. 4, observability and controllability have been well studied in the book by Dáger and Zuazua [4]. Hence, future research should study the relation between their and our results.

Appendix A   Well-Posedness of Ordinary Differential Equations with Measurable Coefficients

We provide the well-posedness theorem underlying the unique existence of the fundamental solutions considered in the main body of the manuscript. The strategy to show well-posedness uses exponentially weighted L_2-spaces. In an o.d.e.-context, this has been employed for instance in [11, Chapter 4] and the references therein. For convenience, we write out the argument in this special situation.

Theorem A.1 *Let* $\mathcal{K} \in L_\infty(a, b)^{d \times d}$. *Then for all* $u_0 \in \mathbb{C}^d$ *there exists a unique* $u \in H^1(a, b)^d$ *such that*

$$u' = \mathcal{K}u, \quad u(a+) = u_0.$$

Proof Without loss of generality, $(a, b) = (0, 1)$. We start off with existence of the solution. For this, let $\rho > 0$ and consider the Hilbert space

$$L_{2,\rho}(0, 1)^d := (L_2(0, 1)^d; (v, w) \mapsto \int_0^1 \langle v(x), w(x) \rangle_{\mathbb{C}^d} e^{-2\rho x} \, dx).$$

Define

$$\Psi : L_{2,\rho}(0, 1)^d \to L_{2,\rho}(0, 1)^d$$

by

$$(\Psi u)(x) := u_0 + \int_0^x \mathcal{K}(s)u(s)\,\mathrm{d}s.$$

Then, clearly Ψ is well-defined. Moreover, for $u, v \in L_{2,\rho}(0,1)^d$ we obtain

$$
\begin{aligned}
\|\Psi u - \Psi v\|_{L_{2,\rho}}^2 &= \int_0^1 \|\int_0^x \mathcal{K}(s)u(s) - \mathcal{K}(s)v(s)\,\mathrm{d}s\|^2 \mathrm{e}^{-2\rho x}\,\mathrm{d}x \\
&\leq \int_0^1 (\int_0^x \|\mathcal{K}(s)\|\,\|u(s) - v(s)\|\,\mathrm{d}s)^2 \mathrm{e}^{-2\rho x}\,\mathrm{d}x \\
&\leq \|\mathcal{K}\|_\infty^2 \int_0^1 (\int_0^x \|u(s) - v(s)\|\mathrm{e}^{-\rho s}\mathrm{e}^{\rho s}\,\mathrm{d}s)^2 \mathrm{e}^{-2\rho x}\,\mathrm{d}x \\
&\leq \|\mathcal{K}\|_\infty^2 \int_0^1 \int_0^x \|u(s) - v(s)\|^2 \mathrm{e}^{-2\rho s}\,\mathrm{d}s \int_0^x \mathrm{e}^{2\rho s}\,\mathrm{d}s\,\mathrm{e}^{-2\rho x}\,\mathrm{d}x \\
&\leq \|\mathcal{K}\|_\infty^2 \|u - v\|_{L_{2,\rho}}^2 \int_0^1 \int_0^x \mathrm{e}^{2\rho s}\,\mathrm{d}s\,\mathrm{e}^{-2\rho x}\,\mathrm{d}x \\
&\leq \|\mathcal{K}\|_\infty^2 \|u - v\|_{L_{2,\rho}}^2 \frac{1}{2\rho}.
\end{aligned}
$$

Choosing $2\rho > \|\mathcal{K}\|_\infty^2$, we get that Ψ is a strict contraction and assumes a unique fixed point u. In consequence,

$$u(x) = \Psi(u)(x) = u_0 + \int_0^x \mathcal{K}(s)u(s)\,\mathrm{d}s,$$

which by Fubini's theorem leads to $u \in H^1(0,1)^d$ with

$$u'(x) = \mathcal{K}(x)u(x), \quad \text{a.e. } x \in (a,b).$$

Moreover, Lebesgue's dominated convergence theorem confirms $u(0+) = u_0$; hence, u is a solution of the desired equation.

To address uniqueness, it suffices to show that the only solution of the differential equation with $u_0 = 0$ is zero. For this, we integrate the desired o.d.e. and obtain

$$u(x) = \int_0^x \mathcal{K}(s)u(s)\,\mathrm{d}s.$$

Hence,

$$\|u(x)\| \leq \|\mathcal{K}\|_\infty \int_0^x \|u(s)\|\,\mathrm{d}s$$

and Gronwall's inequality leads to

$$\|u(t)\| \le 0; \text{ that is, } u = 0.$$

$\square$

Corollary A.2 *Under the assumptions of Theorem A.1, there exists a uniquely determined continuous function* $\Gamma \colon [a, b] \to \mathbb{C}^{d \times d}$ *with* $\Gamma(\cdot)u_0$ *being the unique solution of*

$$u'(x) = \mathcal{K}(x)u(x), \quad u(a) = u_0.$$

It follows that $\Gamma(0) = I_d$.

Proof For uniqueness, let Γ_1 share the properties of Γ. Then for $u_0 \in \mathbb{C}^d$, it follows that $\Gamma_1(\cdot)u_0 = \Gamma(\cdot)u_0$ by the uniqueness result in Theorem A.1. Thus, $\Gamma_1 = \Gamma$. To address existence, we note that the solution operator $S \colon \mathbb{C}^d \to H^1(a, b)^d$ assigning to each initital value u_0 the corresponding solution constructed in Theorem A.1 is linear. By the Sobolev embedding theorem, $A_x \colon u_0 \mapsto (Su_0)(x)$ is well-defined; as it is linear, we infer that for all $x \in [a, b]$ there is a matrix $\Gamma(x)$, which represents A_x. For the continuity of Γ it is enough to see that for all $u_0 \in \mathbb{C}^d$, we have $\Gamma(x)u_0 = A_x u_0 = (Su_0)(x)$ and so $\Gamma(\cdot)u_0 \in H^1(a, b)^d \subseteq C[a, b]^d$, by the Sobolev embedding theorem. The last condition follows from $\Gamma(a)u_0 = u(a) = u_0$.

$\square$

Finally, we present a continuous dependence result in the coefficients, which is a consequence of Gronwall's inequality – it was already observed in an earlier version of [12].

Proposition A.3 *The mapping*

$$L_\infty(a, b)^{d \times d} \ni \mathcal{K} \mapsto \Gamma_{\mathcal{K}} \in C([a, b]; \mathbb{C}^{d \times d}),$$

where $\Gamma_{\mathcal{K}}$ *is the fundamental solution of*

$$u'(x) = \mathcal{K}(x)u(x), \text{ with } \Gamma_{\mathcal{K}}(0) = I_d,$$

is continuous.

Proof Let $\mathcal{K}, \mathcal{L} \in L_\infty(a, b)^{d \times d}$. For $u_0 \in \mathbb{C}^d$ let $u := \Gamma_{\mathcal{K}}(\cdot)u_0$ and $v := \Gamma_{\mathcal{L}}(\cdot)u_0$. Then, we compute for (a.e.) $t \in (a, b)$

$$\partial_t \|u(t) - v(t)\|^2 = \mathrm{Re}\langle u'(t) - v'(t), u(t) - v(t)\rangle$$

$$= \mathrm{Re}\langle (\mathcal{K}u)(t) - (\mathcal{L}v)(t), u(t) - v(t)\rangle$$

$$\le \left(\|\mathcal{K} - \mathcal{L}\|_\infty \|u(t)\| + \|\mathcal{L}\|_\infty \|u(t) - v(t)\| \right) \|u(t) - v(t)\|.$$

Integration over (a, s) yields

$$\|u(s)-v(s)\|^2 \le \|\mathcal{K}-\mathcal{L}\|_\infty \int_a^s \|u(t)\| \|u(t)-v(t)\| dt + \|\mathcal{L}\|_\infty \int_a^s \|u(t)-v(t)\|^2 dt.$$

Thus, by Gronwall's inequality,

$$\|u(s) - v(s)\|^2 \le \|\mathcal{K} - \mathcal{L}\|_\infty \int_a^s \|u(t)\| \|u(t) - v(t)\| dt \cdot e^{\|\mathcal{L}\|_\infty (b-a)};$$

Taking the supremum over s implies

$$\|u - v\|_\infty \le \|\mathcal{K} - \mathcal{L}\|_\infty \int_a^b \|u(t)\| dt \cdot e^{\|\mathcal{L}\|_\infty (b-a)}.$$

So when $\mathcal{L} \to \mathcal{K}$ in $L_\infty(a, b)^{d \times d}$, then $\|u - v\|_\infty = \|(\Gamma_\mathcal{K} - \Gamma_\mathcal{L})u_0\| \to 0$, which shows that the fundamental solution, $\Gamma_\mathcal{K}$, is continuous in $C([a, b]; \mathbb{C}^{d \times d})$. $\qquad\square$

References

1. C.J.K. Batty, T. Duyckaerts, Non-uniform stability for bounded semi-groups on Banach spaces. J. Evol. Equ. **8**(4), 765–780 (2008)
2. R. Chill, D. Seifert, Y. Tomilov, Semi-uniform stability of operator semigroups and energy decay of damped waves Philos. Trans. R. Soc. A. **378**, 20190614 (2020)
3. R.F. Curtain, G. Weiss, Strong stabilization of (almost) impedance passive systems by static output feedback, Math. Control Relat. Fields **9**(4), 643–671 (2019)
4. R. Dáger, E. Zuazua, *Wave Propagation, Observation and Control in 1-d Flexible Multi-Structures.* Mathématiques & Applications (Berlin) [Mathematics & Applications], 50 (Springer, Berlin, 2006)
5. F. Gesztesy, M. Waurick, *The Callias Index Formula Revisited* Lecture Notes in Mathematics 2157 (Springer, Berlin, 2016)
6. T. Eisner, B. Farkas, M. Haase, R. Nagel, *Operator Theoretic Aspects of Ergodic Theory.* Graduate Texts in Mathematics 272 (Springer, Cham, 2015)
7. Y. Le Gorrec, H. Zwart, B. Maschke, Dirac structures and boundary control systems associated with skew-symmetric differential operators. SIAM J. Control Optim. **44**(5), 1864–1892 (2005)
8. B. Jacob, K. Morris, H. Zwart, C_0-semigroups for hyperbolic partial differential equations on a one-dimensional spatial domain. J. Evol. Equ. **15**(2), 493–502 (2015)
9. B. Jacob, H.J. Zwart, *Linear Port-Hamiltonian Systems on Infinite-Dimensional Spaces*, volume 223 of Operator Theory: Advances and Applications. (Birkhäuser/Springer Basel AG, Basel, 2012). Linear Operators and Linear Systems
10. R. Picard, S. Trostorff, B. Watson, M. Waurick, A structural observation on port-Hamiltonian systems. SIAM J. Control Optim. **61**(2), 511–535 (2023)
11. C. Seifert, S. Trostorff, M. Waurick, *Evolutionary Equations – Picard's Theorem for Partial Differential Equations, and Applications* Operator Theory: Advances and Applications 287, 2021
12. S. Trostorff, M. Waurick, Characterisation for Exponential Stability of port-Hamiltonian Systems. Israel J. Math. (2024)
13. J.A. Villegas, H. Zwart, Y. Le Gorrec, B. Maschke, Exponential stability of a class of boundary control systems. IEEE Trans. Autom. Control **54**(1), 142–147 (2009)

Port-Hamiltonian Formulation of Oseen Flows

Timo Reis and Manuel Schaller

1 Introduction

The aim of this work is the representation of Oseen's equations in the port-Hamiltonian context. Hereby, we use recently developed theory of port-Hamiltonian system nodes from [24], which allows for an operator-theoretic description of linear infinite-dimensional port-Hamiltonian systems. The power of this approach is particularly due to the possibility of incorporating boundary control and observation, and it has been shown in [24] that—under a suitable choice of boundary input-output configurations—Maxwell's equations, advection-diffusion equations, and linear hyperbolic systems in one spatial variable fit into the framework of port-Hamiltonian system nodes.

Our setup consists of Oseen's equations with boundary control and observation which is formed by velocity and stress tensor in outward normal direction. Oseen's equations arise in the context of linearizations around steady states of the Navier-Stokes equations, where the latter describe the dynamics of viscous Newtonian fluids. The importance of flow problems for practice and the enormous mathematical challenges in their treatment have led to a huge variety of textbooks and publications on their analysis [11, 16, 19, 22, 27, 32, 33], numerical algorithms and finite-elements [7, 12, 25] and (closed-loop) control [10, 15, 26], without claiming to deliver a complete account. Further, flow problems have also been considered in the

T. Reis
Institute of Mathematics, Faculty of Mathematics and Natural Sciences, Technische Universität Ilmenau, Ilmenau, Germany
e-mail: timo.reis@tu-ilmenau.de

M. Schaller (✉)
Optimization-Based Control Group, Institute of Mathematics, Technische Universität Ilmenau, Ilmenau, Germany
e-mail: manuel.schaller@tu-ilmenau.de

    123
F. L. Schwenninger, M. Waurick (eds.), *Systems Theory and PDEs*,
Trends in Mathematics, https://doi.org/10.1007/978-3-031-64991-2_5

port-Hamiltonian context: In [23], compressible flows with boundary control are considered, where the geometric approach to port-Hamiltonian systems has been applied. In this context, incompressible flows are treated in [14], whereas Euler and shallow water flows are regarded in [30]. Further, an analytic approach to port-Hamiltonian formulation of flows in a one-dimensional spatial domain can be found in [4]. Uncontrolled Oseen equations are considered from a Hamiltonian perspective in [1] for the purpose of stability analysis.

This chapter is organized as follows: the remaining part of this introductory section is devoted to the physical derivation of Oseen's equations. Though this is somehow classical general knowledge, we give a concise presentation in order to make this chapter accessible to a community which is rather dealing with port-Hamiltonian systems than with flow problems. To motivate our configuration of boundary control and observation, we consider the flow through a cylindric domain in Sect. 2. This motivates our imposed boundary conditions, which comprise inflow, normal stress prescription as well as so-called *do nothing* and *no slip conditions*. This part also contains a formal derivation of the energy balance for boundary controlled Oseen's equations. Such an energy balance is typically provided by port-Hamiltonian systems. The analytical part of this chapter starts with Sect. 3, where we briefly review the theory of infinite-dimensional port-Hamiltonian systems from [24]. Thereafter, we present the main part of this chapter in Sect. 4. We consider Oseen's equations on a bounded Lipschitz domain $\Omega \subset \mathbb{R}^d$, $d \geq 2$. The boundary is split into two subdomains which are assumed to be separated by a Lipschitz manifold of dimension $n - 2$. These subdomains are provided with time-dependent functions with values in suitable fractional Sobolev spaces. It is in particular shown that this system fits into the framework of port-Hamiltonian system nodes.

Before starting with the derivation of the model, we declare the notation for differential operators and matrix inner products which are needed for models of flow problems.

For a Lipschitz domain $\Omega \subset \mathbb{R}^d$, $d \geq 2$, with boundary $\Gamma \subset \mathbb{R}^d$, we denote the outward unit normal $n : \Gamma \to \mathbb{R}^d$. For the spatial variable, we use the letter $\xi = (\xi_1, \ldots, \xi_d)^\top \in \mathbb{R}^d$, and $t \in \mathbb{R}$ stands for time. By $\dot{w}$, we denote the (weak) temporal derivative of $w : I \to \mathbb{R}^d$, where $I \subset \mathbb{R}$ is an interval, and by $\nabla w = \left(\frac{\partial w}{\partial \xi_1}, \ldots, \frac{\partial w}{\partial \xi_d} \right)^\top$ the (weak) spatial gradient of $w : \Omega \to \mathbb{R}$. For vector-valued functions $v : \Omega \to \mathbb{R}^d$, the $\mathbb{R}^{d \times d}$-valued function $\nabla v = (\nabla v_1, \ldots, \nabla v_d)^\top$ is the component-wise (weak) gradient, and the $\mathbb{R}^d$-valued function $\delta v = (\delta v_1, \ldots, \delta v_d)^\top$ is the component-wise (weak) Laplace operator. Moreover, the scalar-valued function $\operatorname{div} v = \sum_{i=1}^d \frac{\partial v_i}{\partial \xi_i}$ is the (weak) divergence. For a matrix-valued functions $z = (z_1, \ldots, z_d)^\top : \Omega \to \mathbb{R}^{d \times d}$ with functions $z_i^\top$, $i = 1, \ldots, d$, describing the rows, we denote by $\operatorname{div} z(\xi) = (\operatorname{div} z_1(\xi), \ldots, \operatorname{div} z_d(\xi))^\top$ the row-wise divergence. For vector-valued functions $b, v : \Omega \to \mathbb{R}^d$, we adopt the usual notation $(b \cdot \nabla)v := (b^\top \nabla v_1, \ldots, b^\top \nabla v_d)^\top$, where $v_i : \Omega \to \mathbb{R}$, $i = 1, \ldots, d$ are the components of v. Further, for matrices $A, B \in \mathbb{R}^{d \times d}$, we will use the Hilbert-Schmidt scalar product denoted by $A : B = \operatorname{tr}(A^\top B)$, where tr stands for the trace of a square matrix.

To introduce the model, let $[0, T]$ be the time interval of interest, let $\rho > 0$ be the mass density of the fluid, and let $b : \mathbb{R}^3 \to \mathbb{R}^3$ be the flow-independent stationary convection velocity field along which the flow is linearized. The function $v : [0, T] \times \Omega \to \mathbb{R}^d$ stands for the velocity of the particles, $p : [0, T] \times \Omega \to \mathbb{R}^d$ is the infinitesimal momentum, and $P : [0, T] \times \Omega \to \mathbb{R}$ denotes the pressure. Velocity and infinitesimal momentum are related by the density, i.e., $p = \rho v$.

The force balance for the particles reads

$$\dot{p}(t, \xi) = \operatorname{div} \sigma(v(t, \xi), P(t, \xi)) + \rho \left(b(t, \xi) \cdot \nabla\right) v(t, \xi), \qquad t \in [0, T], \ \xi \in \Omega,$$

(1.1a)

where $\sigma(v(t, \xi), P(t, \xi)) \in \mathbb{R}^{d \times d}$ denotes the stress tensor at (t, ξ). Incompressibility of the fluid means that the velocity field is divergence-free, that is,

$$0 = \operatorname{div} v(t, \xi), \qquad t \in [0, T], \ \xi \in \Omega.$$

(1.1b)

Further, since b is flow-independent stationary convection velocity field, we assume that its divergence as well as its normal trace vanishes, that is, $\operatorname{div} b(\xi) = 0$ for all $\xi \in \Omega$ and $\mathrm{n}(\xi)^\top b(\xi) = 0$ for all $\xi \in \Gamma$. The boundary of Ω is partitioned into three distinct parts Γ_{in}, Γ_{w}, and Γ_{out} with topological conditions which will be concretized in later parts. Boundary conditions on the velocity and normal stress are imposed, that is, for some given $u_v : [0, T] \times \Gamma_{\text{in}} \to \mathbb{R}^d$, $u_\sigma : [0, T] \times \Gamma_{\text{out}} \to \mathbb{R}^d$,

$$u_v(t, \xi) = v(t, \xi) \qquad\qquad t \in [0, T], \ \xi \in \Gamma_{\text{in}} \qquad (1.1c)$$

$$0 = v(t, \xi) \qquad\qquad t \in [0, T], \ \xi \in \Gamma_{\text{w}} \qquad (1.1d)$$

$$u_\sigma(t, \xi) = \sigma(v, P)(t, \xi)\,\mathrm{n}(\xi) \qquad\qquad t \in [0, T], \ \xi \in \Gamma_{\text{out}}, \qquad (1.1e)$$

where we set

$$\sigma(v, P)(t, \xi) = \sigma(v(t, \xi), P(t, \xi)), \qquad t \in [0, T], \ \xi \in \Omega.$$

Further, the infinitesimal momentum is initialized at $t = 0$, i.e.,

$$p(0, \xi) = p_0(\xi), \qquad \xi \in \Omega,$$

(1.1f)

where $p_0 : \mathbb{R}^3 \to \mathbb{R}^3$ is a given initial infinitesimal momentum. The $\mathbb{R}^{d \times d}$-valued stress tensor is defined by

$$\sigma(v, P) := \mu \left(\nabla v + \nabla v^\top\right) - P I_d,$$

where $I_d \in \mathbb{R}^{d \times d}$ is the identity matrix and $\mu > 0$ is the dynamic viscosity of the fluid. Since the velocity is divergence-free, Schwarz's theorem leads to

$$\operatorname{div} \nabla v^\top = \operatorname{div}\left(\nabla v_1 \ \ldots \ \nabla v_d\right) = \begin{pmatrix} \frac{\partial}{\partial \xi_1} \operatorname{div} v \\ \vdots \\ \frac{\partial}{\partial \xi_d} \operatorname{div} v \end{pmatrix} = 0, \tag{1.2}$$

whence the stress tensor slightly simplifies to

$$\operatorname{div} \sigma(v, P) = \operatorname{div}(\mu \nabla v - I_d P).$$

By neglecting the initial and boundary conditions, Oseen's equations read in compact form

$$\begin{pmatrix} \dot{p} \\ 0 \end{pmatrix} = \begin{pmatrix} \mu \delta - \rho b \cdot \nabla & -\nabla \\ -\operatorname{div} & 0 \end{pmatrix} \begin{pmatrix} \frac{1}{\rho} & 0 \\ 0 & I \end{pmatrix} \begin{pmatrix} p \\ P \end{pmatrix}. \tag{1.3}$$

By a formal consideration, the operator in (1.3) is additively composed by skew-adjoint and self-adjoint nonpositive operators, i.e.,

$$\begin{pmatrix} \mu \delta - \rho b \cdot \nabla & -\nabla \\ -\operatorname{div} & 0 \end{pmatrix} = \begin{pmatrix} -\rho b \cdot \nabla & -\nabla \\ -\operatorname{div} & 0 \end{pmatrix} + \begin{pmatrix} \mu \delta & 0 \\ 0 & 0 \end{pmatrix}$$

and, moreover, $\frac{1}{\rho} p$ is the derivative of the quadratic functional

$$\mathcal{H}(p) := \int_\Omega \frac{1}{2\rho} \| p(\xi) \|^2 \, d\xi \tag{1.4}$$

expressing the kinetic energy. Hence, amazingly the system (1.3) resembles the structure of port-Hamiltonian differential-algebraic equations as considered in [6]. In the formulation (1.3), the incompressibility condition can be regarded as an algebraic constraint. In our later—strictly analytical and less formal—considerations, we will regard Oseen's equation rather as an infinite-dimensional ordinary differential equation. Here, this can be done by including the algebraic constraint in the function space on which the infinitesimal momentum evolves. In the case of models for incompressible flows, this is indeed possible as no control enters directly to the incompressibility condition $\operatorname{div} v = 0$.

We note that, whereas finite-dimensional port-Hamiltonian DAEs are very well understood, there are many open questions regarding the infinite-dimensional case. Though first promising results have been published recently [17, 21, 28], the authors think that this area is still in an embryonic stage, and there is a need for a comprehensive theory for infinite-dimensional port-Hamiltonian DAEs. This is, however, beyond the scope of the present chapter.

Moreover, we remark that in the typical formulation of Navier-Stokes (and thus also Oseen's) equations, the velocity rather than the infinitesimal momentum is considered as an unknown. We prefer to use the latter since, for mechanical systems in port-Hamiltonian formulation, the state corresponding to kinetic energy is typically given by momenta [18]. Note that the velocity and momentum in incompressible flow problems only differ by multiplication with the constant mass density of the fluid.

2 Formal Derivation of Energy Balance: Flow Through a Cylinder

Our motivating example is depicted in Fig. 1 and is given by a tube, i.e., a cylindric domain in three dimensions. Inside the tube, there is a flow field with velocity $v : \mathbb{R}^3 \to \mathbb{R}^3$ and pressure $P : \mathbb{R}^3 \to \mathbb{R}$. The boundary is partitioned into three distinct parts, namely the side wall and the two tube ends, i.e.,

$$\Gamma = \Gamma_{\text{in}} \uplus \Gamma_{\text{w}} \uplus \Gamma_{\text{out}}.$$

The left tube end Γ_{in} is the inflow boundary, Γ_{w} is the side wall, and the right tube end Γ_{out} is the outflow boundary.
We consider the boundary conditions

$$
\begin{array}{lll}
v = u_v & \text{on } \Gamma_{\text{in}} & \text{prescribed inflow velocity,} \\[6pt]
v = 0 & \text{on } \Gamma_{\text{w}} & \text{no-slip,} \\[6pt]
\sigma(v, P)\,\mathrm{n} = u_\sigma & \text{on } \Gamma_{\text{out}} & \text{prescribed normal stress.}
\end{array}
$$

In the case where u_σ vanishes, the latter boundary condition is typically referred to as "do-nothing condition."

Next, we deduce an energy balance. Note that all our formal derivations can also be done in arbitrary spatial dimensions $d \geq 2$, and our analytical results in Sect. 4 indeed cover the general case. In this part, we stick to $d = 3$ for illustration purposes.

Fig. 1 Cylindrical domain in $\mathbb{R}^3$ with in- and outflow

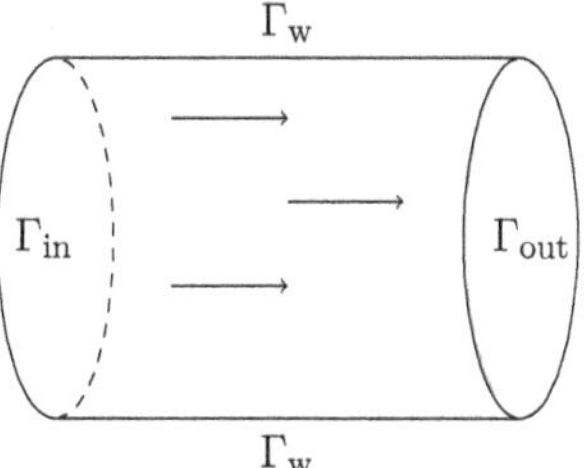

We first derive an auxiliary result with respect the advection term. As b is divergence-free with vanishing normal boundary trace, for a divergence-free function $v : \mathbb{R}^3 \to \mathbb{R}^3$ we have, by using the standard inner product in L^2,

$$\langle (\rho b \cdot \nabla) v, v \rangle_{L^2(\Omega;\mathbb{R}^3)} = \rho \sum_{i=1}^{3} \langle b \cdot \nabla v_i, v_i \rangle_{L^2(\Omega)} = \rho \sum_{i=1}^{3} \langle \nabla v_i, b\, v_i \rangle_{L^2(\Omega;\mathbb{R}^3)}$$

$$= \rho \sum_{i=1}^{3} -\langle v_i, \operatorname{div}(b\, v_i) \rangle_{L^2(\Omega)} + \langle v_i, \mathrm{n}^\top (b\, v_i) \rangle_{L^2(\Gamma)}$$

$$= \rho \sum_{i=1}^{3} -\langle v_i, b \cdot \nabla v_i \rangle_{L^2(\Omega)} = -\langle (\rho b \cdot \nabla) v, v \rangle_{L^2(\Omega;\mathbb{R}^3)},$$

where the latter equality follows by the chain rule of the divergence $\operatorname{div}(bv_i) = \operatorname{div}(b)v_i + b \cdot \nabla v_i$. The above findings in particular imply that

$$\forall v : \operatorname{div} v = 0 : \quad \langle (\rho b \cdot \nabla) v, v \rangle_{L^2(\Omega;\mathbb{R}^3)} = 0. \tag{2.1}$$

Then, by setting $p(t) = p(t, \cdot) : \Omega \to \mathbb{R}^3$ and using the formula (1.4) for the kinetic energy $\mathcal{H}(p(t))$, the infinitesimal momentum in (1.1) formally fulfills the power balance

$$\tfrac{\mathrm{d}}{\mathrm{d}t} \mathcal{H}(p(t)) = \langle \dot{p}(t), v(t) \rangle$$

$$= \langle \operatorname{div} \sigma(v(t), P(t)), v(t) \rangle_{L^2(\Omega;\mathbb{R}^3)} + \langle (\rho b \cdot \nabla) v(t), v(t) \rangle_{L^2(\Omega;\mathbb{R}^3)}$$

$$= -\langle \sigma(v(t), P(t)), \nabla v \rangle_{L^2(\Omega;\mathbb{R}^{3\times3})} + \langle \sigma(v(t), P(t))\mathrm{n}, v(t) \rangle_{L^2(\Gamma,\mathbb{R}^3)}$$

$$= \langle \mu \nabla v(t) - I_3 P(t), \nabla v \rangle_{L^2(\Omega,\mathbb{R}^{3\times3})} + \langle (\mu \nabla v(t) - I_3 P(t))\,\mathrm{n}, v \rangle_{L^2(\Gamma,\mathbb{R}^3)}.$$

Hereby, we note that the inner product $\langle \cdot, \cdot \rangle_{L^2(\Omega,\mathbb{R}^{d\times d})}$ consists of the integral of the pointwise Hilbert-Schmidt inner product of the two arguments.

A consequence of the above computations is that the infinitesimal change of energy is additively expressed by a term on the domain and the boundary. Unsurprisingly, each summand has the physical dimension of power: the first one is the volume integral over the product of force per surface element and infinitesimal change of velocity; the second summand is a surface integral of the the product of force per surface element and velocity.

Now by using that for any divergence-free $v : \Omega \to \mathbb{R}^3$ and $P : \Omega \to \mathbb{R}$,

$$\langle I_3 P, \nabla v \rangle_{L^2(\Omega;\mathbb{R}^{3\times3})} = \int_\Omega \operatorname{tr}\left(\begin{pmatrix} P & & \\ & P & \\ & & P \end{pmatrix} \begin{pmatrix} \frac{\partial v_1}{\partial \xi_1} & \frac{\partial v_1}{\partial \xi_2} & \frac{\partial v_1}{\partial \xi_3} \\ \frac{\partial v_2}{\partial \xi_1} & \frac{\partial v_2}{\partial \xi_2} & \frac{\partial v_2}{\partial \xi_3} \\ \frac{\partial v_3}{\partial \xi_1} & \frac{\partial v_3}{\partial \xi_2} & \frac{\partial v_3}{\partial \xi_3} \end{pmatrix} \right) \mathrm{d}\xi = \int_\Omega P \operatorname{div} v \, \mathrm{d}\xi = 0, \tag{2.2}$$

the above power balance simplifies to

$$\frac{\mathrm{d}}{\mathrm{d}t}\mathcal{H}(p(t)) = -\mu\|\nabla v(t)\|^2_{L^2(\Omega;\mathbb{R}^{d\times d})} + \langle(\mu\nabla v(t) - I_3 P(t))\,\mathrm{n},\, v(t)\rangle_{L^2(\Gamma;\mathbb{R}^d)}.$$

$$(2.3)$$

That is, in the power balance, the pressure only enters on the boundary. The term $\mu\|\nabla v(t)\|^2_{L^2(\Omega;\mathbb{R}^{3\times 3})}$ is called *enstrophy* [19, Chap. I, Sec. 2] and, as can be seen in the above formula (2.4), it is crucial in determining the rate of (interior) kinetic energy decay. By incorporating the split of the boundary into three parts according to Fig. 1, the boundary term reads (for sake of brevity, we omit the time variable)

$$\langle(\mu\nabla v - I_3 P)\,\mathrm{n},\, v\rangle_{L^2(\Gamma,\mathbb{R}^3)}$$

$$= \langle(\mu\nabla v - I_3 P)\,\mathrm{n},\, v\rangle_{L^2(\Gamma_{\mathrm{in}},\mathbb{R}^3)} + \langle(\nu\nabla v - I_3 P)\,\mathrm{n},\, v\rangle_{L^2(\Gamma_{\mathrm{w}},\mathbb{R}^3)}$$

$$+ \langle(\nu\nabla v - I_3 P)\,\mathrm{n},\, v\rangle_{L^2(\Gamma_{\mathrm{out}},\mathbb{R}^d)}$$

$$= \langle(\mu\nabla v - I_3 P)\,\mathrm{n},\, u_v\rangle_{L^2(\Gamma_{\mathrm{in}},\mathbb{R}^3)} + \langle u_\sigma,\, v\rangle_{L^2(\Gamma_{\mathrm{out}},\mathbb{R}^3)},$$

as $v = u_v$ on Γ_{in} (velocity prescription), $v = 0$ on Γ_{w} (no-slip condition), and $(\nu\nabla v - I_3 P)\mathrm{n} = u_\sigma$ on Γ_{out} (normal stress condition).

Using the latter and integrating (2.3) over $[t_0, t_1] \subset [0, T]$, $t_0 \le t_1$, we obtain the energy balance

$$\mathcal{H}(p(t_1)) - \mathcal{H}(p(t_0))$$

$$= -\mu\int_{t_0}^{t_1}\|\nabla v(t)\|^2_{L^2(\Omega;\mathbb{R}^{d\times d})}\mathrm{d}t$$

$$+ \int_{t_0}^{t_1}\langle(\mu\nabla v(t) - I_3 P(t))\,\mathrm{n},\, u_v(t)\rangle_{L^2(\Gamma_{\mathrm{in}},\mathbb{R}^3)}\mathrm{d}t + \int_{t_0}^{t_1}\langle u_\sigma(t),\, v(t)\rangle_{L^2(\Gamma_{\mathrm{out}},\mathbb{R}^3)}\mathrm{d}t$$

$$\le \int_{t_0}^{t_1}\langle(\mu\nabla v(t) - I_3 P(t))\,\mathrm{n},\, u_v(t)\rangle_{L^2(\Gamma_{\mathrm{in}},\mathbb{R}^3)}\mathrm{d}t + \int_{t_0}^{t_1}\langle u_\sigma(t),\, v(t)\rangle_{L^2(\Gamma_{\mathrm{out}},\mathbb{R}^3)}\mathrm{d}t.$$

$$(2.4)$$

Energy loss is expressed by the temporal integral over the enstrophy, whereas the energy exchange with outside is formed by the surface integral over the inner product between the velocity and normal component of the stress tensor on the boundary.

Remark 2.1 One can straightforwardly include velocity-dependent reaction terms of the form $-cv$ in all our considerations, with a constant $c > 0$ on the right-hand side of the dynamics (1.1a). This would lead to an additional energy loss term $-c\int_{t_0}^{t_1}\|v\|^2_{L^2(\Omega;\mathbb{R}^d)}\mathrm{d}t$ in the energy balance (2.4).

3 Port-Hamiltonian System Nodes

In this part, we present the general analytical background for our approach to port-Hamiltonian representation of Oseen's equations.

We first declare some general functional analytic notation. Let $\mathcal{X}$, $\mathcal{Y}$ be Hilbert spaces. Throughout this chapter, all Hilbert spaces are real. The norm in $\mathcal{X}$ will be denoted by $\|\cdot\|_{\mathcal{X}}$ or simply $\|\cdot\|$, if clear from context. The identity mapping in $\mathcal{X}$ is abbreviated by $I_{\mathcal{X}}$ (or just I, if clear from context), and we set $I_n := I_{\mathbb{R}^n}$.

The symbol $\mathcal{X}^*$ stands for the *dual* of $\mathcal{X}$; the duality product is denoted by $\langle\cdot,\cdot\rangle_{\mathcal{X}^*,\mathcal{X}}$, and the inner product in $\mathcal{X}$ by $\langle\cdot,\cdot\rangle_{\mathcal{X}}$. We write $L(\mathcal{X},\mathcal{Y})$ for the space of bounded linear operators from $\mathcal{X}$ to $\mathcal{Y}$. As usual, we abbreviate $L(\mathcal{X}) := L(\mathcal{X},\mathcal{X})$. The adjoint of a densely defined linear operator $A : \mathcal{X} \supset \operatorname{dom} A \to \mathcal{Y}$ is denoted by $A^* : \mathcal{Y}^* \supset \operatorname{dom} A^* \to \mathcal{X}^*$.

We adopt the notation of the book [2] by ADAMS for Lebesgue and Sobolev spaces as well as for the spaces of continuous and continuously differentiable functions. For function spaces with values in a Hilbert space $\mathcal{X}$, we indicate this by denoting "; $\mathcal{X}$" after specifying the (spatial or temporal) domain. For instance, the Lebesgue space of p-integrable $\mathcal{X}$-valued functions on the domain Ω is $L^p(\Omega; \mathcal{X})$. Note that, throughout this chapter, integration of $\mathcal{X}$-valued functions always has to be understood in the Bochner sense [8].

We briefly review the basics of infinite-dimensional port-Hamiltonian systems as presented in [24]. In principle, this approach is based on the formulation

$$
\begin{aligned}
\dot{x}(t) &= (J - R)Hx(t) + (B - P)u(t), \\
y(t) &= (B^\top + P^\top)Hx(t) + (S - N)u(t)
\end{aligned}
\tag{3.1}
$$

of finite-dimensional port-Hamiltonian systems from [6, 20], where $J \in \mathbb{R}^{n \times n}$ and $N \in \mathbb{R}^{m \times m}$ are skew-symmetric, and $H \in \mathbb{R}^{n \times n}$ and $W := \begin{bmatrix} R & P \\ P^\top & S \end{bmatrix} \in \mathbb{R}^{(n+m) \times (n+m)}$ are symmetric positive semi-definite. The total energy of the system is given by the *Hamiltonian* $\mathcal{H}(x) = \frac{1}{2} x^\top H x$. By using (3.1) together with skew-symmetry of J and N, as well as positive semi-definiteness of W and H, for all $t > 0$, $u \in L^2([0, t]; \mathbb{R}^m)$, $x_0 \in \mathbb{R}^n$, the solution of (3.1) with $x(0) = x_0$ fulfills the *dissipation inequality*

$$
\begin{aligned}
\mathcal{H}(x(t)) - \mathcal{H}(x_0) &= -\int_0^t \begin{pmatrix} Hx(\tau) \\ u(\tau) \end{pmatrix}^\top \begin{bmatrix} R & P \\ P^* & S \end{bmatrix} \begin{pmatrix} Hx(\tau) \\ u(\tau) \end{pmatrix} \mathrm{d}\tau + \int_0^t u(\tau)^\top y(\tau)\mathrm{d}\tau \\
&\leq \int_0^t u(\tau)^\top y(\tau)\mathrm{d}\tau,
\end{aligned}
\tag{3.2}
$$

which has the physical interpretation of an energy balance. While $\mathcal{H}(x(t))$ stands for the energy stored at time t, the first integral after the equality sign is the energy dissipated by the system during the time interval $[0, t]$, whereas $u(\tau)^\top y(\tau)$ can be regarded as the external power supply to the system.

In compact form, (3.1) can be rewritten as

$$\begin{pmatrix} \dot{x}(t) \\ y(t) \end{pmatrix} = \begin{bmatrix} I_n & \\ & -I_m \end{bmatrix} M \begin{pmatrix} Hx(t) \\ u(t) \end{pmatrix}, \tag{3.3}$$

where $M := \begin{bmatrix} J-R & B-P \\ -B^\top-P^\top & N-S \end{bmatrix} \in \mathbb{R}^{(n+m)\times(n+m)}$ is a dissipative matrix. The generalization to infinite-dimensional systems as presented in [24] uses exactly this representation by incorporating the theory of *system nodes* by STAFFANS [29]. To allow for partial differential equations, where the dynamics themselves are described by a differential operator, we clearly presume that M is a dissipative operator. However, the incorporation of boundary control requires to allow M to be unbounded and densely defined on some dense subspace of the Cartesian product of the state and input space, whereas the domain is not necessarily a Cartesian product of a dense subspace of the state space and the input space. Note that, in [24], unbounded and non-coercive Hamiltonians are also considered. Since the latter is actually not needed for Oseen equations, we "boil down" our brief introduction of infinite-dimensional port-Hamiltonian systems to those with bounded and coercive Hamiltonians, which is indeed a drastic simplification.

The main ingredient for our concept of infinite-dimensional port-Hamiltonian system is given by *dissipation nodes*. To this end, we denote the canonical projection onto $\mathcal{X}$ and $\mathcal{U}^*$ in $\mathcal{X} \times \mathcal{U}^*$ by $P_\mathcal{X} \in L(\mathcal{X} \times \mathcal{U}^*, \mathcal{X})$ and $P_{\mathcal{U}^*} \in L(\mathcal{X} \times \mathcal{U}^*, \mathcal{U}^*)$, respectively. Further, we can canonically identify $(\mathcal{X} \times \mathcal{U})^* = \mathcal{X}^* \times \mathcal{U}^*$.

Definition 3.1 (Dissipation Node) Let $\mathcal{X}, \mathcal{U}$ be Hilbert spaces, let $\mathcal{U}^*$ be the anti-dual of $\mathcal{U}$, and identify $\mathcal{X}$ with its anti-dual, i.e., $\mathcal{X}^* = \mathcal{X}$. A *dissipation node* on $(\mathcal{X}, \mathcal{U})$ is a (possibly unbounded) linear operator $M : \mathcal{X} \times \mathcal{U} \supset \operatorname{dom} M \to \mathcal{X} \times \mathcal{U}^*$ that satisfies the following:

(a) M is closed and dissipative.
(b) $P_\mathcal{X} M : \mathcal{X} \times \mathcal{U} \supset \operatorname{dom} M \to \mathcal{X}$ is closed.
(c) For all $u \in \mathcal{U}$, there exists some $x' \in \mathcal{X}$ with $\left(\begin{smallmatrix} x' \\ u \end{smallmatrix} \right) \in \operatorname{dom} M$.
(d) For the *main operator* $F : \mathcal{X} \supset \operatorname{dom} F \to \mathcal{X}$ defined by

$$\operatorname{dom} F := \left\{ x' \in \mathcal{X} \,\middle|\, \left(\begin{smallmatrix} x' \\ 0 \end{smallmatrix} \right) \in \operatorname{dom} M \right\}$$

and $Fx' := P_\mathcal{X} M \left(\begin{smallmatrix} x' \\ 0 \end{smallmatrix} \right)$, there exists some $\lambda > 0$ such that $\lambda I - F$ has dense range.

We note that though the theory and results on port-Hamiltonian systems in [24] are formulated and proven for complex Hilbert spaces, the following findings hold as well for the real case, which can be verified by a complexification.

The previous definition justifies the use of the block operator notation

$$M := \begin{bmatrix} F\&G \\ K\&L \end{bmatrix} \tag{3.4}$$

for dissipation nodes, where $K\&L = P_\mathcal{X} M$ and $F\&G = P_{\mathcal{U}^*} M$. The symbol "$\&$" is used to indicate that the domain of M is not necessarily a Cartesian product of subspaces of $\mathcal{X}$ and $\mathcal{U}$. However, it is shown in [24] that, by using the extrapolation space $\mathcal{X}_{-1}$ defined by the completion of $\mathcal{X}$ with respect to the norm $\|x\|_{-1} := \|(\lambda I - F)^{-1}\|_\mathcal{X}$ for some λ in the resolvent of F (the induced topology is indeed independent on λ in the resolvent of F), F extends to an operator $F_{-1} \in L(\mathcal{X}, \mathcal{X}_{-1})$, and we can regard $G \in L(\mathcal{U}, \mathcal{X}_{-1})$; that is,

$$F\&G \left(\begin{smallmatrix} x \\ u \end{smallmatrix}\right) = F_{-1}x + Gu \qquad \forall \left(\begin{smallmatrix} x \\ u \end{smallmatrix}\right) \in \operatorname{dom} M = \operatorname{dom}(F\&G). \tag{3.5}$$

Note that such a splitting is in general not possible for the operator $K\&L$ defining the output.

Definition 3.2 (Port-Hamiltonian System) Let $\mathcal{X}$ and $\mathcal{U}$ be Hilbert spaces. Assume that $H \in L(\mathcal{X})$ is positive, self-adjoint, and has a bounded inverse. Further, let $M : \mathcal{X} \times \mathcal{U} \supset \operatorname{dom} M \to \mathcal{X} \times \mathcal{U}^*$ be a dissipation node on $(\mathcal{X}, \mathcal{U})$. Then we call

$$\begin{pmatrix} \dot{x}(t) \\ y(t) \end{pmatrix} = \begin{bmatrix} F\&G \\ -K\&L \end{bmatrix} \begin{pmatrix} Hx(t) \\ u(t) \end{pmatrix}. \tag{3.6}$$

a *port-Hamiltonian system on* $(\mathcal{X}, \mathcal{U})$. A *classical trajectory* on $[0, T]$ is a triple

$$(y, x, u) \in C([0, T]; \mathcal{U}^*) \times C^1([0, T]; \mathcal{X}) \times C([0, T]; \mathcal{U}),$$

which fulfills (3.6) pointwise on $[0, T]$, and *generalized trajectories* are limits of classical trajectories in the topology of $L^2([0, T]; \mathcal{Y}) \times C([0, T]; \mathcal{X}) \times L^2([0, T]; \mathcal{U})$.

It follows from [24, Prop. 3.6] that $\begin{bmatrix} F\&G \\ -K\&L \end{bmatrix} \begin{bmatrix} H & 0 \\ 0 & I_\mathcal{U} \end{bmatrix}$ defines a system node in the sense of STAFFANS [29], whence this is called a *port-Hamiltonian system node* in [24]. Since H is assumed to be bounded, self-adjoint, positive, and boundedly invertible, the functional $x \mapsto \|x\|_H := (\langle x(t), Hx(t)\rangle)^{1/2}$ is equivalent to the original norm $\|\cdot\|_\mathcal{X}$ on $\mathcal{X}$. It can be moreover seen that FH is a maximally dissipative (see [9] for a definition) on $\mathcal{X}$ endowed with the norm $\|\cdot\|_H$ (and thus with the scalar product $\langle \cdot, H \cdot \rangle$). We collect some properties of port-Hamiltonian systems and their trajectories. Since the result is a specialization of [24, Prop. 3.11], we omit the proof.

Proposition 3.3 (Port-Hamiltonian Systems) *Let* $\mathcal{X}, \mathcal{U}$ *be Hilbert spaces, and let a port-Hamiltonian system (3.6) in the sense of Definition 3.2 on* $(\mathcal{X}, \mathcal{U})$ *be given.*

Then the following holds:

(a) *The main operator* $A = FH$ *of* M *generates a contractive semigroup* $\mathfrak{A} :$ $\mathbb{R}_{\geq 0} \to L(\mathcal{X})$ *in* $(\mathcal{X}, \|\cdot\|_H)$.

(b) *For* $T > 0$, $x_0 \in \mathcal{X}$ *and* $u \in W^{2,1}([0,T]; \mathcal{U})$ *with* $\begin{pmatrix} Hx_0 \\ u(0) \end{pmatrix} \in \mathrm{dom}\, M$, *there exists a unique classical trajectory* (y, x, u) *for* (3.6) *with* $x(0) = x_0$.

(c) *For all* $T > 0$, *the generalized trajectories* $(y, x, u) \in L^2([0,T]; \mathcal{U}^*) \times C([0,T]; \mathcal{X}) \times L^2([0,T]; \mathcal{U})$ *(and thus also the classical solutions) fulfill the dissipation inequality*

$$\forall t \in [0,T]: \qquad \mathcal{H}(x(t)) - \mathcal{H}(x(0))$$

$$= \int_0^t \left\langle M\begin{pmatrix} Hx(\tau) \\ u(\tau) \end{pmatrix}, \begin{pmatrix} Hx(\tau) \\ u(\tau) \end{pmatrix} \right\rangle_{\mathcal{X} \times \mathcal{U}^*, \mathcal{X} \times \mathcal{U}} \mathrm{d}\tau + \int_0^t \langle y(\tau), u(\tau) \rangle_{\mathcal{U}^*, \mathcal{U}} \mathrm{d}\tau$$

$$\leq \int_0^t \langle y(\tau), u(\tau) \rangle_{\mathcal{U}^*, \mathcal{U}} \mathrm{d}\tau, \qquad (3.7)$$

where $\mathcal{H} : \mathcal{X} \to \mathbb{R}$ *with* $\mathcal{H}x := \frac{1}{2}\langle x, Hx \rangle$ *is the Hamiltonian associated to* H.

Though it seems to be exotic in some sense, we emphasize that the input in Proposition 3.3 (b) is truly supposed to be twice weakly differentiable with absolutely integrable second derivative. By using that the norms $\|\cdot\|_H$ and $\|\cdot\|_{\mathcal{X}}$ are equivalent when H is self-adjoint, bounded, and boundedly invertible, the semigroup $\mathfrak{A} : \mathbb{R}_{\geq 0} \to L(\mathcal{X})$ is bounded when $\mathcal{X}$ is equipped with the standard norm $\|\cdot\|_{\mathcal{X}}$. We further give some comments on the solution concept, which follow from the results in [24].

Remark 3.4 (Classical/Generalized Trajectories) Let $T > 0$, let $\mathcal{X}, \mathcal{U}$ be Hilbert spaces, and let a port-Hamiltonian system on $(\mathcal{X}, \mathcal{U})$ in the sense of Definition 3.2 be given.

(a) If (x, u, y) is a classical trajectory, then

$$\begin{pmatrix} x \\ u \end{pmatrix} \in C([0,T]; \mathrm{dom}\, M),$$

where $\mathrm{dom}\, M$ is equipped with the graph norm of M.

(b) For the split of $F \& G$ as in (3.5), consider $F_{-1}H : \mathcal{X}_{-1} \supset \mathcal{X} \to \mathcal{X}_{-1}$. Then $F_{-1}H$ generates a semigroup $\mathfrak{A} : \mathbb{R}_{\geq 0} \to L(\mathcal{X}_{-1})$ on $\mathcal{X}_{-1}$. Then (x, u, y) is a generalized trajectory for (3.6), if $x \in C([0,T]; \mathcal{X})$ and

$$\forall t \in [0,T]: \quad x(t) = \mathfrak{A}(t)x(0) + \int_0^t \mathfrak{A}_{-1}(t - \tau)Gu(\tau)\mathrm{d}\tau, \qquad (3.8)$$

where the latter summand has to be interpreted as an integral in the space $\mathcal{X}_{-1}$.

(c) If (x, u, y) is a generalized trajectory for (3.6), then (3.8) holds. The output evaluation $y(t) = K\&L \left(\begin{smallmatrix} Hx(t) \\ u(t) \end{smallmatrix} \right)$ is—at a glance—not necessarily well-defined for all $t \in [0, T]$. However, it is shown in [29, Lem. 4.7.9] (for the larger context of system nodes) that the second integral of $\left(\begin{smallmatrix} x \\ u \end{smallmatrix} \right)$ is continuous as a mapping from $[0, T]$ to $\mathrm{dom}(K\&L) = \mathrm{dom}\, M$. As a consequence, the output can—in the distributional sense—be defined as the second derivative of $K\&L$ applied to the second integral of $\left(\begin{smallmatrix} Hx \\ u \end{smallmatrix} \right)$. This can be used to show that (x, u, y) is a generalized trajectory for (3.6) if, and only if, (3.8) and

$$y = \left(t \mapsto \frac{\mathrm{d}^2}{\mathrm{d}t^2} K\&L \int_0^t (t - \tau) \left(\begin{smallmatrix} Hx(\tau) \\ u(\tau) \end{smallmatrix} \right) \mathrm{d}\tau \right) \in L^2([0, T]; \mathcal{Y}).$$

4 Oseen Equations as Port-Hamiltonian System

In this part, we formulate the PDE (1.1) by means of a port-Hamiltonian system in the sense of Definition 3.2. In view of the representation (1.3) and the quadratic Hamiltonian energy as in (1.4), we define the Hamiltonian operator given as a simple multiplication operator

$$Hz = \tfrac{1}{\rho}z.$$

such that $\mathcal{H}(p) = \tfrac{1}{2}\langle p, Hp \rangle_{L^2(\Omega; \mathbb{R}^d)}$. Since the density $\rho > 0$ is assumed to be constant, this operator boundedly maps square integrable and divergence-free momenta to square integrable and divergence-free velocities. Having defined the Hamiltonian, it remains to describe the dissipation node incorporating the differential operators in the formulation (1.3) to obtain a port-Hamiltonian system in the form of (3.6). This will be done in the sequel.

4.1 Assumptions on the Domain

The following assumptions are made on the domain Ω and (parts of the) boundary $\Gamma, \Gamma_{\mathrm{in}}, \Gamma_{\mathrm{w}},$ and Γ_{out} throughout the remainder of this work:

- $\Omega \subset \mathbb{R}^d, d \geq 2$ is a Lipschitz domain with boundary Γ.
- $\Gamma_{\mathrm{in}}, \Gamma_{\mathrm{w}},$ and $\Gamma_{\mathrm{out}} \subset \Gamma$ are relatively open and Lipschitz domains of dimension $d - 1$ with boundary.
- Γ_{in} and Γ_{out} are non-empty.
- $\Gamma_{\mathrm{in}}, \Gamma_{\mathrm{w}},$ and Γ_{out} are disjoint, that is,

$$\Gamma_{\mathrm{in}} \cap \Gamma_{\mathrm{w}} = \Gamma_{\mathrm{in}} \cap \Gamma_{\mathrm{out}} = \Gamma_{\mathrm{w}} \cap \Gamma_{\mathrm{out}} = \emptyset.$$

- The closures of Γ_{in}, Γ_{w}, and Γ_{out} unify to the whole boundary, i.e.,

$$\overline{\Gamma_{\mathrm{in}}} \cup \overline{\Gamma_{\mathrm{w}}} \cup \overline{\Gamma_{\mathrm{out}}} = \Gamma.$$

- Γ_{in} and Γ_{out} do not touch, that is,

$$\overline{\Gamma_{\mathrm{in}}} \cap \overline{\Gamma_{\mathrm{out}}} = \emptyset.$$

In view of the tube depicted in Fig. 1, Γ_{w} is the wall of the cylinder as the no-slip boundary with vanishing velocities and Γ_{in} is the inflow boundary with prescribed velocities. Likewise, Γ_{out} will be the part of the boundary in which conditions on the stress tensor $\sigma(v, P)$ in normal direction are imposed. In the example illustrated in Fig. 1, this is the part containing the "do-nothing conditions." In fact, we also allow nonzero conditions on Γ_{out} in our forthcoming analysis.

4.2 Spaces and Traces

Let $H^{1/2}(\Gamma; \mathbb{R}^d)$ be the Sobolev space of fractional order 1/2. Further, for some relatively open subset $\Gamma_i \subset \Gamma$, the space $H_0^{1/2}(\Gamma_i; \mathbb{R}^d)$ consists of elements of $H^{1/2}(\Gamma_i; \mathbb{R}^d)$ whose extension by zero is an element of $H^{1/2}(\Gamma; \mathbb{R}^d)$ [2].

We consider the *trace operator*

$$\gamma : H^1(\Omega; \mathbb{R}^d) \to H^{1/2}(\Gamma; \mathbb{R}^d), \tag{4.1}$$

which is bounded and surjective by [13, Thm. 1.5.1.3]. Further, consider the space

$$H_{\sigma,\mathrm{w}}^1(\Omega; \mathbb{R}^d) = \{v \in H^1(\Omega; \mathbb{R}^d) \,|\, \operatorname{div} v = 0 \text{ on } \Omega \text{ and } \gamma v|_{\Gamma_{\mathrm{w}}} = 0\}.$$

It can be seen that the operator defining the restriction $w|_{\Gamma_{\mathrm{w}}} \in H^{1/2}(\Gamma_{\mathrm{w}}; \mathbb{R}^d)$ of $w \in H^{1/2}(\Gamma; \mathbb{R}^d)$ is bounded. This together with boundedness of the trace operator γ and boundedness of the divergence operator $\operatorname{div} : H^1(\Omega; \mathbb{R}^d) \to L^2(\Omega; \mathbb{R})$ yields that $H_{\sigma,\mathrm{w}}^1(\Omega; \mathbb{R}^d)$ is a closed subspace of $H^1(\Omega; \mathbb{R}^d)$; it is therefore a Hilbert space endowed with the inner product in $H^1(\Omega; \mathbb{R}^d)$.

Since, by our assumptions, Γ_{in} and Γ_{out} do not touch, there is a canonical isomorphism

$$\left\{w \in H^{1/2}(\Gamma; \mathbb{R}^d) \,|\, w|_{\Gamma_{\mathrm{w}}} = 0\right\} \cong H_0^{1/2}(\Gamma_{\mathrm{in}}; \mathbb{R}^d) \times H_0^{1/2}(\Gamma_{\mathrm{out}}; \mathbb{R}^d). \tag{4.2}$$

Now Gauß' divergence theorem yields that γ restricts to

$$\gamma : H^1_{\sigma,\mathrm{w}}(\Omega; \mathbb{R}^d) \to$$
$$\left\{ (w_{\mathrm{in}}, w_{\mathrm{out}}) \in H^{1/2}_0(\Gamma_{\mathrm{in}}; \mathbb{R}^d) \times H^{1/2}_0(\Gamma_{\mathrm{out}}; \mathbb{R}^d) \,\Big|\, \int_{\Gamma_{\mathrm{in}}} w_{\mathrm{in}} \cdot \mathrm{n}\, \mathrm{d}\xi \right.$$
$$\left. + \int_{\Gamma_{\mathrm{out}}} w_{\mathrm{out}} \cdot \mathrm{n}\, \mathrm{d}\xi = 0 \right\}, \qquad (4.3)$$

which is again denoted by γ for sake of brevity. Note that the target space of the latter operator is a closed subspace of $H^{1/2}_0(\Gamma_{\mathrm{in}}; \mathbb{R}^d) \times H^{1/2}_0(\Gamma_{\mathrm{out}}; \mathbb{R}^d)$ with co-dimension one. It is thus a Hilbert space equipped with the inner product in $H^{1/2}_0(\Gamma_{\mathrm{in}}; \mathbb{R}^d) \times H^{1/2}_0(\Gamma_{\mathrm{out}}; \mathbb{R}^d)$. Indeed, it can be concluded from [12, Lem. 2.2] that γ as in (4.3) is surjective.

Now we denote

$$\begin{aligned} \gamma_{\mathrm{in}} : \quad & H^1_{\sigma,\mathrm{w}}(\Omega; \mathbb{R}^d) \to H^{1/2}_0(\Gamma_{\mathrm{in}}; \mathbb{R}^d), \\ \gamma_{\mathrm{out}} : \quad & H^1_{\sigma,\mathrm{w}}(\Omega; \mathbb{R}^d) \to H^{1/2}_0(\Gamma_{\mathrm{out}}; \mathbb{R}^d) \end{aligned} \qquad (4.4)$$

to be the canonical projections of γ onto $H^{1/2}_0(\Gamma_{\mathrm{in}}; \mathbb{R}^d)$ and $H^{1/2}_0(\Gamma_{\mathrm{out}}; \mathbb{R}^d)$, respectively. Since, by our assumption on the domain, Γ_{in} and Γ_{out} are both non-empty, it follows from the surjectivity of γ in (4.3) that both γ_{in} and γ_{out} are surjective.

Moreover, let $H(\mathrm{div}, \Omega; \mathbb{R}^{d \times d})$ be the space of square integrable functions with values in $\mathbb{R}^{d \times d}$ for which the row-wise divergence exists and is square integrable, where the divergence div is defined in a weak sense. That is, for the space $H^1_0(\Omega; \mathbb{R}^d)$ of all elements of $H^1(\Omega; \mathbb{R}^d)$ with vanishing boundary trace,

$$z = \mathrm{div}\, x \quad \Longleftrightarrow \quad \forall \varphi \in H^1_0(\Omega; \mathbb{R}^d) : \ -\langle \nabla\varphi, x \rangle_{L^2(\Omega;\mathbb{R}^{d \times d})} = \langle \varphi, z \rangle_{L^2(\Omega;\mathbb{R}^d)}.$$

Likewise, $H(\mathrm{div}, \Omega; \mathbb{R}^d)$ denotes the space of square integrable $\mathbb{R}^d$-valued functions with square integrable divergence, where the latter is defined via

$$z = \mathrm{div}\, x \quad \Longleftrightarrow \quad \forall \varphi \in H^1_0(\Omega) : \ -\langle \nabla\varphi, x \rangle_{L^2(\Omega;\mathbb{R}^d)} = \langle \varphi, z \rangle_{L^2(\Omega)}.$$

Here we collect the fundamentals to define the normal trace of elements of $H(\mathrm{div}, \Omega; \mathbb{R}^{d \times d})$. First—according to the notation in [2]—we denote, for $\hat{\Gamma} \in \{\Gamma, \Gamma_{\mathrm{in}}, \Gamma_{\mathrm{out}}\}$,

$$H^{-1/2}(\hat{\Gamma}; \mathbb{R}^d) = H^{1/2}_0(\hat{\Gamma}; \mathbb{R}^d)^*.$$

Since the boundary of Γ is empty, we have $H^{1/2}_0(\Gamma; \mathbb{R}^d) = H^{1/2}(\Gamma; \mathbb{R}^d)$.

Using surjectivity of the trace operator γ from $H^1(\Omega; \mathbb{R}^d)$ to $H^{1/2}(\Gamma; \mathbb{R}^d)$, the *normal trace* $\gamma_n \sigma \in H^{-1/2}(\Gamma; \mathbb{R}^d)$ of $\sigma \in H(\mathrm{div}, \Omega; \mathbb{R}^{d \times d})$ is well-defined by

$$\gamma_n \sigma \in H^{-1/2}(\Gamma; \mathbb{R}^d),$$

where

$$\forall z \in H^1(\Omega; \mathbb{R}^d) \;:\; \langle \gamma_n \sigma, \gamma z \rangle_{H^{-1/2}(\Gamma; \mathbb{R}^d), H^{1/2}(\Gamma; \mathbb{R}^d)}$$
$$= \langle \mathrm{div}\, \sigma, z \rangle_{L^2(\Omega; \mathbb{R}^d)} + \langle \sigma, \nabla z \rangle_{L^2(\Omega; \mathbb{R}^{d \times d})}. \tag{4.5}$$

In the case where σ and Γ are smooth, Gauß' divergence theorem yields that $w(\xi) = n(\xi)^\top v(\xi)$ for all $\xi \in \Gamma$, cf. [31, Chap. 16].

The normal traces of $\sigma \in H(\mathrm{div}, \Omega; \mathbb{R}^{d \times d})$ on Γ_{in} and Γ_{out}, which are denoted by

$$\gamma_{n,\mathrm{in}} \sigma \in H^{-1/2}(\Gamma_{\mathrm{in}}; \mathbb{R}^d), \quad \gamma_{n,\mathrm{out}} \sigma \in H^{-1/2}(\Gamma_{\mathrm{out}}; \mathbb{R}^d),$$

can be be defined via testing with weakly differentiable functions vanishing outside Γ_{in} and Γ_{out}, respectively. That is,

$$\forall z \in H^1(\Omega; \mathbb{R}^d) \text{ with } \gamma z|_{\Gamma_w \cup \Gamma_{\mathrm{out}}} = 0 \;:$$
$$\langle \gamma_{n,\mathrm{in}} \sigma, \gamma_{\mathrm{in}} z \rangle_{H^{-1/2}(\Gamma_{\mathrm{in}}; \mathbb{R}^d), H^{1/2}(\Gamma_{\mathrm{in}}; \mathbb{R}^d)} = \langle \mathrm{div}\, \sigma, z \rangle_{L^2(\Omega; \mathbb{R}^d)}$$
$$+ \langle \sigma, \nabla z \rangle_{L^2(\Omega; \mathbb{R}^{d \times d})},$$

$$\forall z \in H^1(\Omega; \mathbb{R}^d) \text{ with } \gamma z|_{\Gamma_w \cup \Gamma_{\mathrm{in}}} = 0 \;:$$
$$\langle \gamma_{n,\mathrm{out}} \sigma, \gamma_{\mathrm{out}} z \rangle_{H^{-1/2}(\Gamma_{\mathrm{out}}; \mathbb{R}^d), H^{1/2}(\Gamma_{\mathrm{out}}; \mathbb{R}^d)} = \langle \mathrm{div}\, \sigma, z \rangle_{L^2(\Omega; \mathbb{R}^d)}$$
$$+ \langle \sigma, \nabla z \rangle_{L^2(\Omega; \mathbb{R}^{d \times d})}.$$

Note that in the above definition, γ_{in} and γ_{out} denote the surrogates of the restricted trace as defined in (4.4) for functions $H^1(\Omega; \mathbb{R}^3)$, mapping surjectively onto $H_0^{1/2}(\Gamma_{\mathrm{in}}; \mathbb{R}^3)$ and $H_0^{1/2}(\Gamma_{\mathrm{out}}; \mathbb{R}^3)$, respectively.

A consequence is the following Green's identity, which will play a central role in the remainder:

$$\forall z \in H^1_{\sigma,w}(\Omega; \mathbb{R}^d), \, \sigma \in H(\mathrm{div}, \Omega; \mathbb{R}^{d \times d}) \;:$$

$$\langle \gamma_{n,\mathrm{in}} \sigma, \gamma_{\mathrm{in}} z \rangle_{H^{-1/2}(\Gamma_{\mathrm{in}}; \mathbb{R}^d), H^{1/2}(\Gamma_{\mathrm{in}}; \mathbb{R}^d)} + \langle \gamma_{n,\mathrm{out}} \sigma, \gamma_{\mathrm{out}} z \rangle_{H^{-1/2}(\Gamma_{\mathrm{out}}; \mathbb{R}^d), H^{1/2}(\Gamma_{\mathrm{out}}; \mathbb{R}^d)}$$

$$= \langle \mathrm{div}\, \sigma, z \rangle_{L^2(\Omega; \mathbb{R}^d)} + \langle \sigma, \nabla z \rangle_{L^2(\Omega; \mathbb{R}^{d \times d})}. \tag{4.6}$$

4.3 Dissipation Node Corresponding to the Oseen System

We now introduce the system node corresponding to the Oseen problem (1.1).
Hereby, we assume that $\Omega \subset \mathbb{R}^d$, $d \geq 2$, with boundary Γ, and boundary parts
$\Gamma_{\text{in}}, \Gamma_{\text{w}}, \Gamma_{\text{out}}$ having the properties defined in Sect. 4.1. Further, we assume that
$b \in L^\infty(\Omega; \mathbb{R}^d) \cap H(\text{div}, \Omega; \mathbb{R}^d)$ with $\text{div}\, b = 0$ and trivial normal trace, i.e.,
$\gamma_{\text{n}} b = 0$. Further, $\mu, \rho \in \mathbb{R}$ are positive constants for the dynamic viscosity and the
mass density.

The remainder of this section is devoted to prove that the following is a dissipa-
tion node on $(\mathcal{X}, \mathcal{U}) = \left(L_\sigma^2(\Omega; \mathbb{R}^3), H_0^{1/2}(\Gamma_{\text{in}}; \mathbb{R}^d) \times H_0^{-1/2}(\Gamma_{\text{out}}; \mathbb{R}^d) \right)$ with the
state space $L_\sigma^2(\Omega; \mathbb{R}^d) = \{v \in H(\text{div}, \Omega; \mathbb{R}^d) \mid \text{div}\, v = 0\}$:

$$\text{dom}\, M = \text{dom}\, F\&G := \left\{ \begin{pmatrix} v \\ u_v \\ u_\sigma \end{pmatrix} \in H_{\sigma,\text{w}}^1(\Omega; \mathbb{R}^d) \times H_0^{1/2}(\Gamma_{\text{in}}; \mathbb{R}^d) \times H_0^{-1/2}(\Gamma_{\text{out}}; \mathbb{R}^d) \right.$$

$$\exists P \in L^2(\Omega) \text{ s.t. } \mu \nabla v - P I_d \in H(\text{div}, \Omega; \mathbb{R}^{d\times d})$$

$$\left. \wedge\ \gamma_{\text{in}} v = u_v \ \wedge\ \gamma_{\text{n,out}}(\mu \nabla v - P I_d) = u_\sigma \right\}$$

$$\tag{4.7a}$$

and

$$\forall \begin{pmatrix} v \\ u_v \\ u_\sigma \end{pmatrix} \in \text{dom}\, M : \quad F\&G \begin{pmatrix} v \\ u_v \\ u_\sigma \end{pmatrix} = \text{div}(\mu \nabla v - P I_d) + \rho (b \cdot \nabla) v, \tag{4.7b}$$

$$K\&L \begin{pmatrix} v \\ u_v \\ u_\sigma \end{pmatrix} = - \begin{pmatrix} \gamma_{\text{n,in}}(\mu \nabla v - P I_d) \\ \gamma_{\text{out}} v \end{pmatrix}, \quad M = \begin{bmatrix} F\&G \\ K\&L \end{bmatrix}. \tag{4.7c}$$

Crucial in showing that (4.7) is a dissipation node is the following simplified
and weak version of De Rham's theorem characterizing the orthogonal complement
of divergence-free functions as gradient fields, cf. [12, Thm. 2.3] or [33, Rem. 1.9].
This result ensures well definition of $F\&G$ as it implies uniqueness of the pressure
gradient in the definition (4.7b), cf. also [5].

Lemma 4.1 *Let* $\Omega \subset \mathbb{R}^d$ *be a bounded Lipschitz domain, and let* $f \in$
$H^{-1}(\Omega; \mathbb{R}^d) = H_0^1(\Omega; \mathbb{R}^d)^*$, *such that* $\langle f, v \rangle_{L^2(\Omega; \mathbb{R}^d)} = 0$ *for all* $v \in H_0^1(\Omega; \mathbb{R}^d)$
with $\text{div}\, v = 0$. *Then there exists some* $P \in L^2(\Omega)$ *such that* $f = \nabla P$ *holds in the*
distributional sense. That is,

$$\forall z \in H_0^1(\Omega; \mathbb{R}^d) : \quad \langle f, z \rangle_{H^{-1}(\Omega; \mathbb{R}^d), H_0^1(\Omega; \mathbb{R}^d)} = -\langle P, \text{div}\, z \rangle_{L^2(\Omega)}.$$

Further, P is unique up to addition of a constant function.

We further advance a lemma which basically states that the drift term $\rho(b \cdot \nabla) v$ does
not contribute to dissipation. It is the weak form of eq. (2.1).

Lemma 4.2 *Let $\Omega \subset \mathbb{R}^d$ be a bounded Lipschitz domain, and let $b \in L^\infty(\Omega) \cap H(\mathrm{div}, \Omega)$ with $\mathrm{div}\, b = 0$ and trivial normal trace, that is, $\gamma_n b = 0$. Then for all $v \in H^1(\Omega; \mathbb{R}^d)$,*

$$\langle v, \rho(b \cdot \nabla)v \rangle_{L^2(\Omega;\mathbb{R}^d)} = 0.$$

Proof By boundedness of Ω and the fact that smooth functions are contained in both $L^1(\Omega; \mathbb{R}^d)$ and $L^2(\Omega; \mathbb{R}^d)$, we have that $L^2(\Omega; \mathbb{R}^d)$ is dense in $L^1(\Omega; \mathbb{R}^d)$. As a consequence, our condition on b is equivalent to

$$b \in L^\infty(\Omega; \mathbb{R}^d) \ \wedge \ \forall z \in W^{1,1}(\Omega; \mathbb{R}^d): \ \langle \nabla z, b \rangle_{L^1(\Omega;\mathbb{R}^d), L^\infty(\Omega;\mathbb{R}^d)} = 0, \quad (4.8)$$

where the latter stands for the canonical duality product of L^1 and $L^\infty \cong (L^1)^*$. Let $v \in H^1(\Omega; \mathbb{R}^d)$. The product rule for weak derivatives [3, Thm. 4.25] yields that the pointwise inner product $v \cdot v \in W^{1,1}(\Omega)$ with

$$\tfrac{1}{2}\nabla(v \cdot v) = v \cdot \nabla v \in L^1(\Omega; \mathbb{R}^d),$$

and thus

$$\langle v, (b \cdot \nabla)v \rangle_{L^2(\Omega;\mathbb{R}^d)} = \langle b, v \cdot \nabla v \rangle_{L^2(\Omega;\mathbb{R}^d)}$$

$$= \tfrac{1}{2}\langle b, \nabla(v \cdot v) \rangle_{L^\infty(\Omega;\mathbb{R}^d), L^1(\Omega;\mathbb{R}^d)} \overset{(4.8)}{=} 0.$$

$\square$

Now we are able to show that M as in (4.7) is a dissipation node.

Theorem 4.3 (Dissipation Node for Oseen Equations) *Let $\Omega \subset \mathbb{R}^d$, $d \geq 2$, with boundary Γ, and $\Gamma_{\mathrm{in}}, \Gamma_{\mathrm{w}}, \Gamma_{\mathrm{out}} \subset \Gamma$ satisfying the properties as constituted in Sect. 4.1. Further, let $b \in L^\infty(\Omega; \mathbb{R}^d) \cap H(\mathrm{div}, \Omega; \mathbb{R}^d)$ with $\mathrm{div}\, b = 0$ and $\gamma_n b = 0$, and let $\mu, \rho \in \mathbb{R}_{>0}$.*

Then the operator M as in (4.7) is a dissipation node on $(\mathcal{X}, \mathcal{U})$ with $\mathcal{X} = L^2(\Omega; \mathbb{R}^d)$ and $\mathcal{U} = H_0^{1/2}(\Gamma_{\mathrm{in}}; \mathbb{R}^d) \times H^{-1/2}(\Gamma_{\mathrm{out}}; \mathbb{R}^d)$.

Proof We successively show that M fulfills (a)–(3.1) in Definition 3.1.

(a) *Step 1:* Let $\begin{pmatrix} v \\ u_v \\ u_\sigma \end{pmatrix} \in \mathrm{dom}\, M$. We first note that—for obvious reasons— (2.2) holds for weakly differentiable and divergence-free $\mathbb{R}^d$-valued functions as well, whence

$$\forall w \in H^1_{\sigma,\mathrm{w}}(\Omega; \mathbb{R}^d): \quad \langle \nabla w, P I_d \rangle_{L^2(\Omega;\mathbb{R}^{d \times d})} = 0. \quad (4.9)$$

Now dissipativity of M follows from (for sake of brevity, we neglect the subindices indicating the spaces specifying the duality product)

$$\left\langle \begin{pmatrix} v \\ u_v \\ u_\sigma \end{pmatrix}, M \begin{pmatrix} v \\ u_v \\ u_\sigma \end{pmatrix} \right\rangle$$

$$= \left\langle \begin{pmatrix} v \\ u_v \\ u_\sigma \end{pmatrix}, \begin{pmatrix} \operatorname{div}(\mu\nabla - PI_d) + \rho(b\cdot\nabla)v \\ -\mu\gamma_{\mathrm{n,in}}(\mu\nabla v - PI_d) \\ -\gamma_{\Gamma_{\mathrm{out}}} v \end{pmatrix} \right\rangle$$

$$= \langle v, \operatorname{div}(\mu\nabla v - PI_d)\rangle + \underbrace{\langle v, \rho(b\cdot\nabla)v\rangle}_{=0 \text{ by Lemma 4.2}} - \langle u_v, \mu\gamma_{\mathrm{n,in}}(\mu\nabla v - PI_d)\rangle$$

$$\quad - \langle u_\sigma, \gamma_{\mathrm{n,out}} v\rangle$$

$$= \langle v, \operatorname{div}(\mu\nabla v - PI_d)\rangle - \langle \gamma_{\mathrm{in}} v, \gamma_{\mathrm{n,in}}(\mu\nabla v - PI_d)\rangle$$

$$\quad - \langle \gamma_{\mathrm{out}} v, \gamma_{\mathrm{n,out}}(\mu\nabla v - PI_d)\rangle$$

$$\overset{(4.6)}{=} -\mu\langle\nabla v, \mu\nabla v - PI_d\rangle$$

$$\overset{(4.9)}{=} -\mu\langle\nabla v, \nabla v\rangle \le 0.$$

Step 2: We show that $M - \mathcal{R}$ is onto, where $\mathcal{R} : \mathcal{X} \times \mathcal{U} \to \mathcal{X} \times \mathcal{U}^*$ and in a matrix notation, $\mathcal{R} = \operatorname{diag}(I_{\mathcal{X}}, \mathcal{R}_{\mathrm{in}}, \mathcal{R}_{\mathrm{out}}^{-1})$ with the Riesz isomorphisms

$$\mathcal{R}_{\mathrm{in}} : H_0^{1/2}(\Gamma_{\mathrm{in}}; \mathbb{R}^d) \to H^{-1/2}(\Gamma_{\mathrm{in}}; \mathbb{R}^d),$$
$$\mathcal{R}_{\mathrm{out}} : H_0^{1/2}(\Gamma_{\mathrm{out}}; \mathbb{R}^d) \to H^{-1/2}(\Gamma_{\mathrm{out}}; \mathbb{R}^d).$$

Let $z \in \mathcal{X} = L^2(\Omega; \mathbb{R}^d)$ and $(w_1, w_2) \in H^{-1/2}(\Gamma_{\mathrm{in}}; \mathbb{R}^d) \times H_0^{1/2}(\Gamma_{\mathrm{out}}; \mathbb{R}^d)$. We have to find $(x, u_v, u_\sigma) \in \operatorname{dom} M$ such that

$$(M - \mathcal{R})\begin{pmatrix} v \\ u_v \\ u_\sigma \end{pmatrix} = \begin{pmatrix} z \\ w_1 \\ w_2 \end{pmatrix}. \tag{4.10}$$

Defining the bilinear form and the linear functional

$$b(v, \varphi) := \mu\langle\nabla\varphi, \nabla v\rangle_{L^2(\Omega;\mathbb{R}^{d\times d})} + \langle\varphi, v\rangle_{L^2(\Omega;\mathbb{R}^d)} - \langle\varphi, \rho(b\cdot\nabla)v\rangle_{L^2(\Omega;\mathbb{R}^d)}$$

$$+ \langle\gamma_{\mathrm{in}}\varphi, \gamma_{\mathrm{in}} v\rangle_{H_0^{1/2}(\Gamma_{\mathrm{in}};\mathbb{R}^d)} + \langle\gamma_{\mathrm{out}}\varphi, \gamma_{\mathrm{out}} v\rangle_{H_0^{1/2}(\Gamma_{\mathrm{out}};\mathbb{R}^d)},$$

$$F_{z,w_1,w_2}(\varphi) := -\langle\varphi, z\rangle_{L^2(\Omega;\mathbb{R}^d)} - \langle\gamma_{\mathrm{in}}\varphi, w_1\rangle_{H_0^{1/2}(\Gamma_{\mathrm{in}};\mathbb{R}^d), H^{-1/2}(\Gamma_{\mathrm{in}};\mathbb{R}^d)}$$

$$- \langle\gamma_{\mathrm{out}}\varphi, w_2\rangle_{H_0^{1/2}(\Gamma_{\mathrm{out}};\mathbb{R}^d)},$$

we can conclude from the Lax-Milgram lemma [13, Lem. 2.2.1.1] that there exists some $v \in H_{\sigma,\mathrm{w}}^1(\Omega; \mathbb{R}^d)$, such that

$$b(v, \varphi) = F_{z,w_1,w_2}(\varphi) \tag{4.11}$$

for all $\varphi \in H^1_{\sigma,\mathrm{w}}(\Omega; \mathbb{R}^d)$. In particular, for all test functions $\varphi \in H^1_0(\Omega; \mathbb{R}^d)$ with $\operatorname{div} \varphi = 0$, we have

$$\langle \operatorname{div} \mu \nabla v - v + \rho(b \cdot \nabla)v - z, \varphi \rangle_{H^{-1}(\Omega;\mathbb{R}^d), H^1_0(\Omega;\mathbb{R}^d)} = 0.$$

An application of Lemma 4.1 gives rise to the existence of some $P \in L^2(\Omega)$ with $\operatorname{div} \mu \nabla v - v + \rho(b \cdot \nabla)v - z = \nabla P$. This means that $\operatorname{div}(\mu \nabla v - P I_d) = v - \rho(b \cdot \nabla v) + z \in L^2(\Omega)$, i.e.,

$$\operatorname{div}(\mu \nabla v - P I_d) + \rho(b \cdot \nabla)v - v = z. \tag{4.12}$$

By abbreviating $\sigma := \mu \nabla v - P I_d$ and by setting $u_v = \gamma_{\mathrm{in}} v$ and $u_\sigma = \gamma_{\mathrm{n,out}} \sigma$, we compute

$$\langle \gamma_{\mathrm{in}} \varphi, u_v \rangle_{H^{1/2}_0(\Gamma_{\mathrm{in}};\mathbb{R}^d)} + \langle \gamma_{\mathrm{out}} \varphi, \gamma_{\mathrm{out}} v \rangle_{H^{1/2}_0(\Gamma_{\mathrm{out}};\mathbb{R}^d)}$$

$$\overset{(4.9)}{=} \mu \langle \nabla \varphi, \nabla v \rangle_{L^2(\Omega;\mathbb{R}^{d\times d})} - \langle \nabla \varphi, \sigma \rangle_{L^2(\Omega;\mathbb{R}^{d\times d})}$$

$$+ \langle \gamma_{\mathrm{in}} \varphi, u_v \rangle_{H^{1/2}_0(\Gamma_{\mathrm{in}};\mathbb{R}^d)} + \langle \gamma_{\mathrm{out}} \varphi, \gamma_{\mathrm{out}} v \rangle_{H^{1/2}_0(\Gamma_{\mathrm{out}};\mathbb{R}^d)}$$

$$\overset{\substack{(4.11)\\ \& (4.12)}}{=} -\langle \varphi, \operatorname{div} \sigma + \rho(b \cdot \nabla)v - v \rangle_{L^2(\Omega;\mathbb{R}^d)} - \langle \nabla \varphi, \sigma \rangle_{L^2(\Omega;\mathbb{R}^{d\times d})}$$

$$- \langle \varphi, v \rangle_{L^2(\Omega;\mathbb{R}^d)} + \langle \varphi, \rho(b \cdot \nabla)v \rangle_{L^2(\Omega;\mathbb{R}^d)}$$

$$- \langle \gamma_{\mathrm{in}} \varphi, w_1 \rangle_{H^{1/2}_0(\Gamma_{\mathrm{in}};\mathbb{R}^d), H^{-1/2}(\Gamma_{\mathrm{in}};\mathbb{R}^d)} - \langle \gamma_{\mathrm{out}} \varphi, w_2 \rangle_{H^{1/2}_0(\Gamma_{\mathrm{out}};\mathbb{R}^d)}$$

$$= -\langle \varphi, \operatorname{div} \sigma \rangle_{L^2(\Omega;\mathbb{R}^d)} - \langle \nabla \varphi, \sigma \rangle_{L^2(\Omega;\mathbb{R}^{d\times d})}$$

$$- \langle \gamma_{\mathrm{in}} \varphi, w_1 \rangle_{H^{1/2}_0(\Gamma_{\mathrm{in}};\mathbb{R}^d), H^{-1/2}(\Gamma_{\mathrm{in}};\mathbb{R}^d)} - \langle \gamma_{\mathrm{out}} \varphi, w_2 \rangle_{H^{1/2}_0(\Gamma_{\mathrm{out}};\mathbb{R}^d)}$$

$$\overset{(4.6)}{=} -\langle \gamma_{\mathrm{n,in}} \sigma, \gamma_{\mathrm{in}} \varphi \rangle_{H^{-1/2}(\Gamma_{\mathrm{in}};\mathbb{R}^d), H^{1/2}(\Gamma_{\mathrm{in}};\mathbb{R}^d)}$$

$$- \langle u_\sigma, \gamma_{\mathrm{out}} \varphi \rangle_{H^{-1/2}(\Gamma_{\mathrm{out}};\mathbb{R}^d), H^{1/2}(\Gamma_{\mathrm{out}};\mathbb{R}^d)}$$

$$- \langle \gamma_{\mathrm{in}} \varphi, w_1 \rangle_{H^{1/2}_0(\Gamma_{\mathrm{in}};\mathbb{R}^d), H^{-1/2}(\Gamma_{\mathrm{in}};\mathbb{R}^d)} - \langle \gamma_{\mathrm{out}} \varphi, w_2 \rangle_{H^{1/2}_0(\Gamma_{\mathrm{out}};\mathbb{R}^d)}$$

$$= -\langle \gamma_{\mathrm{in}} \varphi, \mathcal{R}^{-1}_{\mathrm{in}} \gamma_{\mathrm{n,in}} \sigma + \mathcal{R}^{-1}_{\mathrm{in}} w_1 \rangle_{H^{1/2}(\Gamma_{\mathrm{in}};\mathbb{R}^d)}$$

$$- \langle \gamma_{\mathrm{out}} \varphi, w_2 + \mathcal{R}^{-1}_{\mathrm{out}} u_\sigma \rangle_{H^{1/2}(\Gamma_{\mathrm{out}};\mathbb{R}^d)}.$$

Since the latter holds for all $\varphi \in H^1_{\sigma,\mathrm{w}}(\Omega; \mathbb{R}^d)$, we obtain that

$$w_1 = -\gamma_{\mathrm{in}} \sigma - \mathcal{R}_{\mathrm{in}} u_v,$$

$$w_2 = -\gamma_{\mathrm{out}} v - \mathcal{R}^{-1}_{\mathrm{out}} u_\sigma.$$

These two equations together with (4.12) yield $(v, u_v, u_\sigma) \in \operatorname{dom} M$ satisfies (4.10), which shows the claim.

Step 3: We show that M is closed.

By steps 1 and 2, M is dissipative, and $M - \mathcal{R}$ is onto. Since this implies that $M - \mathcal{R}$ has closed range, [9, Prop. 3.14] implies that M is closed.

(b) We show that $P_\chi M = F\&G$ is closed. As M is closed by (a), and in view of [24, Lem. 2.3], we may equivalently show there exists some $C > 0$, such that for all $\begin{pmatrix} v \\ u_v \\ u_\sigma \end{pmatrix} \in \operatorname{dom} M$

$$\left\| K\&L \begin{pmatrix} v \\ u_v \\ u_\sigma \end{pmatrix} \right\|_{H^{-1/2}(\Gamma_{\mathrm{in}};\mathbb{R}^d) \times H_0^{1/2}(\Gamma_{\mathrm{out}};\mathbb{R}^d)}$$

$$\leq C \left(\left\| \begin{pmatrix} v \\ u_v \\ u_\sigma \end{pmatrix} \right\|_{L^2(\Omega;\mathbb{R}^d) \times H_0^{1/2}(\Gamma_{\mathrm{in}};\mathbb{R}^d) \times H^{-1/2}(\Gamma_{\mathrm{out}};\mathbb{R}^d)} + \left\| F\&G \begin{pmatrix} v \\ u_v \\ u_\sigma \end{pmatrix} \right\|_{L^2(\Omega;\mathbb{R}^d)} \right).$$

To this end, consider a sequence

$$\left(\begin{pmatrix} v_n \\ u_{v,n} \\ u_{\sigma,n} \end{pmatrix} \right) \quad \text{bounded in } L^2(\Omega;\mathbb{R}^d) \times H_0^{1/2}(\Gamma_{\mathrm{in}};\mathbb{R}^d) \times H^{-1/2}(\Gamma_{\mathrm{out}};\mathbb{R}^d),$$

$$\left(F\&G \begin{pmatrix} v_n \\ u_{v,n} \\ u_{\sigma,n} \end{pmatrix} \right) \quad \text{bounded in } L^2(\Omega;\mathbb{R}^d).$$

$$(4.13)$$

Step 1: We show that (v_n) is a bounded sequence in $H^1_{\sigma,\mathrm{w}}(\Omega;\mathbb{R}^d)$. Since the trace operator $\gamma_{\mathrm{in}} : H^1_{\sigma,\mathrm{w}}(\Omega;\mathbb{R}^d) \to H_0^{1/2}(\Gamma_{\mathrm{in}};\mathbb{R}^d)$ is bounded and surjective, it possesses a bounded right inverse

$$\gamma_{\mathrm{in}}^- \in L(H_0^{1/2}(\Gamma_{\mathrm{in}};\mathbb{R}^d), H^1_{\sigma,\mathrm{w}}(\Omega;\mathbb{R}^d))$$

with

$$\gamma_{\mathrm{in}}\gamma_{\mathrm{in}}^- = I_{H_0^{1/2}(\Gamma_{\mathrm{in}};\mathbb{R}^d)}.$$

This together with the bounded emdedding $H^1_{\sigma,\mathrm{w}}(\Omega;\mathbb{R}^d) \subset L^2(\Omega;\mathbb{R}^d)$ implies that the sequence $(v_n - \gamma_{\mathrm{in}}^- u_{v,n})$ is bounded in $L^2(\Omega;\mathbb{R}^d)$. As $(v_n, u_{v,n}, u_{\sigma,n}) \in \operatorname{dom} M$, we have that $\gamma_{\mathrm{in}}(v_n - \gamma_{\mathrm{in}}^- u_{v,n}) = 0$ for all $n \in \mathbb{N}$. Further, let (P_n) be such that

$$\mu \nabla v_n - P_n I_d \in H(\operatorname{div}, \Omega; \mathbb{R}^{d \times d}).$$

Then

$$\langle v_n - \gamma_{\text{in}}^- u_{v,n}, F\&G\left(\begin{smallmatrix} v_n \\ u_{v,n} \\ u_{\sigma,n} \end{smallmatrix}\right)\rangle_{L^2(\Omega;\mathbb{R}^d)}$$

$$= \langle v_n - \gamma_{\text{in}}^- u_{v,n}, \operatorname{div}(\mu\nabla v_n - P_n I_d)\rangle_{L^2(\Omega;\mathbb{R}^d)} + \langle v_n - \gamma_{\text{in}}^- u_{v,n}, \rho(b\cdot\nabla)v\rangle_{L^2(\Omega;\mathbb{R}^d)}$$

$$\overset{\substack{(4.6)\\ \&(4.9)}}{=} -\mu\langle\nabla(v_n - \gamma_{\text{in}}^- u_{v,n}), \nabla v_n\rangle_{L^2(\Omega;\mathbb{R}^{d\times d})} + \langle v_n - \gamma_{\text{D},\Gamma_{\text{D}}}^- u_{\text{D},n}, \rho(b\cdot\nabla)v_n\rangle_{L^2(\Omega;\mathbb{R}^d)}$$

$$+ \underbrace{\langle\gamma_{\text{out}}(v_n - \gamma_{\text{in}}^- u_{v,n}), \gamma_{n,\text{out}}(\mu\nabla v_n - P_n I_d)\rangle_{H_0^{1/2}(\Gamma_{\text{out}};\mathbb{R}^d)\times H^{-1/2}(\Gamma_{\text{out}};\mathbb{R}^d)}}$$

$$+ \underbrace{\langle\gamma_{\text{in}}(v_n - \gamma_{\text{in}}^- u_{v,n}), \gamma_{n,\text{in}}(\mu\nabla v_n - P_n I_d)\rangle_{H_0^{1/2}(\Gamma_{\text{in}};\mathbb{R}^d)\times H^{-1/2}(\Gamma_{\text{in}};\mathbb{R}^d)}}_{=0}$$

$$= -\mu\langle\nabla v_n, \nabla v_n\rangle_{L^2(\Omega;\mathbb{R}^{d\times d})} + \mu\langle\nabla\gamma_{\text{in}}^- u_{v,n}, \nabla v_n\rangle_{L^2(\Omega;\mathbb{R}^{d\times d})} + \langle v_n, \rho(b\cdot\nabla)v_n\rangle_{L^2(\Omega;\mathbb{R}^d)}$$

$$- \langle\gamma_{\text{in}}^- u_{v,n}, \rho(b\cdot\nabla)v_n\rangle_{L^2(\Omega;\mathbb{R}^d)} + \langle\gamma_{\text{out}}v_n, u_{\sigma,n}\rangle_{H_0^{1/2}(\Gamma_{\text{out}};\mathbb{R}^d), H^{-1/2}(\Gamma_{\text{out}};\mathbb{R}^d)}$$

$$- \langle\gamma_{\text{out}}\gamma_{\text{in}}^- u_{v,n}, u_{\sigma,n}\rangle_{H_0^{1/2}(\Gamma_{\text{out}};\mathbb{R}^d), H^{-1/2}(\Gamma_{\text{out}};\mathbb{R}^d)}$$

$$\leq -\mu\|\nabla v_n\|_{L^2(\Omega;\mathbb{R}^{d\times d})}^2 + \mu\|\nabla\gamma_{\text{in}}^- u_{v,n}\|_{L^2(\Omega;\mathbb{R}^{d\times d})}\|\nabla v_n\|_{L^2(\Omega;\mathbb{R}^{d\times d})}$$

$$+ \rho\|v_n\|_{L^2(\Omega;\mathbb{R}^d)}\|b\|_{L^\infty(\Omega;\mathbb{R}^d)}\|\nabla v_n\|_{L^2(\Omega;\mathbb{R}^{d\times d})}$$

$$+ \rho\|\gamma_{\text{in}}^- u_{v,n}\|_{L^2(\Omega;\mathbb{R}^d)}\|b\|_{L^\infty(\Omega;\mathbb{R}^d)}\|\nabla v_n\|_{L^2(\Omega;\mathbb{R}^{d\times d})}$$

$$+ \|\gamma_{\text{out}}v_n\|_{H_0^{1/2}(\Gamma_{\text{out}};\mathbb{R}^d)}\|u_{\sigma,n}\|_{H^{-1/2}(\Gamma_{\text{out}};\mathbb{R}^d)}$$

$$+ \|\gamma_{\text{out}}\gamma_{\text{in}}^- u_{v,n}\|_{H_0^{1/2}(\Gamma_{\text{out}};\mathbb{R}^d)}\|u_{\sigma,n}\|_{H^{-1/2}(\Gamma_{\text{out}};\mathbb{R}^d)}$$

Rearranging the terms in the above inequality, invoking (4.13) and continuity of the trace operators, there are constants $c_1, c_2 > 0$, such that, for all $n \in \mathbb{N}$,

$$\|\nabla v_n\|_{L^2(\Omega;\mathbb{R}^{d\times d})}^2 \leq c_1 + c_2\|v_n\|_{H^1(\Omega;\mathbb{R}^d)}.$$

Thus, as (v_n) is bounded in $L^2(\Omega;\mathbb{R}^d)$, there is $c_3 \geq 0$ such that, for all $n \in \mathbb{N}$,

$$\|v_n\|_{H^1(\Omega;\mathbb{R}^d)}^2 \leq c_3 + c_2\|v_n\|_{H^1(\Omega;\mathbb{R}^d)}.$$

Applying Young's inequality, this shows that (v_n) is a bounded sequence in $H_{\sigma,w}^1(\Omega;\mathbb{R}^d)$.

Step 2: We show that

$$\left(K\&L\left(\begin{smallmatrix} v_n \\ u_{v,n} \\ u_{\sigma,n} \end{smallmatrix}\right)\right) \quad \text{is bounded in } H_0^{1/2}(\Gamma_{\text{in}};\mathbb{R}^d) \times H^{-1/2}(\Gamma_{\text{out}};\mathbb{R}^d).$$

By using the Banach-Steinhaus theorem [3, Thm. 7.3], it suffices to show that, for all $y = (y_1, y_2) \in H_0^{1/2}(\Gamma_{\text{in}};\mathbb{R}^d) \times H^{-1/2}(\Gamma_{\text{out}};\mathbb{R}^d)$, the scalar sequence

$$\left(\left\langle y,\; K\&L\begin{pmatrix} v_n \\ u_{v,n} \\ u_{\sigma,n} \end{pmatrix}\right\rangle_{H_0^{1/2}(\Gamma_{\mathrm{in}};\mathbb{R}^d)\times H^{-1/2}(\Gamma_{\mathrm{out}};\mathbb{R}^d),\, H^{-1/2}(\Gamma_{\mathrm{out}};\mathbb{R}^d)\times H_0^{1/2}(\Gamma_{\mathrm{in}};\mathbb{R}^d)}\right)$$

$$= \left(\langle y_1,\, \gamma_{\mathrm{n,in}}(\mu\nabla v_n - P_n I_d)\rangle_{H_0^{1/2}(\Gamma_{\mathrm{in}};\mathbb{R}^d),\, H^{-1/2}(\Gamma_{\mathrm{in}};\mathbb{R}^d)}\right.$$

$$\left. + \langle y_2,\, \gamma_{\mathrm{out}} v_n\rangle_{H^{-1/2}(\Gamma_{\mathrm{out}};\mathbb{R}^d),\, H_0^{1/2}(\Gamma_{\mathrm{out}};\mathbb{R}^d)}\right)$$

is bounded, where, again, (P_n) is such that $\mu\nabla v_n - P_n I_d \in H(\mathrm{div},\Omega;\mathbb{R}^{d\times d})$. As (v_n) is bounded in $H_{\sigma,\mathrm{w}}^1(\Omega;\mathbb{R}^d)$ by Step 1, the sequence

$$\left(\langle y_2,\, \gamma_{\mathrm{out}} v_n\rangle_{H^{-1/2}(\Gamma_{\mathrm{out}};\mathbb{R}^d),\, H_0^{1/2}(\Gamma_{\mathrm{out}};\mathbb{R}^d)}\right)$$

is bounded by continuity of the trace operator $\gamma_{\mathrm{out}}\; :\; H_{\mathrm{w}}^1(\Omega;\mathbb{R}^d)\;\to\; H_0^{1/2}(\Gamma_{\mathrm{out}};\mathbb{R}^d)$ as defined in (4.4).

As $y_1 \in H_0^{1/2}(\Gamma_{\mathrm{in}};\mathbb{R}^d)$, we define $w = \gamma_{\mathrm{in}}^- y_1 \in H_{\sigma,\mathrm{w}}^1(\Omega;\mathbb{R}^d)$, where γ_{in}^- is the right-inverse of the trace operator as defined in the beginning of step 1. Then

$$\left\langle w,\; F\&G\begin{pmatrix} v_n \\ u_{v,n} \\ u_{\sigma,n} \end{pmatrix}\right\rangle_{L^2(\Omega;\mathbb{R}^d)}$$

$$= -\mu\langle\nabla w,\nabla v_n\rangle_{L^2(\Omega;\mathbb{R}^{d\times d})} + \langle y_1,\, \gamma_{\mathrm{n,in}}(\mu\nabla v_n - P_n I_d)\rangle_{H_0^{1/2}(\Gamma_{\mathrm{in}};\mathbb{R}^d),\, H^{-1/2}(\Gamma_{\mathrm{in}};\mathbb{R}^d)}$$

$$+ \langle\gamma_{\mathrm{out}} w,\, \underbrace{\gamma_{\mathrm{n,out}}(\mu\nabla v_n - P_n I_d)}_{=u_{\sigma,n}}\rangle_{H_0^{1/2}(\Gamma_{\mathrm{in}};\mathbb{R}^d),\, H^{-1/2}(\Gamma_{\mathrm{in}};\mathbb{R}^d)}$$

$$+ \langle w,\, \rho(b\cdot\nabla)v_n\rangle_{L^2(\Omega;\mathbb{R}^d)}.$$

Now, the scalar sequence $\langle y_1,\, \gamma_{\mathrm{n,in}}(\mu\nabla v_n - P_n I_d)\rangle_{H_0^{1/2}(\Gamma_{\mathrm{in}};\mathbb{R}^d),\, H^{-1/2}(\Gamma_{\mathrm{in}};\mathbb{R}^d)}$ is bounded, as all other terms in the above identity are bounded.

(c) We show that for all $(u_v, u_\sigma) \in H_0^{1/2}(\Gamma_{\mathrm{in}};\mathbb{R}^d) \times H^{-1/2}(\Gamma_{\mathrm{out}};\mathbb{R}^d)$, there exists some $v \in H_{\sigma,\mathrm{w}}^1(\Omega;\mathbb{R}^d)$ with $\begin{pmatrix} v \\ u_v \\ u_\sigma \end{pmatrix} \in \mathrm{dom}\, M$. Let $(u_v, u_\sigma) \in H_0^{1/2}(\Gamma_{\mathrm{in}};\mathbb{R}^d) \times H^{-1/2}(\Gamma_{\mathrm{out}};\mathbb{R}^d)$. Set $v_{\mathrm{in}} := \gamma_{\mathrm{in}}^- u_v \in H_{\sigma,\mathrm{w}}^1(\Omega;\mathbb{R}^d)$, where γ_{in}^- is the right-inverse defined in step 1 of (b). Due to the Lax-Milgram lemma, there exists some $v_0 \in H_{\sigma,\mathrm{w}}^1(\Omega;\mathbb{R}^d)$ with $\gamma_{\mathrm{in}} v = 0$ such that, for all $\varphi \in H_{\sigma,\mathrm{w}}^1(\Omega;\mathbb{R}^d)$ with $\gamma_{\mathrm{in}}\varphi = 0$,

$$\mu\langle\nabla\varphi,\nabla v_0\rangle_{L^2(\Omega;\mathbb{R}^{d\times d})} + \langle\varphi, v_0\rangle_{L^2(\Omega;\mathbb{R}^d)} - \langle\varphi, \rho(b\cdot\nabla)v_0\rangle_{L^2(\Omega;\mathbb{R}^d)}$$

$$= -\mu\langle\nabla\varphi,\nabla v_{\mathrm{in}}\rangle_{L^2(\Omega;\mathbb{R}^{d\times d})} - \langle\varphi, v_{\mathrm{in}}\rangle_{L^2(\Omega;\mathbb{R}^d)} + \langle\varphi, \rho(b\cdot\nabla)v_{\mathrm{in}}\rangle_{L^2(\Omega;\mathbb{R}^d)}$$

$$- \mu\langle\gamma_{\mathrm{out}}\varphi, u_\sigma\rangle_{H_0^{1/2}(\Gamma_{\mathrm{out}};\mathbb{R}^d),\, H^{-1/2}(\Gamma_{\mathrm{out}};\mathbb{R}^d)}.$$

By linearity, $v := v_0 + v_{\text{in}}$ fulfills $\gamma_{\text{in}} v = u_v$, $\gamma_{\text{n,out}}(\mu \nabla v - P I_d) = u_\sigma$, and, for all $\varphi \in H_0^1(\Omega; \mathbb{R}^d)$ with $\operatorname{div} \varphi = 0$,

$$-\mu \langle \nabla \varphi, \nabla v \rangle_{L^2(\Omega; \mathbb{R}^{d \times d})} + \langle \varphi, \rho(b \cdot \nabla) v \rangle_{L^2(\Omega)} = \langle \varphi, v \rangle_{L^2(\Omega; \mathbb{R}^d)}.$$

Again, by Lemma 4.1, there exists some $P \in L^2(\Omega)$ with $\operatorname{div} \mu \nabla v + \rho(b \cdot \nabla) v - v = \nabla P$ and thus

$$\operatorname{div}(\mu \nabla v - P I_d) = -\rho(b \cdot \nabla) v + v \in L^2(\Omega; \mathbb{R}^3).$$

This gives $\mu \nabla v - P I_d \in H(\operatorname{div}, \Omega; \mathbb{R}^{d \times d})$. Hence, we conclude that $\begin{pmatrix} v \\ u_v \\ u_\sigma \end{pmatrix} \in$ dom $F \& G$.

(d) We show that $F - I$ is surjective, i.e., in particular has dense range. Let $z \in L^2(\Omega; \mathbb{R}^3)$. Again using the Lax-Milgram lemma, there exists some $v \in H_{\sigma,\text{w}}^1(\Omega; \mathbb{R}^d)$ with $\gamma_{\text{in}} v = 0$, such that for all $\varphi \in H_{\sigma,\text{w}}^1(\Omega; \mathbb{R}^d)$ with $\gamma_{\text{in}} \varphi = 0$,

$$\mu \langle \nabla \varphi, \nabla v \rangle_{L^2(\Omega; \mathbb{R}^{d \times d})} + \langle \varphi, v \rangle_{L^2(\Omega; \mathbb{R}^d)} - \langle \varphi, \rho(b \cdot \nabla) v \rangle_{L^2(\Omega; \mathbb{R}^d)}$$
$$= -\langle \varphi, z \rangle_{L^2(\Omega; \mathbb{R}^d)}. \tag{4.14}$$

Again invoking Lemma 4.1, there exists some $P \in L^2(\Omega)$ with

$$\operatorname{div} \mu \nabla v - v + \rho(b \cdot \nabla) v - z = \nabla P,$$

which gives rise to

$$\operatorname{div}(\mu \nabla v - P I_d) = -\rho(b \cdot \nabla) v + z \in L^2(\Omega; \mathbb{R}^d). \tag{4.15}$$

Moreover, by $\gamma_{\text{in}} v = 0$, Green's identity (4.6) together with (4.14) and (4.15), we obtain that

$$\gamma_{\text{n,out}}(\mu \nabla v - P I_d) = 0.$$

This gives $v \in \operatorname{dom} F$ with $(F - I) v = z$.

$\square$

By having shown that $M = \begin{bmatrix} F \& G \\ K \& L \end{bmatrix}$ as in (4.7) defines a dissipation node, and by further incorporating the (extremely simple) Hamiltonian as in (1.4), we have brought the Oseen equations

$$\dot{p}(t) = \mu \delta v(t) + \rho(b \cdot \nabla) v + \nabla P(t),$$
$$\operatorname{div} v(t) = 0,$$
$$v|_{\Gamma_{\text{w}}} = 0,$$

$$v|_{\Gamma_{\text{in}}} = u_v(t), \qquad\qquad \sigma(t)\mathbf{n}|_{\Gamma_{\text{out}}} = u_\sigma(t),$$

$$y_\sigma(t) = \sigma(t)\mathbf{n}|_{\Gamma_{\text{in}}}, \qquad\qquad y_v(t) = v|_{\Gamma_{\text{out}}},$$

$$p(0) = p_0$$

with $v(t) = \frac{1}{\rho}p(t) \in L^2(\Omega; \mathbb{R}^d)$ and $\sigma(t) = \mu\left(\nabla v(t) + \nabla v(t)^\top\right) - P(t)I_d \in L^2(\Omega; \mathbb{R}^{d\times d})$ into the framework of port-Hamiltonian system nodes. The input consists of velocities at the inflow part Γ_{in} of the boundary together with the stress tensor in normal direction at the outflow part Γ_{out} of the boundary. The output is composed of the stress tensor in normal direction at Γ_{in} and the velocity trace at Γ_{in}. We have shown that this system is—under the assumptions on the domain as specified in Sect. 4.1, together with μ, $\rho > 0$ and essential boundedness, divergence-freeness and trivial normal boundary trace of b—port-Hamiltonian in the sense of Definition 3.2. This allows to apply the results known for port-Hamiltonian systems of this type, such as, e.g., Proposition 3.3: For instance, we can conclude that the free dynamics of the above system (i.e., $u_v = 0$ and $u_\sigma = 0$) is described by a contractive semigroup. We can further conclude from Proposition 3.3 that the above system has a classical solution, if

$$u_v \in W^{2,1}([0, T]; H_0^{1/2}(\Gamma_{\text{in}}; \mathbb{R}^d)),$$

$$u_\sigma \in W^{2,1}([0, T]; H^{-1/2}(\Gamma_{\text{out}}; \mathbb{R}^d)),$$

and the initial infinitesimal momentum $p_0 \in H^1_{\sigma,\text{w}}(\Omega; \mathbb{R}^d)$ has the property that there exists some $P_0 \in H^1_0(\Omega)$ with $\mu\nabla v - P_0 I_d \in H(\text{div}, \Omega; \mathbb{R}^{d\times d})$, joint with the compatibility conditions

$$\frac{1}{\rho}p_0|_{\Gamma_{\text{in}}} = u_v(0),$$

$$\mu\left(\left(\nabla p_0 + \nabla p_0^\top\right) - P_0 I_d\right)\mathbf{n}|_{\Gamma_{\text{out}}} = u_\sigma(0).$$

Moreover, for the Hamiltonian $\mathcal{H}$ as in (1.4), and $0 \le t_0 \le t_1 \le T$, the weak (and thus also the classical) solutions on the interval $[0, T]$ fulfill the energy balance

$$\mathcal{H}(p(t_1)) - \mathcal{H}(p(t_0))$$

$$= -\mu \int_{t_0}^{t_1} \|\nabla v(t)\|^2_{L^2(\Omega; \mathbb{R}^{d\times d})}\,dt$$

$$+ \int_{t_0}^{t_1} \langle y_\sigma, u_v\rangle_{H^{-1/2}(\Gamma_{\text{in}};\mathbb{R}^d), H_0^{1/2}(\Gamma_{\text{in}};\mathbb{R}^d)}\,dt$$

$$+ \int_{t_0}^{t_1} \langle y_v, u_\sigma\rangle_{H^{-1/2}(\Gamma_{\text{out}};\mathbb{R}^d), H_0^{1/2}(\Gamma_{\text{out}};\mathbb{R}^d)}\,dt.$$

5 Conclusion

We have formulated Oseen flows by means of port-Hamiltonian system nodes, providing a functional analytic framework for a system-theoretic and energy-based modeling of boundary controlled linear flow problems. A system node corresponding to the Oseen system has been introduced, and it has been shown that this indeed defines a dissipation node, such that, together with the corresponding kinetic energy Hamiltonian, we obtained an energy balance, linking the change of energy to the enstrophy and the inner product between input and output. Further, we have provided an application to a flow in a tube.

Acknowledgments The authors would like to thank Volker Mehrmann for pointing out the problem of port-Hamiltonian formulation of Oseen flows.

References

1. F. Achleitner, A. Arnold, V. Mehrmann, Hypocoercivity in algebraically constrained partial differential equations with application to Oseen equations (2022). preprint arXiv:2212.06631
2. R.A. Adams, J.J. Fournier, *Sobolev Spaces*, volume 140 of Pure and Applied Mathematics, 2nd ed. (Academic Press, Amsterdam, San Diego, Oxford, London, 2003)
3. H.W. Alt, *Linear Functional Analysis, An Application-Oriented Introduction*. Universitext (Springer, London, 2016)
4. R. Altmann, P. Schulze, A port-Hamiltonian formulation of the Navier–Stokes equations for reactive flows. Systems Control Lett. **100**, 51–55 (2017)
5. W. Arendt, A. ter Elst, From forms to semigroups. Spectral Theory Math. Syst. Theory Evol. Equ. Differ. Difference Equ. **221**, 47–69 (2012)
6. C. Beattie, V. Mehrmann, H. Xu, H. Zwart, Linear port-Hamiltonian descriptor systems. Math. Control Signals Syst. **30**(4), 17 (2018)
7. M. Braack, E. Burman, V. John, G. Lube, Stabilized finite element methods for the generalized Oseen problem. Comput. Methods Appl. Mech. Eng. **196**(4–6), 853–866 (2007)
8. J. Diestel, J. Uhl, *Vector Measures*, volume 15 of Mathematical Surveys and Monographs (American Mathematical Society, Providence, 1977)
9. K.-J. Engel, R. Nagel, *One-Parameter Semigroups for Linear Evolution Equations*, volume 194 (Springer, New York, 2000)
10. E. Fernández-Cara, On the control of the Navier-Stokes equations and related systems, in *Recent Advances in Pure and Applied Mathematics*, ed. by F. Ortegón Gallego, J.I. García García (Springer, Cham, 2020), pp. 1–20
11. Y. Giga, A. Novotný, *Handbook of Mathematical Analysis in Mechanics of Viscous Fluids* (Springer, Berlin, 2018)
12. V. Girault, P.-A. Raviart, *Finite Element Methods for Navier-Stokes Equations: Theory and Algorithms*, volume 5 (Springer, Berlin, 2012)
13. P. Grisvard, *Elliptic Problems in Nonsmooth Domains*, volume 24 of Monographs and Studies in Mathematics (Pitman Advanced Publishing Program, Boston, London, Melbourne, 1985)
14. G. Haine, D. Matignon, Incompressible Navier-Stokes equation as port-Hamiltonian systems: velocity formulation versus vorticity formulation. IFAC-PapersOnLine **54**(19), 161–166 (2021)
15. J. Heiland, Convergence of coprime factor perturbations for robust stabilization of Oseen systems. Math. Control Relat. Fields **12**(3), 747–761 (2022)

16. M. Hieber, On operator semigroups arising in the study of incompressible viscous fluid flows. Philos. Trans. Roy. Soc. A **378**(2185), 20190618 (2020)
17. B. Jacob, K. Morris, On solvability of dissipative partial differential-algebraic equations. IEEE Control Syst. Lett. **6**, 3188–3193 (2022)
18. D. Jeltsema, A. van der Schaft, Port-Hamiltonian systems theory: An introductory overview. Found. Trends Syst. Control **1**(2–3), 173–387 (2014)
19. M. Marion, R. Temam, Navier-Stokes equations: theory and approximation. Handbook Numer. Anal. **6**, 503–689 (1998)
20. V. Mehrmann, R. Morandin, Structure-preserving discretization for port-Hamiltonian descriptor systems, in *2019 IEEE 58th Conference on Decision and Control (CDC)* (2019), pp. 6863–6868
21. V. Mehrmann, H. Zwart, Abstract dissipative Hamiltonian differential-algebraic equations are everywhere (2023). preprint arXiv:2311.03091
22. S. Monniaux, Navier-Stokes equations in arbitrary domains: the Fujita-Kato scheme. Math. Res. Lett. **13**(3), 455–461 (2006)
23. L.A. Mora, Y. Le Gorrec, D. Matignon, H. Ramirez, J.I. Yuz, About dissipative and pseudo port-Hamiltonian formulations of irreversible Newtonian compressible flows. IFAC-PapersOnLine **53**(2), 11521–11526 (2020)
24. F. Philipp, T. Reis, M. Schaller, Infinite-dimensional port-Hamiltonian systems—a system node approach (2023). preprint arXiv:2302.05168
25. A. Quarteroni, A. Valli, *Numerical Approximation of Partial Differential Equations*, vol. 23 (Springer, Berlin, 2008)
26. J.-P. Raymond, Feedback boundary stabilization of the two-dimensional Navier–Stokes equations. SIAM J. Control Optim. **45**(3), 790–828 (2006)
27. J.-P. Raymond, Stokes and Navier-Stokes equations with nonhomogeneous boundary conditions. Annales de l'IHP Analyse non linéaire **24**(6), 921–951 (2007)
28. T. Reis, Some notes on port-Hamiltonian systems on banach spaces. IFAC-PapersOnLine **54**(19), 223–229 (2021)
29. O. Staffans, *Well-Posed Linear Systems*, volume 103 of Encyclopedia of Mathematics and its Applications (Cambridge University Press, Cambridge, 2005)
30. G.E. Swaters, *Introduction to Hamiltonian Fluid Dynamics and Stability Theory*, volume 102 of Monographs and Surveys in Pure and Applied Mathematics (CRC Press, Boca Raton, London, New York, Washington, 1999)
31. L. Tartar, *An Introduction to Sobolev Spaces and Interpolation Spaces*. Lecture Notes of the Unione Matematica Italiana (Springer, Berlin, Heidelberg, 2007)
32. R. Temam, *Navier–Stokes Equations and Nonlinear Functional Analysis* (SIAM, 1995)
33. R. Temam, *Navier-Stokes Equations: Theory and Numerical Analysis*, vol. 343 (American Mathematical Society, 2001)

On the Equivalence of Geometric and Descriptor Representations of Linear Port-Hamiltonian Systems

Hannes Gernandt, Friedrich M. Philipp (iD), Till Preuster, and Manuel Schaller

1 Introduction

Development and operation of modern technologies require a deep understanding and control of complex dynamical systems. The class of port-Hamiltonian (pH) systems represents such an elegant mathematical framework for modeling and analysis of multi-physics systems. Due to their inherent energy-based structure, these systems are very well suited to describe the energy flows, energy conservation, and interconnection of physical systems in a wide range of applications. From a modeling perspective, they offer the additional benefit of coupling capability. Port-Hamiltonian systems have found numerous applications in physical domains such as robotics, renewable energy systems, and mechatronics [19, 26].

From a mathematical point of view, there exist different approaches to pH systems fertilized by different areas of mathematics and mathematical physics. On the one hand, one can describe this class of systems by geometrical structures [26],

Funding: HG acknowledges funding within the BMBF project EIZ - Project number 03SF0693A. FP was funded by the Carl Zeiss Foundation within the project *DeepTurb–Deep Learning in and from Turbulence* and by the free state of Thuringia and the German Federal Ministry for Education and Research (BMBF) within the project *THInKI–Thüringer Hochschulinitiative für KI im Studium*. E-Mail: hannes.gernandt@ieg.fraunhofer.de, {friedrich.philipp, till.preuster, manuel.schaller}@tu-ilmenau.de.

H. Gernandt
Fraunhofer Research Institution for Energy Infrastructures and Geothermal Systems IEG Cottbus, Cottbus, Germany

F. M. Philipp · T. Preuster · M. Schaller (✉)
Optimization-Based Control Group, Institute of Mathematics, Technische Universität Ilmenau, Ilmenau, Germany
e-mail: manuel.schaller@tu-ilmenau.de

F. L. Schwenninger, M. Waurick (eds.), *Systems Theory and PDEs*,
Trends in Mathematics, https://doi.org/10.1007/978-3-031-64991-2_6

leading to the concept of Dirac structures. Moreover, the total energy of the system is given by the Hamiltonian density, which can be generalized by so-called Lagrangian subspaces.

In the language of system and control theory, pH systems can be characterized as descriptor systems with the physical structure of the system being inscribed in the algebraic properties of the coefficient matrices. This perspective allows the application of numerical methods as well as many results from simulation and solution theory and interprets pH systems as open Hamiltonian systems interacting with their environment by means of inputs and outputs [3, 19]. For example, the pH structure implies certain restrictions on the Kronecker canonical form of the underlying matrix pencil [16] and also provides robustness of the eigenvalues under structured perturbations [17].

Eventually, there is also a functional analytical approach to pH systems theory. This point of view allows the extension of the description of energy-based physical systems on infinite-dimensional state spaces in terms of partial-differential equations and boundary control systems, cf. [13] for one-dimensional state domains and [21, 24] for recent approaches to higher-dimensional state domains.

In this chapter, our aim is to reveal a connection between the geometric pH formulation by means of Lagrange structures, Dirac structures, and resistive structures and the system theoretic formulation in finite dimensions by means of input-state-output systems given as a differential-algebraic equation (DAE) of the form

$$\begin{bmatrix} \frac{\mathrm{d}}{\mathrm{d}t} Ez(t) \\ y(t) \end{bmatrix} = \begin{bmatrix} J - R & B - P \\ (B + P)^* & S + N \end{bmatrix} \begin{bmatrix} Qz(t) \\ u(t) \end{bmatrix}, \quad t \geq 0 \tag{1}$$

with $\mathbb{K}^m$-valued input u and output y, $\mathbb{K}^n$-valued state z, and matrices $E, J, R, Q \in \mathbb{K}^{n \times n}$, $B, P \in \mathbb{K}^{n \times m}$, $S, N \in \mathbb{K}^{m \times m}$ having additional structural properties, cf. Definition 13.

In [28], a first link between the geometric modeling of pH systems outlined in [26] and the state-space representations for DAE-systems from [3] was established, where the authors only considered Lagrange and Dirac structures without any dissipation or external ports. This results in a state space system of the form

$$K \frac{d}{dt} Pz(t) = LSz(t), \quad t \geq 0, \tag{2}$$

where the matrices $K, L \in \mathbb{K}^{n \times n}$ are given by the kernel representation of the Dirac structure and $P, S \in \mathbb{K}^{n \times n}$ are given by a range representation of the Lagrange structure.

The case with dissipation was considered in [10] for pH descriptor systems. To this end, the Dirac structure was replaced by a dissipative subspace and for structural results on the underlying matrix pencils nonnegativity of the Lagrange structure was assumed.

The geometric setting in [10] was further generalized in [20] where, contrary to the dissipative subspace in [10], in addition to the Dirac and Lagrange structures, a resistive structure was used to model the dissipation. The relation to state space systems of the form (2) was studied, and the index as well as the Kronecker canonical form of (2) was investigated. However, no external port variables were considered.

Recently, in [30], a geometric description of dissipative pH descriptor systems including port variables was given. It was shown that the previously used geometric definition of pH systems, either via a separate resistive structure or via a dissipative structure (called *monotone* in [30]), is in fact equivalent. This was used to obtain a state space formulation (1) from the geometric description using a monotone structure.

The main contribution of the present note is to also provide a converse result, i.e., for pH descriptor systems (1) satisfying $\ker E \cap \ker Q = \{0\}$, we derive an equivalent geometric formulation. This extends previous results from [15], where no additional Lagrange structure was considered, leading to a one-to-one correspondence of the two formulations in the behavioral sense. This means that for each solution of the geometric pH descriptor system, there is a corresponding solution of the state-space formulation and vice versa.

Furthermore, in comparison to [30], we show that geometric pH systems have a state space formulation (1), where Q equals the identity, which is often assumed in pH literature [19]. Incorporating the converse direction, it follows that each pH descriptor system is equivalent to another one in a possibly larger state-space with $Q = I$.

The chapter is organized as follows: In Sect. 2, we recall notations and well-known facts from multivalued linear algebra. After presenting both the geometric and the descriptor formulation of pH descriptor systems in Sect. 3, the one-to-one correspondence between these two formulations is shown in Sect. 4. We conclude the chapter and discuss open problems in Sect. 5.

2 Preliminaries from Multivalued Linear Algebra

Notation $\mathbb{K}$ denotes either $\mathbb{C}$ or $\mathbb{R}$—consistently throughout this chapter. The graph $\{(x, Ax) : x \in \mathbb{K}^n\}$ of a linear map $A : \mathbb{K}^n \to \mathbb{K}^m$ is denoted by $\operatorname{gr} A$. Its inverse (as a linear relation) is given and denoted by $\operatorname{gr}^{-1} A = \{(Ax, x) : x \in \mathbb{K}^n\}$. For $A \in \mathbb{K}^{n \times n}$, we write $A^* := \overline{A}^\top$ where $\overline{A}$ is the entry-wise complex conjugate of A, i.e., if $A \in \mathbb{R}^{n \times n}$ we have $A^* = A^\top$. The Euclidean inner product in $\mathbb{K}^n$ will be denoted by $\langle x, y \rangle := y^* x$ for all $x, y \in \mathbb{K}^n$ with the resulting Euclidean norm $\|x\|^2 := \langle x, x \rangle$.

Recall the notions of kernel, domain, multivalued part, and range of a linear subspace of a product space (also called *linear relation*).

Definition 1 The *kernel, domain, multivalued part,* and *range* of a linear subspace $\mathcal{A} \subset \mathbb{K}^n \times \mathbb{K}^m$ are defined by

$$\ker \mathcal{A} := \{ f \in \mathbb{K}^n : (f, 0) \in \mathcal{A} \},$$

$$\operatorname{dom} \mathcal{A} := \{ f \in \mathbb{K}^n : \exists\, e \in \mathbb{K}^m \text{ s.t. } (f, e) \in \mathcal{A} \},$$

$$\operatorname{mul} \mathcal{A} := \{ e \in \mathbb{K}^m : (0, e) \in \mathcal{A} \},$$

$$\operatorname{ran} \mathcal{A} := \{ e \in \mathbb{K}^m : \exists\, f \in \mathbb{K}^n \text{ s.t. } (f, e) \in \mathcal{A} \},$$

respectively. The *inverse* $\mathcal{A}^{-1}$, the *adjoint* $\mathcal{A}^*$, and *scalar multiples* $\alpha\mathcal{A}$ of $\mathcal{A}$ are defined by

$$\mathcal{A}^{-1} := \{ (e, f) : (f, e) \in \mathcal{A} \},$$

$$\mathcal{A}^* := \left\{ (e', f') : \langle f', e \rangle = \langle e', f \rangle \; \forall (e, f) \in \mathcal{A} \right\},$$

$$\alpha\mathcal{A} := \{ (e, \alpha f) : (e, f) \in \mathcal{A} \}, \quad \alpha \in \mathbb{K}.$$

In particular, non-invertible matrices A can be inverted in the sense of linear relations by considering $\operatorname{gr}^{-1} A$ which might then be multivalued if A is not injective or not-everywhere defined if A is not surjective.

In the following, we collect some notions for subspaces which have additional structural properties.

Definition 2 Let $\mathcal{D}, \mathcal{L}, \mathcal{M},$ and $\mathcal{R}$ be subspaces of $\mathbb{K}^{2n}$.

(i) $\mathcal{L}$ is called a *Lagrange structure* if $\mathcal{L} = \mathcal{L}^*$.
(ii) $\mathcal{D}$ is called a *Dirac structure* if $\mathcal{D} = -\mathcal{D}^*$.
(iii) $\mathcal{R}$ is called a *(maximal) resistive structure* if $\mathcal{R} \subset \mathcal{R}^*$,

$$\langle e, f \rangle \leq 0 \quad \text{for all } \begin{bmatrix} e \\ f \end{bmatrix} \in \mathcal{R} \quad (\text{and } \dim \mathcal{R} = n).$$

(iv) $\mathcal{M}$ is called a *(maximal) monotone structure* if

$$\operatorname{Re} \langle e, f \rangle \geq 0 \quad \text{for all } \begin{bmatrix} e \\ f \end{bmatrix} \in \mathcal{M} \quad (\text{and } \dim \mathcal{M} = n).$$

Remark 3

(a) In the language of linear relations slightly different nomenclature is used. Dirac, Lagrange, and (maximal) monotone structures are called skew-adjoint, self-adjoint, and (maximal) accretive, respectively, cf. [2]. A resistive structure would be called a nonpositive symmetric relation.

(b) $\mathcal{L}$ is a Lagrange structure if and only if $\mathcal{L} \subset \mathcal{L}^*$ and $\dim \mathcal{L} = n$. In particular, a maximal resistive structure is also a Lagrange structure. Similarly, $\mathcal{D}$ is a Dirac structure if and only if $\mathcal{D} \subset -\mathcal{D}^*$ and $\dim \mathcal{D} = n$.

Remark 4 In the case $\mathbb{K} = \mathbb{C}$, resistive structures always have maximal resistive extensions. This follows directly from [2, Theorem 5.3.1]. Similarly, monotone structures always have maximal monotone extensions. Indeed, if $\mathcal{M}$ is a monotone structure in $\mathbb{C}^{2n}$, then its Cayley transform is a linear contraction $V : \text{dom } V \to \mathbb{C}^n$ with $\text{dom } V \subset \mathbb{C}^n$, cf. [2, Proposition 1.6.6]. Let $\tilde{V} : \mathbb{C}^n \to \mathbb{C}^n$ be a contractive extension of V. Then the inverse Cayley transform of $\tilde{V}$ is an extension as desired.

There are two common representations which will be referred to in this chapter as *kernel* and *image representation*; see [4, Theorem 3.3] and also [2].

Proposition 5 *Let $\mathcal{A}$ be a subspace of $\mathbb{K}^{2n}$ of dimension d. Then there exists matrices $K, L \in \mathbb{K}^{(2n-d)\times n}$ and $F, G \in \mathbb{K}^{n\times d}$ such that the following holds*

$$\mathcal{M} = \ker[K, L] = \text{ran}\begin{bmatrix} F \\ G \end{bmatrix}. \tag{3}$$

If $\mathcal{D} = \ker[K, L]$ is a Dirac structure for some $K, L \in \mathbb{K}^{n\times n}$, then [28] used the notion *Dirac algebraic constraint* if K is not invertible. This is equivalent to the existence of $(z, 0) \in \mathcal{D}$ with $z \neq 0$, or in the language of linear relations $\ker \mathcal{D} \neq \{0\}$. A special case of such constraints are kinematic constraints (see Example 2.7 in [28]). Analogously, for a Lagrange structure $\mathcal{L} = \text{ran}\begin{bmatrix} P \\ S \end{bmatrix}$ for some $P, S \in \mathbb{K}^{n\times n}$, there are said to be *Lagrange algebraic constraints* if P is not invertible. These can be used to model algebraic state constraints.

The matrices in the kernel and range representations (3) can be used to characterize the structural properties from Definition 2. In the next proposition, we restrict ourselves to the range representation.

Proposition 6 *Let $\mathcal{A}$ be a subspace of $\mathbb{K}^{2n}$ which is given by $\mathcal{A} = \text{ran}\begin{bmatrix} P \\ S \end{bmatrix}$ for some $P, S \in \mathbb{K}^{n\times m}$. Then the following equivalences hold:*

(i) $\mathcal{A} = \text{ran}\begin{bmatrix} P \\ S \end{bmatrix}$ *is a Lagrange structure if and only if $S^*P = P^*S$ and* $\text{rank}\begin{bmatrix} P \\ S \end{bmatrix} = n$.

(ii) $\mathcal{A} = \text{ran}\begin{bmatrix} P \\ S \end{bmatrix}$ *is a Dirac structure if and only if $S^*P = -P^*S$ and* $\text{rank}\begin{bmatrix} P \\ S \end{bmatrix} = n$.

(iii) $\mathcal{A} = \text{ran}\begin{bmatrix} P \\ S \end{bmatrix}$ *is (maximal) monotone if and only if $S^*P + P^*S \geq 0$ (and* $\text{rank}\begin{bmatrix} P \\ S \end{bmatrix} = n$*).*

(iv) $\mathcal{A} = \text{ran}\begin{bmatrix} P \\ S \end{bmatrix}$ *is (maximal) resistive if and only if $S^*P = P^*S \leq 0$ (and* $\text{rank}\begin{bmatrix} P \\ S \end{bmatrix} = n$*).*

Remark 7 Similar characterizations as in Proposition 6 can also be derived for subspaces $\mathcal{A}$ given in kernel representations $\mathcal{A} = \ker[K, L]$ for some $K, L \in \mathbb{K}^{(2n-d)\times n}$. Then the adjoint relation $\mathcal{A}^*$ is given in range representation

$$\mathcal{A}^* = \text{ran}\begin{bmatrix} L^* \\ -K^* \end{bmatrix}.$$

Furthermore, $\mathcal{A}$ is Lagrange (resp. Dirac, maximal monotone, maximal resistive) if and only if $\mathcal{A}^*$ has this property. Indeed, Lagrange and maximal resistive structures satisfy $\mathcal{R} = \mathcal{R}^*$, Dirac structures satisfy $\mathcal{D} = -\mathcal{D}^*$, see, e.g., [10], and it was shown in [2, Proposition 1.6.7] that $\mathcal{M}$ is maximal monotone if and only if $\mathcal{M}^*$ is maximal monotone.

Therefore, we can apply Proposition 6 to $\mathcal{A}^*$ which implies that $\mathcal{A} = \ker[K, L]$ is Lagrange (resp. Dirac, maximal monotone, maximal resistive) if and only if $(KL^* = LK^*, KL^* = -LK^*, KL^* + LK^* \leq 0, KL^* = LK^* \geq 0)$ and $\mathrm{rank}[K, L] = n$.

The following result is the key to rewrite maximal subspaces as graphs of matrices in a larger subspace.

Proposition 8 *Let $\mathcal{A} \subset \mathbb{K}^N \times \mathbb{K}^N$ be a Dirac (resp. Lagrange, maximal resistive, maximal monotone) structure and let $l := \dim \ker \mathcal{A}$. Then there exist matrices $G \in \mathbb{K}^{N \times l}$ with $\ker G = \{0\}$ and $A \in \mathbb{K}^{N \times N}$ satisfying $A = -A^*$ (resp. $A = A^*$, $A = A^* \leq 0$, $A + A^* \geq 0$) and $A \ker G^* \subset \ker G^*$ such that*

$$\mathcal{A} = \left\{ \begin{bmatrix} Ae - G\lambda \\ e \end{bmatrix} : G^*e = 0, \ \lambda \in \mathbb{K}^l \right\}. \tag{4}$$

Proof The claim has been proven for Dirac structures in [30, Proposition 3.8] (see also [7, Theorem 3.1]), for Lagrange structures in [28, Proposition 5.3] and for maximal resistive and monotone subspaces in [10]. $\qquad\qquad\square$

We conclude this section with a remark.

Remark 9

(i) If $\begin{bmatrix} x \\ y \end{bmatrix} \in \mathcal{A}$, where $\mathcal{A}$ is as in (4), then both e and λ are uniquely determined: $e = y$ and $\lambda = G^\dagger(Ay - x)$, where $G^\dagger$ is any left-inverse of G.

(ii) Since the inverse relation $\mathcal{A}$ of Dirac, Lagrange, maximal resistive, and maximal monotone structures inherits the particular property, we can apply Proposition 8 to $\mathcal{A}^{-1}$ and obtain the existence of $\hat{A}$ and an injective $\hat{G}$ such that

$$\mathcal{A} = \left\{ \begin{bmatrix} e \\ \hat{A}e - \hat{G}\lambda \end{bmatrix} : \hat{G}^*e = 0, \ \lambda \in \mathbb{K}^{\hat{l}} \right\}.$$

3 Two Formulations of pH Systems

In this part, we introduce the two formulations of pH systems we will consider in the remainder of this work. In the upcoming Sect. 3.1 we introduce the geometric representation, whereas in Sect. 3.2, we recall the formulation by means of a differential algebraic descriptor system.

3.1 Geometric Representation of pH Systems

The following geometric description of pH systems was recently introduced in [30] and extends the geometric formulation from [28] by incorporating resistive variables, inputs, and outputs.

Definition 10 ([30]) A *geometric representation* of a pH system (in short, a *geometric pH system*) with state space $\mathbb{K}^n$ and external dimension m is given by a triple $(\mathcal{D}, \mathcal{L}, \mathcal{R})$ consisting of

- A Dirac structure $\mathcal{D} \subseteq \mathbb{K}^{n+r+m} \times \mathbb{K}^{n+r+m}$,
- A Lagrange structure $\mathcal{L} \subset \mathbb{K}^n \times \mathbb{K}^n$, and
- A maximal resistive structure $\mathcal{R} \subset \mathbb{K}^r \times \mathbb{K}^r$.

By a *solution* of the pH system $(\mathcal{D}, \mathcal{L}, \mathcal{R})$, we understand an input-state-output trajectory $(u, x, y) \in C([0, \infty); \mathbb{K}^m) \times C^1([0, \infty); \mathbb{K}^n) \times C([0, \infty); \mathbb{K}^m)$ for which there exist continuous functions $f_R, e_R,$ and e_L such that for all $t \geq 0$, we have

$$\big(-\dot{x}(t), f_R(t), y(t), e_L(t), e_R(t), u(t) \big) \in \mathcal{D}, \quad (x(t), e_L(t)) \in \mathcal{L},$$

$$(f_R(t), e_R(t)) \in \mathcal{R}. \tag{5}$$

The functions f_R and e_R are called the *resistive flow and effort variables*, respectively, and e_L is the *Lagrangian effort*.

We briefly comment on this definition in view of previous works and generalizations.

Remark 11 In a more general setting, the structures in Definition 10 might also depend on time t and state x, cf. [26]. However, here we only consider stationary structures. Furthermore, the variables u and y, usually denoting inputs and outputs in systems and control theory frameworks, were called f_P and e_P in [30], respectively. Last, we note that in, e.g., [30, Definition 2.1] or in [20, Defintion 14], negated maximal resistive structures are called nonnegative Lagrange structures. To avoid confusion with the Lagrange structure $\mathcal{L}$, we will utilize the notion maximal resistive structure for $\mathcal{R}$.

Solutions of geometric pH systems obey a *power-balance*, as the following elementary result shows:

Lemma 12 Let $(u, x, y) \in C([0, \infty); \mathbb{K}^m) \times C^1([0, \infty); \mathbb{K}^n) \times C([0, \infty); \mathbb{K}^m)$ be a solution of the geometric pH system $(\mathcal{D}, \mathcal{L}, \mathcal{R})$. Then, for all $t \geq 0$, the following power balance holds:

$$\mathrm{Re}\,\langle \dot{x}(t), e_L(t) \rangle = \mathrm{Re}\,\langle f_R(t), e_R(t) \rangle + \mathrm{Re}\,\langle y(t), u(t) \rangle \leq \mathrm{Re}\,\langle y(t), u(t) \rangle.$$

Proof Since $\mathcal{D}$ is a Dirac structure as defined in Definition 2(ii), and due to (5), we compute for all $t \geq 0$

$$0 = \mathrm{Re} \left\langle \begin{bmatrix} -\dot{x}(t) \\ f_R(t) \\ y(t) \end{bmatrix}, \begin{bmatrix} e_L(t) \\ e_R(t) \\ u(t) \end{bmatrix} \right\rangle = -\mathrm{Re}\,\langle \dot{x}(t), e_L(t) \rangle + \mathrm{Re}\,\langle f_R(t), e_R(t) \rangle \tag{6}$$

$$+ \mathrm{Re}\,\langle y(t), u(t) \rangle.$$

We have $\mathrm{Re}\,\langle f_R(t), e_R(t) \rangle \leq 0$ as $(f_R(t), e_R(t)) \in \mathcal{R}$ and $\mathcal{R}$ is resistive, cf. Definition 2(iii). $\qquad\square$

3.2 Port-Hamiltonian Descriptor Systems

A second formulation of linear pH systems is given in a somewhat more explicit form, involving a differential-algebraic equation (DAE), see, e.g., [19, Definition 4.9], see also [3, 15].

Definition 13 ([19]) A pH descriptor system is a DAE with inputs and outputs of the form

$$\begin{bmatrix} \frac{\mathrm{d}}{\mathrm{d}t} E z(t) \\ y(t) \end{bmatrix} = \begin{bmatrix} J - R & B - P \\ (B + P)^* & S + N \end{bmatrix} \begin{bmatrix} Q z(t) \\ u(t) \end{bmatrix} \tag{7}$$

with $\mathbb{K}^m$-valued input u and output y, $\mathbb{K}^n$-valued state z, matrices $E, J, R, Q \in \mathbb{K}^{n \times n}$, $B, P \in \mathbb{K}^{n \times m}$, $S, N \in \mathbb{K}^{m \times m}$ satisfying

$$E^*Q = Q^*E, \quad J = -J^*, \quad N = -N^*, \quad R = R^*, \quad S = S^*$$

such that

$$W := \begin{bmatrix} Q^* & 0 \\ 0 & I \end{bmatrix} \begin{bmatrix} R & P \\ P^* & S \end{bmatrix} \begin{bmatrix} Q & 0 \\ 0 & I \end{bmatrix} \geq 0. \tag{8}$$

The *Hamiltonian* of the system is defined as $H(z) = z^*Q^*Ez$. A solution of (7) is an input-state-output trajectory $(u, z, y) \in C([0, \infty); \mathbb{K}^{n+2m})$ with $Ez \in C^1([0, \infty); \mathbb{K}^n)$ such that (7) is satisfied for all $t \geq 0$.

Note that one could also generalize the above definition to inputs $u \in L^1_{\mathrm{loc}}((0, \infty); \mathbb{K}^m)$ when considering $Ez \in W^{1,1}_{\mathrm{loc}}([0, \infty); \mathbb{K}^n)$. In this case, our results concluded in the following (in particular Theorem 15 and Theorem 18) remain to hold when requiring the inclusion (5) in Definition 10 to only hold almost everywhere in time.

The following result yields a regularity result of the Hamiltonian along solutions and power balance for the DAE system (7). Its proof follows by straightforward modifications of [8, Lemma 2.2], where a similar result was shown for solutions in $W_{\text{loc}}^{1,1}([0, \infty); \mathbb{K}^m)$ and we state it here for completeness.

Lemma 14 *If (u, z, y) is a solution of (7), then $H \circ z \in C^1([0, \infty); \mathbb{K}^n)$, and the following power balance holds:*

$$\frac{\mathrm{d}}{\mathrm{d}t} H(z(t)) = \mathrm{Re}\langle u(t), y(t)\rangle - \left\| W^{\frac{1}{2}} \begin{bmatrix} z(t) \\ u(t) \end{bmatrix} \right\|^2. \tag{9}$$

Proof Let $\mathcal{P}$ denote the orthogonal projection onto $\mathrm{ran}\, E^*$, i.e., $I - \mathcal{P}$ maps onto $(\mathrm{ran}\, E^*)^\perp = \ker E$. Hence, we have $E = E\mathcal{P} + E(I - \mathcal{P}) = E\mathcal{P}$. Let $E^\dagger$ denote the Moore-Penrose inverse of E. Then $\mathcal{P} = E^\dagger E$ and therefore

$$Ez \in C^1([0, \infty), \mathbb{K}^n) \quad \Longleftrightarrow \quad \mathcal{P}z \in C^1([0, \infty), \mathbb{K}^n).$$

Since $E^*Q = Q^*E$, we have $H(z) = \frac{1}{2}z^*\mathcal{P}E^*Qz = \frac{1}{2}(\mathcal{P}z)^*Q^*E(\mathcal{P}z)$ and thus $H \circ z \in C^1([0, \infty); K^n)$. Consequently, and as $\mathrm{Re}(z^*Q^*JQz) = \mathrm{Re}(u^*Nu) = 0$, we obtain

$$\frac{\mathrm{d}}{\mathrm{d}t}(H \circ z) = \mathrm{Re}\left[\left(\frac{\mathrm{d}}{\mathrm{d}t}\mathcal{P}z\right)^* Q^*E\mathcal{P}z\right] = \mathrm{Re}\left[\left(\frac{\mathrm{d}}{\mathrm{d}t}\mathcal{P}z\right)^* E^*Qz\right] = \mathrm{Re}\left[\left(\frac{\mathrm{d}}{\mathrm{d}t}Ez\right)^* Qz\right]$$

$$= \mathrm{Re}\left[(J - R)Qz + (B - P)u\right]^* Qz = -z^*Q^*RQz + \mathrm{Re}(u^*(B - P)^*Qz)$$

$$= \mathrm{Re}\left[u^*\left((B + P)^*Qz + (S + N)u\right) - 2u^*P^*Qz - u^*Su\right] - z^*Q^*RQz$$

$$= \mathrm{Re}\left[u^*y\right] - \begin{bmatrix} z^* & u^* \end{bmatrix} W \begin{bmatrix} z \\ u \end{bmatrix},$$

which is the claimed power balance. $\qquad\square$

4 Equivalence of the Two Formulations

In this section, we associate a pH descriptor system in the sense of Definition 13 with a geometric pH system as defined in Definition 10 and vice versa. This shows that the two formulations introduced in Sect. 3 are equivalent.

4.1 From Geometric pH to Descriptor pH

The next theorem shows that geometric pH systems $(\mathcal{D}, \mathcal{L}, \mathcal{R})$ can be associated with particular pH descriptor systems such that solutions of the geometric pH system are uniquely determined parts of solutions of the descriptor system and vice versa.

Theorem 15 *Let a geometric pH system $(\mathcal{D}, \mathcal{L}, \mathcal{R})$ be given as in Definition 10 and set $p = \dim \ker \mathcal{D} + \dim \ker \mathcal{R} + \dim \ker \mathcal{L}$. Then there exists a pH descriptor system of the form*

$$\begin{bmatrix} \frac{d}{dt} Ez(t) \\ y(t) \end{bmatrix} = \begin{bmatrix} J - R & B \\ B^* & N \end{bmatrix} \begin{bmatrix} z(t) \\ u(t) \end{bmatrix} \tag{10}$$

as in (7) with $Q = I$, $P = 0$, $S = 0$ with the state $z \in \mathbb{K}^{n+r+p}$ such that the following hold:

(i) *If (u, x, y) is a solution of $(\mathcal{D}, \mathcal{L}, \mathcal{R})$ then there exists z such that (u, z, y) solves (10).*

(ii) *If (u, z, y) is a solution of (10), then for every $(x_0, e_L(0)) \in \mathcal{L}$ there exists x such that (u, x, y) solves $(\mathcal{D}, \mathcal{L}, \mathcal{R})$ with $x(0) = x_0$.*

Furthermore, $E = E^*$ holds and if $-\mathcal{L}$ is resistive, then $E = E^* \geq 0$ holds.

Proof Let $d = \dim \ker \mathcal{D}$, $k = \dim \ker \mathcal{R}$, $l = \dim \ker \mathcal{L}$, and $N = n + r + m$. By Proposition 8, there exist an injective $G \in \mathbb{K}^{N \times d}$ and a skew-adjoint $\tilde{J} \in \mathbb{K}^{N \times N}$ such that

$$\mathcal{D} = \left\{ \begin{bmatrix} \tilde{J}e - G\lambda \\ e \end{bmatrix} : G^*e = 0, \ \lambda \in \mathbb{K}^d, \ e \in \mathbb{K}^N \right\}. \tag{11}$$

Let (u, x, y) be a solution of $(\mathcal{D}, \mathcal{L}, \mathcal{R})$ with f_R, e_R, and e_L as in (5), i.e.,

$$\left(-\dot{x}(t), f_R(t), y(t), e_L(t), e_R(t), u(t) \right) \in \mathcal{D},$$
$$(x(t), e_L(t)) \in \mathcal{L}, \quad (f_R(t), e_R(t)) \in \mathcal{R}. \tag{12}$$

Hence, we find that there exists $\lambda : [0, \infty) \to \mathbb{K}^d$ such that

$$\begin{bmatrix} -\dot{x}(t) \\ f_R(t) \\ y(t) \\ 0 \end{bmatrix} = \begin{bmatrix} \tilde{J} & -G \\ G^* & 0 \end{bmatrix} \begin{bmatrix} e_L(t) \\ e_R(t) \\ u(t) \\ \lambda(t) \end{bmatrix} = \begin{bmatrix} J_{11} & J_{12} & J_{13} & -G_1 \\ -J_{12}^* & J_{22} & J_{23} & -G_2 \\ -J_{13}^* & -J_{23}^* & J_{33} & -G_3 \\ G_1^* & G_2^* & G_3^* & 0 \end{bmatrix} \begin{bmatrix} e_L(t) \\ e_R(t) \\ u(t) \\ \lambda(t) \end{bmatrix}. \tag{13}$$

Making use of Proposition 8 again, we find that the maximal resistive structure $\mathcal{R}$ has a representation

$$\mathcal{R} = \left\{ \begin{bmatrix} -\tilde{R}e_R + G_R\lambda_R \\ e_R \end{bmatrix} : G_R^* e_R = 0, \ \lambda_R \in \mathbb{K}^k, \ e_R \in \mathbb{K}^r \right\}, \tag{14}$$

where $G_R \in \mathbb{K}^{r \times k}$ is injective and $\widetilde{R} \in \mathbb{K}^{r \times r}$ is a positive semi-definite Hermitian matrix. Hence, (14) implies

$$f_R(t) = -\widetilde{R} e_R(t) + G_R \lambda_R(t), \qquad G_R^* e_R(t) = 0.$$

Therefore, (13) can be equivalently rewritten as

$$
\begin{bmatrix} \dot{x}(t) \\ 0 \\ y(t) \\ 0 \\ 0 \end{bmatrix}
=
\begin{bmatrix}
-J_{11} & -J_{12} & -J_{13} & G_1 & 0 \\
J_{12}^* & -J_{22} - \widetilde{R} & -J_{23} & G_2 & G_R \\
-J_{13}^* & -J_{23}^* & J_{33} & -G_3 & 0 \\
-G_1^* & -G_2^* & -G_3^* & 0 & 0 \\
0 & -G_R^* & 0 & 0 & 0
\end{bmatrix}
\begin{bmatrix} e_L(t) \\ e_R(t) \\ u(t) \\ \lambda(t) \\ \lambda_R(t) \end{bmatrix}
$$

and after an additional permutation of the rows and columns, we obtain

$$
\begin{bmatrix} \dot{x}(t) \\ 0 \\ 0 \\ 0 \\ y(t) \end{bmatrix}
=
\begin{bmatrix}
-J_{11} & -J_{12} & G_1 & 0 & -J_{13} \\
J_{12}^* & -J_{22} - \widetilde{R} & G_2 & G_R & -J_{23} \\
-G_1^* & -G_2^* & 0 & 0 & -G_3^* \\
0 & -G_R^* & 0 & 0 & 0 \\
-J_{13}^* & -J_{23}^* & -G_3 & 0 & J_{33}
\end{bmatrix}
\begin{bmatrix} e_L(t) \\ e_R(t) \\ \lambda(t) \\ \lambda_R(t) \\ u(t) \end{bmatrix} .
\tag{15}
$$

Leveraging Proposition 8 one more time, we may express the Lagrange structure $\mathcal{L}$ as

$$
\mathcal{L} = \left\{ \begin{bmatrix} L e_L - G_L \lambda_L \\ e_L \end{bmatrix} : G_L^* e_L = 0, \ \lambda_L \in \mathbb{K}^l, \ e_L \in \mathbb{K}^n \right\},
\tag{16}
$$

where $G_L \in \mathbb{K}^{n \times l}$ is injective and $L \in \mathbb{K}^{n \times n}$ is a Hermitian matrix. Hence, for given $e_L(t) \in \mathbb{K}^n$ there exists a unique $\lambda_L(t) \in \mathbb{K}^l$ satisfying

$$
x(t) = L e_L(t) - G_L \lambda_L(t), \qquad G_L^* e_L = 0
\tag{17}
$$

By Proposition 8, $L e_L(t)$ is orthogonal to $G_L \lambda_L(t)$ for all $t \geq 0$ and therefore $x \in C^1([0, \infty), \mathbb{K}^n)$ holds if and only if $L e_L, G_L \lambda_L \in C^1([0, \infty), \mathbb{K}^n)$ holds. Moreover, as G_L is injective, we have $G_L^\dagger G_L = I_l$ and so $G_L \lambda_L \in C^1([0, \infty), \mathbb{K}^n)$ is equivalent to $\lambda_L \in C^1([0, \infty), \mathbb{K}^l)$. Using (17), the system (15) is equivalent to

$$
\begin{bmatrix} \frac{d}{dt} L e_L(t) - G_L \dot{\lambda}_L(t) \\ 0 \\ 0 \\ 0 \\ y(t) \end{bmatrix}
=
\begin{bmatrix}
-J_{11} & -J_{12} & G_1 & 0 & -J_{13} \\
J_{12}^* & -J_{22} - R & G_2 & G_R & -J_{23} \\
-G_1^* & -G_2^* & 0 & 0 & -G_3^* \\
0 & -G_R^* & 0 & 0 & 0 \\
-J_{13}^* & -J_{23}^* & -G_3 & 0 & J_{33}
\end{bmatrix}
\begin{bmatrix} e_L(t) \\ e_R(t) \\ \lambda(t) \\ \lambda_R(t) \\ u(t) \end{bmatrix}
$$

which can be rewritten as

$$
\begin{bmatrix} \frac{d}{dt} L e_L(t) \\ 0 \\ 0 \\ 0 \\ 0 \\ y(t) \end{bmatrix} = \begin{bmatrix} -J_{11} & -J_{12} & G_1 & 0 & G_L & -J_{13} \\ J_{12}^* & -J_{22} - \widetilde{R} & G_2 & G_R & 0 & -J_{23} \\ -G_1^* & -G_2^* & 0 & 0 & 0 & -G_3^* \\ 0 & -G_R^* & 0 & 0 & 0 & 0 \\ -G_L^* & 0 & 0 & 0 & 0 & 0 \\ -J_{13}^* & -J_{23}^* & -G_3 & 0 & 0 & J_{33} \end{bmatrix} \begin{bmatrix} e_L(t) \\ e_R(t) \\ \lambda(t) \\ \lambda_R(t) \\ \dot{\lambda}_L(t) \\ u(t) \end{bmatrix}.
\tag{18}
$$

We now define

$$
J := \begin{bmatrix} -J_{11} & -J_{12} & G_1 & 0 & G_L \\ J_{12}^* & -J_{22} & G_2 & G_R & 0 \\ -G_1^* & -G_2^* & 0 & 0 & 0 \\ 0 & -G_R^* & 0 & 0 & 0 \\ -G_L^* & 0 & 0 & 0 & 0 \end{bmatrix}, \quad R := \begin{bmatrix} 0 & 0 & 0 & 0 & 0 \\ 0 & \widetilde{R} & 0 & 0 & 0 \\ 0 & 0 & 0 & 0 & 0 \\ 0 & 0 & 0 & 0 & 0 \\ 0 & 0 & 0 & 0 & 0 \end{bmatrix}, \quad E = \begin{bmatrix} L & 0 & 0 & 0 & 0 \\ 0 & 0 & 0 & 0 & 0 \\ 0 & 0 & 0 & 0 & 0 \\ 0 & 0 & 0 & 0 & 0 \\ 0 & 0 & 0 & 0 & 0 \end{bmatrix},
$$

$$
B := \begin{bmatrix} -J_{13} \\ -J_{23} \\ -G_3^* \\ 0 \\ 0 \end{bmatrix}, \quad N = J_{33}, \quad S = 0, \quad Q = I, \quad P = 0.
\tag{19}
$$

If (u, x, y) solves $(\mathcal{D}, \mathcal{L}, \mathcal{R})$, then $L e_L \in C^1([0, \infty), \mathbb{K}^n)$ holds and (u, z, y) with $z = (e_L, e_R, \lambda, \lambda_R, \dot{\lambda}_L)$ solves (18). This proves (i). To show (ii), let (u, z, y) with $z = (e_L, e_R, \lambda, \lambda_R, \mu_L)$ solve (18). Then for $(x_0, e_L(0)) \in \mathcal{L}$, there exists unique $\lambda_L^0 \in \mathbb{K}^l$ such that $x_0 = L e_L(0) + G_L \lambda_L^0$ holds. We set $\lambda_L(t) = \lambda_L^0 + \int_0^t \mu_L(s)\, ds$ and define $x(t) = L e_L(t) - G_L \lambda_L(t)$. Since $L e_L(0)$ is given, we have $x(0) = L e_L(0) + G_L \lambda_L^0 = x_0$. Furthermore, $f_R(t) = -\widetilde{R} e_R(t) + G_R \lambda_R(t)$ fulfills $(f_R(t), e_R(t)) \in \mathcal{R}$ and consequently (5) is satisfied. Therefore, (u, x, y) is a solution of $(\mathcal{D}, \mathcal{L}, \mathcal{R})$ which proves (ii).

If $-\mathcal{L}$ is resistive, then Proposition 8 implies that L as in (16) fulfills $L = L^* \geq 0$ and therefore E given by (19) satisfies $E = E^* \geq 0$. $\qquad \square$

Remark 16 Assume that for the Lagrange structure $\mathcal{L}$ in Definition 10, the subspace $-\mathcal{L}$ is resistive. Theorem 15 yields $E \geq 0$. Therefore, the pencil $(E, J - R)$ is regular,[1] if and only if $\ker E \cap \ker(J - R) = \{0\}$ holds, see, e.g., [1, Lemma 6.1.4]. Furthermore, it was shown in [1] that the following holds:

$$
(sE - (J - R))^{-1} + (sE^* - (J - R)^*)^{-1} \geq 0 \quad \text{for } \operatorname{Re}(s) > 0.
$$

---

[1] i.e., $\lambda E - (J - R)$ is invertible for some $\lambda \in \mathbb{C}$.

Therefore, the resulting descriptor system is positive real, i.e., that the transfer function $G(s) = B^*(sE - (J - R))^{-1}B + N$ fulfills

$$G(s) + G(s)^* = B^*((sE - (J - R))^{-1} + (sE^* - (J - R)^*)^{-1})B \geq 0 \quad \text{for } \text{Re}(s) > 0.$$

More details on the relation of positive real and pH descriptor systems, as well as their relation to passive descriptor systems, can be found in [5].

Remark 17 In the proof of Theorem 15, we applied Proposition 8 to each of the subspaces $\mathcal{D}$, $\mathcal{L}$, and $\mathcal{R}$. Since Proposition 8 holds for monotone subspaces as well, one might generalize and assume that $\mathcal{L}$ is monotone. In this case, the corresponding pH descriptor system can be derived as in Theorem 15 and $E + E^* \geq 0$ holds. Then, the resulting matrix pair $(E, J - R)$ is said to have positive Hermitian part. In [18], the spectral properties as well as the regularity and the Kronecker canonical form of these pencils were further analyzed.

4.2 *From Descriptor pH to Geometric pH*

Next, we show how to associate a geometric pH system with a given pH descriptor system such that there is a one-to-one correspondence between the solutions of the two systems. The following result is a slight extension of [15, Theorem 3] where no additional Lagrange structure was considered.

Theorem 18 *Let a pH descriptor system as in Definition 13 be given with*

$$\ker E \cap \ker Q = \{0\}.$$

Let $W = \begin{bmatrix} R & P \\ P^* & S \end{bmatrix}$ *as in* (8), $\Gamma := \begin{bmatrix} J & B \\ -B^* & -N \end{bmatrix}$, *set* $r = n + m$, *and define*

$$\mathcal{L} := \text{ran} \begin{bmatrix} E \\ Q \end{bmatrix} \subset \mathbb{K}^n \times \mathbb{K}^n, \qquad \mathcal{R} := \text{gr}(-W) = \text{ran} \begin{bmatrix} I_r \\ -W \end{bmatrix} \subset \mathbb{K}^r \times \mathbb{K}^r.$$

Further, with the matrices

$$U := \begin{bmatrix} I_n & 0 & 0 \\ 0 & 0 & I_r \\ 0 & I_m & 0 \end{bmatrix} \in \mathbb{K}^{(n+r+m)\times(n+r+m)} \qquad and$$

$$\tilde{D} := \begin{bmatrix} -\Gamma & -I_r \\ I_r & 0 \end{bmatrix} \in \mathbb{K}^{(n+r+m)\times(n+r+m)}$$

define the subspace

$$\mathcal{D} := \text{gr}^{-1}(U\tilde{D}U^*).$$

Then $(\mathcal{D}, \mathcal{L}, \mathcal{R})$ is a geometric pH system. Moreover, the following hold:

(i) *If (u, z, y) solves the DAE (7), then (u, Ez, y) solves the geometric pH system $(\mathcal{D}, \mathcal{L}, \mathcal{R})$.*

(ii) *If (u, x, y) solves the geometric pH system $(\mathcal{D}, \mathcal{L}, \mathcal{R})$ with f_R, e_R, and e_L as in Definition 10, then (u, z, y) solves the DAE (7), where z is the unique function satisfying $x = Ez$ and $e_L = Qz$.*

Proof We first show that $(\mathcal{D}, \mathcal{L}, \mathcal{R})$ is a geometric pH system in the sense of Definition 10. To this end, note that $U\tilde{D}U^*$ is skew-symmetric and therefore $\mathcal{D}$ defines a Dirac structure. By Definition 13, one has $W \geq 0$ and therefore $\mathcal{R}$ is maximal resistive by Proposition 6 (iv). Moreover, Definition 13 yields $Q^*E = E^*Q$ and the assumption $\ker E \cap \ker Q = \{0\}$ implies $\mathrm{rank}\begin{bmatrix} E \\ Q \end{bmatrix} = n$. Using Proposition 6 (i), it follows that $\mathcal{L}$ is a Lagrange structure. Assume that (u, z, y) solves (7). Then, setting

$$x := Ez, \quad e_L := Qz, \quad f_R := \begin{bmatrix} Qz \\ u \end{bmatrix}, \quad e_R := -Wf_R,$$

we have

$$\begin{bmatrix} -\dot{x} \\ y \\ f_R \end{bmatrix} = \begin{bmatrix} -\frac{d}{dt}Ez \\ y \\ Qz \\ u \end{bmatrix} = \begin{bmatrix} (-J+R)Qz + (-B+P)u \\ (B+P)^*Qz + (S+N)u \\ Qz \\ u \end{bmatrix} = \begin{bmatrix} (-\Gamma+W)f_R \\ f_R \end{bmatrix}$$

$$= \begin{bmatrix} -\Gamma f_R - e_R \\ f_R \end{bmatrix} = \tilde{D}\begin{bmatrix} Qz \\ u \\ e_R \end{bmatrix}.$$

Now applying U from the left to this equation shows that $(-\dot{x}, f_R, y, e_L, e_R, u) \in \mathcal{D}$.

Conversely, if (u, x, y) solves the geometric pH system $(\mathcal{D}, \mathcal{L}, \mathcal{R})$, then there exist functions f_R, e_L, e_R such that $(-\dot{x}, f_R, y, e_L, e_R, u) \in \mathcal{D}$, $(x, e_L) \in \mathcal{L}$, and $(f_R, e_R) \in \mathcal{R}$. By the definition of $\mathcal{L}$ and $\mathcal{R}$, there exists a unique function z such that $x = Ez$ and $e_L = Qz$. Moreover, $e_R = -Wf_R$, and we obtain

$$\begin{bmatrix} -\frac{d}{dt}Ez \\ y \\ f_R \end{bmatrix} = \begin{bmatrix} -\dot{x} \\ y \\ f_R \end{bmatrix} = \tilde{D}\begin{bmatrix} e_L \\ u \\ e_R \end{bmatrix} = \begin{bmatrix} -\Gamma & -I_r \\ I_r & 0 \end{bmatrix}\begin{bmatrix} Qz \\ u \\ -Wf_R \end{bmatrix}$$

$$= \begin{bmatrix} -\Gamma\begin{bmatrix} Qz \\ u \end{bmatrix} + Wf_R \\ \begin{bmatrix} Qz \\ u \end{bmatrix} \end{bmatrix}.$$

This implies that $f_R = \begin{bmatrix} Qz \\ u \end{bmatrix}$ and thus

$$\begin{bmatrix} -\frac{d}{dt} Ez \\ y \end{bmatrix} = (W - \Gamma) \begin{bmatrix} Qz \\ u \end{bmatrix},$$

which means that (u, z, y) solves (7). $\qquad\square$

We briefly comment on this result. For pH descriptor system as in Definition 13 satisfying $\ker E \cap \ker Q = \{0\}$, we obtain a geometric pH system from Theorem 18. Applying Theorem 15 to this geometric system leads back to a pH descriptor system as in Definition (13), which fulfills $Q = Id$. Hence, we obtain a pH descriptor system with invertible Q that is equivalent to the original descriptor system, but with a larger state space dimension; see also [17, 19] for alternative methods on achieving invertibility of Q. Furthermore, note that the condition $\ker E \cap \ker Q = \{0\}$ is a natural assumption for pH descriptor systems (10) and, as explained in Remark 16, easily follows from the typically assumed regularity of the underlying matrix pair $(E, (J - R)Q)$.

5 Conclusion, Extensions, and Open Problems

We have shown that the geometric formulation of port-Hamiltonian systems is equivalent to the state-space representation by means of differential algebraic equations. To this end, we utilized tools from multilinear algebra and provided constructive proofs to transfer either of the formulation to the other. The main assumption to derive a geometric representation from a pH-DAE was that the matrices in the Hamiltonian $\mathcal{H}(x) = \frac{1}{2} x^* Q^* E x$, $x \in \mathbb{K}^n$, satisfy $\ker E \cap \ker Q = \{0\}$.

Concerning future research, a first extension could be the investigation of the case of a nontrivial kernel intersection. In this case, one cannot directly define a Lagrange subspace by means of these two matrices as it was done in the proof of Theorem 18. One possible remedy could be to isolate the common kernel by means of a common singular value decomposition of E and Q, similarly as in [11, Lemma 3.6]. Furthermore, the geometric pH representation obtained in Theorem 18 could not be optimal in the sense that the dimensions of constructed Dirac and the maximal resistive structure, which is equal to $2(n + m)$, might be further reduced.

In view of an extension to the infinite-dimensional case, a first step would be to define an infinite-dimensional differential algebraic formulation of pH systems. Whereas infinite-dimensional DAEs are a very delicate issue [9, 22, 23, 25], a definition for closed systems was given in [12] by incorporating the pH structure, i.e., the dissipativity of the main operator. From a geometric point of view, infinite-dimensional pH systems give rise to (Stokes)-Dirac structures [27] including also boundary port variables. An analytical viewpoint on Dirac structures for skew-symmetric differential operators was provided in [14, 29].

In the present note, we focused on continuous time pH systems. Recently, in [6], a definition for discrete-time pH descriptor systems was given. Therein, the discrete time pH system was obtained from a Cayley transformation of continuous time systems. Furthermore, the Cayley transform was applied to the underlying Dirac structures which result in contractive subspaces as discrete-time counterparts of Dirac structures. It remains an open problem to compare the geometric pH formulation and the descriptor pH formulation for discrete-time systems.

References

1. T. Berger. *On differential-algebraic control systems*. Ph.D. Thesis, TU Ilmenau, 2013
2. J. Behrndt, S. Hassi, H. De Snoo, *Boundary Value Problems, Weyl Functions, and Differential Operators* (Birkhäuser, Cham, 2020)
3. C.A. Beattie, V. Mehrmann, H. Xu, H.J. Zwart, Linear port-Hamiltonian descriptor systems. Math. Control Signals Syst. **30**, 1–27 (2018)
4. T. Berger, C. Trunk, H. Winkler, Linear relations and the Kronecker canonical form. Linear Algebra Appl. **488**, 13–44 (2016)
5. K. Cherifi, H. Gernandt, D. Hinsen, The difference between port-Hamiltonian, passive and positive real descriptor systems. Math. Control Signals Syst. **36**(2), 451–482 (2024)
6. K. Cherifi, H. Gernandt, D. Hinsen, V. Mehrmann, On discrete-time dissipative port-Hamiltonian (descriptor) systems. Math. Control Signals Syst. (2023)
7. M. Dalsmo, A.J. van der Schaft, On representations and integrability of mathematical structures in energy-conserving physical systems. SIAM J. Control Optim. **37**(1), 54–91 (1998)
8. T. Faulwasser, B. Maschke, F. Philipp, M. Schaller, K. Worthmann, Optimal control of port-Hamiltonian descriptor systems with minimal energy supply. SIAM J. Control Optim. **60**(4), 2132–2158 (2022)
9. A. Favini, A. Yagi, *Degenerate Differential Equations in Banach Spaces* (CRC Press, Boca Raton, 1998)
10. H. Gernandt, F.E. Haller, T. Reis, A linear relations approach to port-Hamiltonian differential-algebraic equations. SIAM J. Matrix Anal. Appl. **42**(2), 1011–1044 (2021)
11. A. Ilchmann, J. Kirchhoff, M. Schaller, Port-Hamiltonian descriptor systems are relative generically controllable and stabilizable (2023). preprint arXiv:2302.05156
12. B. Jacob, K. Morris, On solvability of dissipative partial differential-algebraic equations. IEEE Control Syst. Lett. **6**, 3188–3193 (2022)
13. B. Jacob, H. Zwart, *Linear Port-Hamiltonian Systems on Infinite-Dimensional Spaces*, volume 223 of Operator Theory: Advances and Applications (Birkhäuser, Basel, 2012)
14. Y. Le Gorrec, H. Zwart, B. Maschke, Dirac structures and boundary control systems associated with skew-symmetric differential operators. SIAM J. Control Optim. **44**(5), 1864 (2005)
15. V. Mehrmann, R. Morandin, Structure-preserving discretization for port-Hamiltonian descriptor systems, n *2019 IEEE 58th Conference on Decision and Control (CDC)* (2019), pp. 6863–6868
16. C. Mehl, V. Mehrmann, M. Wojtylak, Linear algebra properties of dissipative Hamiltonian descriptor systems. SIAM J. Matrix Anal. Appl. **39**, 1489–1519 (2018)
17. C. Mehl, V. Mehrmann, M. Wojtylak, Distance problems for dissipative Hamiltonian systems and related matrix polynomials. Linear Algebra Appl. **623**, 335–366 (2021)
18. C. Mehl, V. Mehrmann, M. Wojtylak, Matrix pencils with coefficients that have positive semidefinite Hermitian parts. SIAM J. Matrix Anal. Appl. **43**(3), 1186–1212 (2022)
19. V. Mehrmann, B. Unger, Control of port-Hamiltonian differential-algebraic systems and applications. Acta Numerica **32**, 395–515 (2023)

20. V. Mehrmann, A.J. van der Schaft, Differential–algebraic systems with dissipative Hamiltonian structure. Math. Control Signals Syst. **35**(3), 541–584 (2023)
21. F. Philipp, T. Reis, M. Schaller, Infinite-dimensional port-Hamiltonian systems - a system node approach (2023). preprint arXiv:2302.05168
22. T. Reis, *Systems theoretic aspects of PDAEs and applications to electrical circuits.* Ph.D. Thesis, University of Kaiserslautern, 2006
23. G.A. Sviridyuk, V.E. Fedorov, *Linear Sobolev Type Equations and Degenerate Semigroups of Operators* (De Gruyter, Berlin, Boston, 2003)
24. N. Skrepek, *Linear port-Hamiltonian Systems on Multidimensional Spatial Domains.* Ph.D. Thesis, University of Wuppertal, 2021
25. C. Seifert, S. Trostorff, M. Waurick, *Evolutionary Equations: Picard's Theorem for Partial Differential Equations, and Applications* (Springer, Berlin, 2022).
26. A.J. van der Schaft, D. Jeltsema, Port-Hamiltonian systems theory: an introductory overview. Found. Trends® Syst. Control **1**(2–3), 173–378 (2014)
27. A.J. van der Schaft, B. Maschke, Hamiltonian formulation of distributed-parameter systems with boundary energy flow. J. Geom. Phys. **42**(1–2), 166–194 (2002)
28. A.J. van der Schaft, B. Maschke, Generalized port-Hamiltonian DAE systems. Syst. Control Lett. **121**, 31–37 (2018)
29. J.A. Villegas, *A port-Hamiltonian approach to distributed parameter systems.* Ph.D. Thesis, University of Twente, 2007
30. A. van der Schaft, V. Mehrmann. Linear port-Hamiltonian DAE systems revisited. Syst. Control Lett. **177**, 105564 (2023)

On Differential-Algebraic Equations with Bounded Spectrum in Banach Spaces

Friedrich M. Philipp (iD)

1 Introduction

In this note we are interested in solving homogeneous differential-algebraic equations (DAEs) with an initial value condition of the form

$$\frac{d}{dt}Ex = Ax, \qquad Ex(0) = Ex_0, \tag{1.1}$$

where E and A are linear operators on a Banach space $\mathcal{X}$. We shall assume throughout that E is *not* boundedly invertible and that the corresponding pencil $sE - A$ is regular.

Problem (1.1) can obviously be regarded as a generalized Cauchy problem. On the other hand, every Cauchy problem of the form $\dot{z} = Tz$, $z(0) = z_0 \in \operatorname{dom} T$, with an operator T with nonempty resolvent set $\rho(T)$ can be recast in the form (1.1) by setting $E = (T - \mu)^{-1}$, $A = T(T - \mu)^{-1}$, and $x = (T - \mu)z$ for some $\mu \in \rho(T)$.

In finite dimensions, the solution theory for this problem is well understood, see, e.g., [5, 12] or [7, Appendix A]. It can be shown that there always exist invertible matrices U and V such that UEV and UAV are of the form

$$UEV = \begin{bmatrix} I & 0 \\ 0 & N \end{bmatrix} \quad \text{and} \quad UAV = \begin{bmatrix} M & 0 \\ 0 & I \end{bmatrix},$$

The author was funded by the Carl Zeiss Foundation within the project *DeepTurb–Deep Learning in and from Turbulence* and by the free state of Thuringia and the German Federal Ministry for Education and Research (BMBF) within the project *THInKI–Thüringer Hochschulinitiative für KI im Studium*.

F. M. Philipp (✉)
Institute of Mathematics, Technische Universität Ilmenau, Ilmenau, Germany
e-mail: friedrich.philipp@tu-ilmenau.de

© The Author(s), under exclusive license to Springer Nature Switzerland AG 2024
F. L. Schwenninger, M. Waurick (eds.), *Systems Theory and PDEs*,
Trends in Mathematics, https://doi.org/10.1007/978-3-031-64991-2_7

where N is nilpotent and I denotes the identity matrix. Obviously, this decouples the DAE in (1.1) into an ordinary differential equation (ODE) and the particularly simple DAE of the form $\frac{d}{dt} Nz = z$, which—due to the nilpotency of N—has the only solution $z = 0$.

The matrices U and V can be obtained with the help of the so-called *Wong sequences* (see, e.g., [3, 5, 18]). These are two sequences of nested subspaces N_k and R_k, where the N_k are ascending and the R_k are descending. The minimal $m \in \mathbb{N}$ such that $N_{m+1} = N_m$ is called the *index* of the problem and can be shown to coincide with the maximal length of Jordan chains at the eigenvalue ∞ of the pencil $sE - A$. We then also have $R_{m+1} = R_m$ and $\mathcal{X} = N_m \oplus R_m$, where $\oplus$ denotes the direct sum of subspaces. If P_0 denotes the projection onto N_m with respect to the latter decomposition and $S = E(I - P_0) + AP_0$, then S is invertible and

$$E = S(I_{R_m} \oplus N') \qquad \text{and} \qquad A = S(M' \oplus I_{N_m})$$

with linear maps $N' : N_m \to N_m$ and $M' : R_m \to R_m$, where N' is nilpotent. The matrices U and V are now easy to obtain.

However, in infinite dimensions—even in the case of bounded coefficient operators E and A—it is not clear at all whether the ascending and descending Wong sequences N_k and R_k become stationary at some point or whether the Wong subspaces R_k are closed. One of the main results of this chapter is that the two Wong sequences become stationary if and only if the spectrum of the pencil (E, A) is bounded and the resolvent $(sE - A)^{-1}$ has polynomial growth as $s \to \infty$. We say that the pencil has finite index in this case. Several equivalent conditions are provided in Theorem 3.7. The solution of the initial value problem (1.1) can then be found analogously to the finite-dimensional situation, cf. Theorem 3.13.

The situation is drastically different when the spectrum of (E, A) is bounded, but the resolvent growth at ∞ is *not* of polynomial type. In this case, the problem reduces to an ODE and a DAE of the form $\frac{d}{dt} Tx = x$, where the spectrum of T merely consists of zero. Such operators are called quasi-nilpotent. We give two examples of quasi-nilpotent, but non-nilpotent operators T in one of which the only solution to the corresponding DAE is the trivial one, but in the other nontrivial solutions also exist. We leave open the question for which quasi-nilpotent operators T the solution of the DAE $\frac{d}{dt} Tx = x$ is unique (and thus trivial). However, we are able to characterize the existence of L^∞-solutions on the half-axis (Theorem 4.5), L^2-solutions on compact time intervals (Theorem 4.8), and analytic solutions (Proposition 4.12).

This chapter is organized as follows. In Sect. 2 we describe the setting of the paper and perform the first reductions: from unbounded to bounded coefficients E and A and from pencils (E, A) to operators. We close with a generalization of the Riesz–Dunford spectral projection and show that

$$P_\sigma := \frac{1}{2\pi i} \int_C (zE - A)^{-1} E \, dz$$

can also be regarded as a spectral projection for the pencil (E, A). Here, C denotes a closed curve enclosing a spectral set σ of (E, A). In Sect. 3 we consider pencils of finite index and prove the main result, Theorem 3.7. After that, we provide some conditions which guarantee the closedness of the Wong subspaces R_k. The section ends with the above-mentioned result Theorem 3.13 on the solution of the initial value problem (1.1) in the finite-index case.

Section 4 is devoted to the study of the DAE $\frac{d}{dt}Tx = x$, where T is quasi-nilpotent, but not nilpotent. We provide the above-mentioned two examples, which show that the solution behavior of the DAE is not determined alone by the fact that T is quasi-nilpotent. Here, we provide the above-mentioned necessary and sufficient conditions for the existence of L^∞-solutions on $[0, \infty)$, L^2-solutions on a compact time interval, and analytic solutions.

2 Setting

In this chapter, we consider initial value problems (IVP) of the form

$$\tfrac{d}{dt}Ex = Ax, \qquad Ex(0) = Ex_0, \tag{2.1}$$

in a reflexive or separable Banach space $\mathcal{X}$. Letting $L(\mathcal{X})$ denote the space of all bounded linear operators from $\mathcal{X}$ into itself, we assume that either $E \in L(\mathcal{X})$ and $A : \mathcal{X} \supset \operatorname{dom} A \to \mathcal{X}$ is closed or $E : \mathcal{X} \supset \operatorname{dom} E \to \mathcal{X}$ is closed and $A \in L(\mathcal{X})$. We set $\operatorname{dom}(E, A) := \operatorname{dom} A$ in the first case and $\operatorname{dom}(E, A) := \operatorname{dom} E$ in the second, i.e., $\operatorname{dom}(E, A) = \operatorname{dom} E \cap \operatorname{dom} A$.

A *solution* of (2.1) is a trajectory $x \in C(\mathbb{R}_0^+, \mathcal{X})$ with $x(t) \in \operatorname{dom}(E, A)$ for all $t \geq 0$, which satisfies $Ex \in C^1(\mathbb{R}_0^+, \mathcal{X})$ such that $\frac{d}{dt}(Ex)(t) = Ax(t)$ for all $t \geq 0$ and $Ex(0) = Ex_0$.

We define the *resolvent set* of the pencil (E, A) as

$$\rho(E, A) := \{s \in \mathbb{C} \mid sE - A : \operatorname{dom}(E, A) \to \mathcal{X} \text{ is bijective}\}.$$

The *spectrum* of (E, A) is defined by $\sigma(E, A) := \mathbb{C} \setminus \rho(E, A)$, where we have intentionally excluded the point ∞.

We assume here and throughout the chapter that the DAE in (2.1) and the pencil (E, A) are *regular*, i.e., that $\rho(E, A) \neq \varnothing$.

2.1 Reduction to Bounded Coefficient Operators

We fix an arbitrary $\mu \in \rho(E, A)$ and define the bounded operators

$$F := E(\mu E - A)^{-1} \qquad \text{and} \qquad B := A(\mu E - A)^{-1}.$$

It is easy to see that if x solves (2.1), then $z := (\mu E - A)x = \mu Ex - \frac{d}{dt}Ex$ is continuous and satisfies

$$\frac{d}{dt}Fz = Bz, \qquad Fz(0) = Fz_0, \tag{2.2}$$

where $z_0 := (\mu E - A)x_0$. Conversely, each solution z of (2.2) transforms to the solution $x = (\mu E - A)^{-1}z$ of (2.1).

Therefore, we shall assume without loss of generality from Sect. 2.2 on that both operators E and A are bounded. We add here that the relation

$$(sF - B)(\mu E - A) = sE - A, \qquad s \in \mathbb{C}$$

immediately yields

$$\sigma(E, A) = \sigma(F, B).$$

In particular, the pencil (F, B) is regular with $\mu F - B = I$.

It will be shown below (cf. Lemma 2.2) that $\sigma(F, B)$ is bounded if and only if zero is an isolated point of the spectrum of the operator $(\mu F - B)^{-1}F = F = E(\mu E - A)^{-1}$:

Example 2.1

(a) If $K := A(\mu E - A)^{-1}$ is compact for some $\mu \neq 0$, we have $\mu F = I + K$, and it follows from the spectral theory for compact operators that zero is an isolated point of $\sigma(F)$. Since it is also a pole of the resolvent of F, the pencil (F, B) has finite index, cf. Lemma 3.2, which is the subject of Sect. 3.

(b) If E is a bounded finite-rank operator, then so is $F = E(\mu E - A)^{-1}$ so that zero is an isolated spectral point of F.

2.2 Reduction to Operators

Consider the IVP (2.1) with $E, A \in L(\mathcal{X})$. Next, we shall reduce the spectral properties of the pencil (E, A) to those of a special operator. Fixing an arbitrary $\mu \in \rho(E, A)$, we apply the resolvent $(\mu E - A)^{-1}$ to both the DAE and the initial value condition and obtain the equivalent IVP

$$\frac{d}{dt}T_\mu x = (\mu T_\mu - I)x, \qquad T_\mu x(0) = T_\mu x_0, \tag{2.3}$$

where

$$T_\mu := (\mu E - A)^{-1}E.$$

Indeed, we have

$$(\mu E - A)^{-1}A = (\mu E - A)^{-1}(\mu E - (\mu E - A)) = \mu T_\mu - I.$$

Note that for $s \neq \mu$,

$$sE - A = (s - \mu)(\mu E - A)(T_\mu - \tau_\mu(s)), \tag{2.4}$$

where

$$\tau_\mu(z) := \frac{1}{\mu - z}, \qquad z \in \mathbb{C} \setminus \{\mu\}.$$

Therefore, for $s \neq \mu$,

$$s \in \sigma(E, A) \quad \Longleftrightarrow \quad \tau_\mu(s) \in \sigma(T_\mu). \tag{2.5}$$

Moreover, it is easily seen that for $s, z \in \rho(E, A)$ we have

$$T_z T_s = -\frac{T_s - T_z}{s - z}. \tag{2.6}$$

In particular, the operators T_z, $z \in \rho(E, A)$, all commute with one another.

Note that $0 \in \sigma(T_\mu)$ as E is assumed to be noninvertible. What we now would like to assume is that *zero is an isolated point of the spectrum of* T_μ. In this case, we may apply the Riesz–Dunford spectral projector[1]

$$P_0 = -\frac{1}{2\pi i} \int_C (T_\mu - \lambda)^{-1} \, d\lambda \tag{2.7}$$

and the complementary projection $I - P_0$ to (2.3), respectively, to obtain the two decoupled IVPs:

$$\frac{d}{dt} S_0 z_0 = (\mu S_0 - I)z_0, \qquad S_0 z_0(0) = S_0 P_0 x_0,$$
$$\frac{d}{dt} S_1 z_1 = (\mu S_1 - I)z_1, \qquad S_1 z_1(0) = S_1(I - P_0)x_0, \tag{2.8}$$

where $S_0 = T_\mu|_{P_0 \mathcal{X}} \in L(P_0 \mathcal{X})$ and $S_1 = T_\mu|_{(I - P_0)\mathcal{X}} \in L((I - P_0)\mathcal{X})$. Then S_1 is boundedly invertible and $\sigma(S_0) = \{0\}$. Hence, the set of IVPs (2.8) is equivalent to

$$\frac{d}{dt} T z_0 = z_0, \qquad T z_0(0) = T P_0 x_0, \tag{2.9}$$
$$\dot{z}_1 = (\mu - S_1^{-1})z_1, \qquad z_1(0) = (I - P_0)x_0, \tag{2.10}$$

where $T = (\mu S_0 - I)^{-1} S_0$. Note also that $\sigma(T) = \{0\}$.

---

[1] Where $C = \partial B_r(0)$, $r > 0$ sufficiently small, is a positively oriented circle line around the zero point.

Clearly, the IVP (2.10) has the unique solution $z_1(t) = e^{(\mu - S_1^{-1})t}(I - P_0)x_0$ so that it remains to solve the IVP (2.9). If z_0 is a solution, then the original problem (2.3) has the solution $x = z_0 + z_1$.

The next lemma immediately follows from (2.5).

Lemma 2.2 *Zero is an isolated point of $\sigma(T_\mu)$ if and only if $\sigma(E, A)$ is bounded. In this case, the IVP (2.1) has a solution if and only if the IVP (2.9) has a solution.*

2.3 Spectral Sets

A set $\sigma \subset \sigma(E, A)$ will be called a *spectral set* for the pencil (E, A), if it is both open and closed in $\sigma(E, A)$. If σ is a bounded spectral set for (E, A), then we find a bounded open neighborhood U of σ whose boundary C consists of a finite number of rectifiable Jordan curves, oriented in the positive sense, and such that $\overline{U} \cap \sigma(E, A) = \sigma$. We then define

$$P_\sigma := \frac{1}{2\pi i} \int_C (zE - A)^{-1} E \, dz = \frac{1}{2\pi i} \int_C T_z \, dz.$$

It is evident that this definition does not depend on the set U and its boundary, but only on σ.

Parts of the following proposition have been proved in [13, Theorem 5.1] for the special case of isolated points of $\sigma(E, A)$, i.e., $\sigma = \{\lambda_0\}$ with $\lambda_0 \in \mathbb{C}$. Its proof can be found in the appendix of this paper.

Proposition 2.3 *Let σ be a bounded spectral set of (E, A), let $\mu \in \rho(E, A)$ be arbitrary, and choose C as above such that μ is in the exterior of C. Then zero is in the exterior of $\tau_\mu(C)$, $\tau_\mu(C)$ is positively oriented, and $\tau_\mu(\sigma)$ is a spectral set for the operator T_μ which is contained in the interior of $\tau_\mu(C)$. Moreover,*

$$P_\sigma = -\frac{1}{2\pi i} \int_{\tau_\mu(C)} (T_\mu - \lambda)^{-1} \, d\lambda, \tag{2.11}$$

i.e., P_σ is the spectral projection of T_μ corresponding to the spectral set $\tau_\mu(\sigma)$.

3 DAEs of Finite Index

Next, we carry over the notion of the index of a DAE or pencil from the finite-dimensional case to the infinite-dimensional situation.

Definition 3.1 The regular DAE $\frac{d}{dt} Ex = Ax$ (or the regular pencil (E, A)) is said to have *finite index* $m \in \mathbb{N}$ if $\sigma(E, A)$ is bounded and $\|(sE - A)^{-1}\| = O(|s|^{m-1})$ as $|s| \to \infty$, where m is minimal, i.e., $\|(sE - A)^{-1}\| \neq O(|s|^{m-2})$ as $|s| \to \infty$.

Recall the operators $T_\mu = (\mu E - A)^{-1} E$, where $\mu \in \rho(E, A)$. The next lemma is an immediate consequence of (2.4).

Lemma 3.2 *The regular pencil* (E, A) *has finite index* m *if and only if zero is a pole of the resolvent of* T_μ *of order* m.

Let $T : \mathcal{X} \to \mathcal{X}$ be a linear operator. Note that for all $k \in \mathbb{N}$ we have $\ker T^k \subset \ker T^{k+1}$ and that $\ker T^k = \ker T^{k+1}$ implies that $\ker T^{k+j} = \ker T^k$ for all $j \in \mathbb{N}$. Similarly, $\operatorname{ran} T^{k+1} \subset \operatorname{ran} T^k$, and equality implies $\operatorname{ran} T^{k+j} = \operatorname{ran} T^k$ for all $j \in \mathbb{N}$. Hence, the closed subspaces $\ker T^k$ are ascending, and the subspaces $\operatorname{ran} T^k$ are descending. The *ascent* and the *descent* of T are defined as

$$\alpha(T) := \min\{n \in \mathbb{N} : \ker T^{n+1} = \ker T^n\} \quad \text{and}$$

$$\delta(T) := \min\{n \in \mathbb{N} : \operatorname{ran} T^{n+1} = \operatorname{ran} T^n\},$$

respectively. If the minimum does not exist, we put $\alpha(T) = \infty$ or $\delta(T) = \infty$, accordingly.

Theorem 3.3 ([14, Corollary 20.5]) *Let* $T \in L(\mathcal{X})$ *such that* $\alpha(T), \delta(T) < \infty$. *Then* $\alpha(T) = \delta(T) =: n$, $\operatorname{ran} T^n$ *is closed, and* $\mathcal{X} = \ker T^n \oplus \operatorname{ran} T^n$.

The next proposition shows in particular that the condition in Theorem 3.3 means that zero is a pole of the resolvent of T. In the proof we shall frequently make use of the relation

$$(T - \lambda)^{-1}(T^n - \lambda^n) = \sum_{k=0}^{n-1} \lambda^k T^{n-k-1}, \tag{3.1}$$

which holds for $T \in L(\mathcal{X})$, $n \in \mathbb{N}_{\geq 1}$, and $\lambda \in \rho(T)$.

Proposition 3.4 *For* $T \in L(\mathcal{X})$ *and* $m \in \mathbb{N}_{\geq 1}$, *the following statements are equivalent:*

  (i) *Zero is a pole of the resolvent of* T *of order* m.
  (ii) *Zero is an isolated point of* $\sigma(T)$ *and* $T^m \int_C (T - \lambda)^{-1} \, d\lambda = 0$ *(m minimal), where* $C = \partial B_r(0)$ *such that* $\overline{B_r(0)} \cap \sigma(T) = \{0\}$.
  (iii) *Zero is an isolated point of* $\sigma(T)$ *and* $\delta(T) = m$.
  (iv) $\mathcal{X} = \ker T^m \oplus \operatorname{ran} T^m$ *(m minimal).*
  (v) $\alpha(T) = \delta(T) = m$.
  (vi) *There exists a sequence* $(\lambda_n) \subset \rho(T)$ *such that* $\lambda_n \to 0$ *and*

$$\|(T - \lambda_n)^{-1}\| = O(|\lambda_n|^{-m}) \text{ as } n \to \infty \quad (m \text{ minimal }), \tag{3.2}$$

and $\operatorname{ran} T^m$ *is closed.*

(vii) *There exists a sequence $(\lambda_n) \subset \rho(T)$ with $\lambda_n \to 0$ as in (3.2), and $\delta(T) < \infty$.*

Proof Assume that all items except (iii) are equivalent. Then (ii) and (v) certainly imply (iii). Conversely, if (iii) holds, denote the spectral projection for T with respect to the spectral set $\sigma = \{0\}$ by P_0. Then $T_0 := T|_{P_0 \mathcal{X}}$ is quasi-nilpotent (i.e., $\sigma(T_0) = \{0\}$) and $\delta(T_0) = m$. Hence, [8, Theorem 2] implies that T_0 is nilpotent. Consequently, we have $\alpha(T) = \delta(T) = m$ and thus (v). This shows that we may neglect (iii) in the following.

(i)$\Leftrightarrow$(ii) This is the statement of [6, Theorem VII.3.18].

(ii)$\Rightarrow$(iv) Let $P_0 = -\frac{1}{2\pi i} \int_C (T - \lambda)^{-1} d\lambda$ be the spectral projection of T with respect to the spectral set $\{0\}$. Then $T^m P_0 = 0$ means that $P_0 \mathcal{X} = \ker T^m$. Hence $\mathcal{X} = \ker T^m \oplus (I - P_0)\mathcal{X}$ implies that $\operatorname{ran} T^m = T^m(I - P_0)\mathcal{X} = (I - P_0)\mathcal{X}$ as $T|_{(I-P_0)\mathcal{X}}$ is boundedly invertible. As to the minimality of m, assume that $\mathcal{X} = \ker T^k \oplus \operatorname{ran} T^k$ for some $1 \le k \le m$. Then $\operatorname{ran} T^k \subset \operatorname{ran} T^{2k}$, and hence $\operatorname{ran} T^k \subset \operatorname{ran} T^{jk}$ for all $j \in \mathbb{N}$. Choose $j \in \mathbb{N}$ such that $jk \ge m$. Then from $T^k P_0 \mathcal{X} = P_0 T^k \mathcal{X} \subset P_0 T^{jk} \mathcal{X} = T^{jk} P_0 \mathcal{X} = \{0\}$ and the minimality in (ii), we obtain $k = m$.

(iv)$\Rightarrow$(v) Let $x \in \ker T^{m+1}$. Then also $x \in \ker T^{2m}$ so that $T^m x \in \ker T^m \cap \operatorname{ran} T^m = \{0\}$, i.e., $x \in \ker T^m$. It follows that $\alpha(T) \le m$. If $y \in \operatorname{ran} T^m$, $y = T^m x$ with some $x \in \mathcal{X}$, then $x = u + T^m z$ with some $u \in \ker T^m$ and $z \in \mathcal{X}$. Thus, $y = T^{2m} z \in \operatorname{ran} T^{m+1}$. This implies $\delta(T) \le m$. Now, (v) follows from Theorem 3.3 and the minimality in (iv).

(v)$\Rightarrow$(i) By Theorem 3.3, $\operatorname{ran} T^m$ is closed and $\mathcal{X} = \ker T^m \oplus \operatorname{ran} T^m$. With respect to this decomposition, T decomposes as $T = N \oplus S$, where $N \in L(\ker T^m)$ is nilpotent ($N^m = 0$, $N^{m-1} \ne 0$), and $S \in L(\operatorname{ran} T^m)$ is boundedly invertible. In particular, $\sigma(T) = \sigma(N) \cup \sigma(S) = \{0\} \cup \sigma(S)$, which shows that zero is an isolated point of $\sigma(T)$. Now, (i) follows from the representation $(N - \lambda)^{-1} = -\sum_{j=-m}^{-1} \lambda^j N^{-j-1}$ of the resolvent of the nilpotent operator N, which can be deduced from (3.1).

Before we turn to the proof of (i)$\Rightarrow$(vi)$\Rightarrow$(vii)$\Rightarrow$(i), let us show that the existence of a sequence (λ_n) as in (vi) and (vii) implies $\alpha(T) \le m$. Indeed, (3.1) with $n = m + 1$ yields for $\lambda \in \rho(T)$ that

$$(T - \lambda)^{-1}(T^{m+1} - \lambda^{m+1}) = \sum_{k=0}^{m} \lambda^k T^{m-k}.$$

Applying this to $x \in \ker T^{m+1}$ with $\lambda = \lambda_n$ and letting $n \to \infty$ in fact yield $T^m x = 0$, i.e., $x \in \ker T^m$.

(i)$\Rightarrow$(vi) We can choose any sequence (λ_n) such that $\lambda_n \to 0$ to satisfy the requirements in (vi). The closedness of $\operatorname{ran} T^m$ follows from (i)$\Leftrightarrow$(iv) and Theorem 3.3.

(vi)$\Rightarrow$(vii) The following argumentation is valid if $\mathcal{X}$ is separable: Since $\operatorname{ran} T^m$ is closed, so is $\operatorname{ran}(T')^m$. Here, $T' : \mathcal{X}' \to \mathcal{X}'$ denotes the adjoint of T. Note that $\sigma(T') = \sigma(T)$ and $\|(T' - \lambda)^{-1}\| = \|(T - \lambda)^{-1}\|$ for $\lambda \in \rho(T)$. Hence, by the intermediate part of this proof, $\alpha(T') \le m$. Hence, we only have to prove that $\operatorname{ran}(T')^{m+1} = \operatorname{ran}(T')^m$, since then $\operatorname{ran}(T')^{m+1}$ is closed and consequently ran $T^{m+1} = {}^\perp\mathrm{ker}(T')^{m+1} = {}^\perp\mathrm{ker}(T')^m = \mathrm{ran}\, T^m$. For this, let $x'_j \in \mathrm{ran}(T')^j$ for $j = 0, \ldots, m$ such that $T'x'_j = x'_{j+1}$, $j = 0, \ldots, m-1$. It follows from (3.1) that for $\lambda \in \rho(T')$ we have

$$(T' - \lambda)^{-1}x'_m = \sum_{j=0}^{m-1} \lambda^j x'_{m-1-j} + \lambda^m (T' - \lambda)^{-1}x'_0.$$

Therefore, due to the property of (λ_n), the sequence (u'_n) with $u'_n = (T' - \lambda_n)^{-1}x'_m \in \mathrm{ran}(T')^m$ is bounded. By the sequential Banach–Alaoglu theorem (see, e.g., [15, Theorem 3.17]) (u'_n) has a weak*-convergent subsequence. Without loss of generality, we may assume that this subsequence is (u'_n) itself. Set $u' := \mathrm{w}^* - \lim_{n\to\infty} u'_n$. As $\mathrm{ran}(T')^m$ is weak*-closed (see [15, Theorem 4.14]), we have $u' \in \mathrm{ran}(T')^m$. Moreover,

$$T'u' = \mathrm{w}^* - \lim_{n\to\infty} T'(T' - \lambda_n)^{-1}x'_m = x'_m + \mathrm{w}^* - \lim_{n\to\infty} \lambda_n u'_n = x'_m,$$

and thus $x'_m \in \mathrm{ran}(T')^{m+1}$.

If $\mathcal{X}$ is reflexive, we can prove $\mathrm{ran}\, T^{m+1} = \mathrm{ran}\, T^m$ by proceeding similarly as in the above proof with T instead of T' and by using the fact that bounded sequences in $\mathrm{ran}\, T$ have a weakly convergent subsequence with weak limit in $\mathrm{ran}\, T$.

(vii)$\Rightarrow$(i) By the intermediate part of the proof and Theorem 3.3, we have that $k := \alpha(T) = \delta(T) \leq m$. Hence, due to the already proved equivalence of (v) and (i), it follows that zero is a pole of the resolvent of T of order k. But m in (vii) is minimal so that $k = m$. $\qquad\square$

The so-called *Wong sequences* (N_k) and (R_k) (see, e.g., [3–5, 18]) for the pencil (E, A) are defined by $N_0 = \{0\}$, $R_0 = \mathcal{X}$, and

$$N_{k+1} = E^{-1}AN_k \qquad \text{and} \qquad R_{k+1} = A^{-1}ER_k$$

for $k = 0, 1, \ldots$. It is easily seen that the N_k are ascending and the R_k are descending.

The statement of the next lemma holds in full generality—even for singular pencils (E, A). It was proved as [5, Lemma 2.3] for the special case of finite-dimensional regular pencils and only for $s \in \rho(E, A)$.

Lemma 3.5 *For each $s \in \mathbb{C}$ we have*

$$R_{k+1} = (sE - A)^{-1}ER_k \qquad and \qquad N_{k+1} = E^{-1}(sE - A)N_k.$$

Proof Let $s \in \mathbb{C}$. If $x \in R_{k+1}$, then $Ax \in ER_k$ and also $x \in R_k$. Hence, we have $(sE - A)x = sEx - Ax \in ER_k$, which gives $x \in (sE - A)^{-1}ER_k$. For the opposite direction, let $x \in (sE - A)^{-1}ER_k$, i.e., $(sE - A)x \in ER_k \subset ER_\ell$ for all $\ell = 0, \ldots, k$. From $(sE - A)x \in ER_0$ we conclude that $Ax \in ER_0$ and thus

$x \in R_1$. Now, an inductive argument leads to $x \in R_k$ and thus $Ax \in ER_k$, which means $x \in R_{k+1}$.

If $x \in E^{-1}(sE - A)N_k$, then $Ex = (sE - A)u$ with some $u \in N_k$. Hence, $E(su - x) = Au \in AN_k$, i.e., $su - x \in N_{k+1}$. But also $u \in N_{k+1}$ and so $x \in N_{k+1}$. The converse inclusion is proved via induction. Clearly, $N_1 = \ker E = E^{-1}(sE - A)N_0$. Now, if $N_{k+1} \subset E^{-1}(sE - A)N_k$ holds for some k, let $x \in N_{k+2}$. Then $Ex = Au$ with $u \in N_{k+1}$, thus $Eu = (sE - A)v$ with $v \in N_k \subset N_{k+1}$, and we obtain

$$(sE - A)(sv - u) = s(sE - A)v - sEu + Au = Au = Ex.$$

Since $sv - u \in N_{k+1}$, it follows that $x \in E^{-1}(sE - A)N_{k+1}$, as desired. $\qquad\square$

The next corollary follows directly from the definitions of the R_k and N_k and the fact that $\ker ST = T^{-1} \ker S$ for all $S, T \in L(\mathcal{X})$.

Corollary 3.6 *For any $\mu \in \rho(E, A)$ and $k \in \mathbb{N}_0$, we have*

$$R_k = \operatorname{ran} T_\mu^k \qquad and \qquad N_k = \ker T_\mu^k.$$

Combining Corollary 3.6 and Proposition 3.4 immediately yields the following theorem characterizing the finite-index property of the regular pencil (E, A).

Theorem 3.7 *Let $m \in \mathbb{N}$. Then the following statements are equivalent:*

(i) *The pencil (E, A) has finite index m.*
(ii) *$\sigma(E, A)$ is bounded and $R_{m+1} = R_m$ (m minimal).*
(iii) *$\mathcal{X} = N_m \oplus R_m$ (m minimal).*
(iv) *$R_{m+1} = R_m$ and $N_{m+1} = N_m$ (m minimal).*
(v) *There exists a sequence $(s_n) \subset \rho(E, A)$ such that $|s_n| \to \infty$ and*

$$\|(s_n E - A)^{-1}\| = O(|s_n|^{m-1}) \text{ as } n \to \infty \quad (m\ minimal), \tag{3.3}$$

and R_m is closed.
(vi) *There exists a sequence $(s_n) \subset \rho(E, A)$ with $|s_n| \to \infty$ as in (3.3), and $R_{k+1} = R_k$ for some $k \in \mathbb{N}$.*

Corollary 3.8 *The pencil (E, A) has finite index if and only if $R_{k+1} = R_k$ and $N_{j+1} = N_j$ for some $k, j \in \mathbb{N}$.*

Remark 3.9 In [16, Problem 1.2] it was asked as to whether the closedness of all R_k and the index of (E, A) being m imply $\min\{k : R_k = R_{k+1}\} = m$. The implication (v)$\Rightarrow$(iv) in Theorem 3.7 answers this question in the affirmative.

The following example shows that the Wong subspaces $R_1, \ldots, R_{m-1}$ do not necessarily have to be closed.

Example 3.10 Assume that the equivalent conditions in Theorem 3.7 are satisfied and let $\mu \in \rho(E, A)$. Then T_μ decomposes as $T_\mu = N \oplus S$ with respect to the decomposition $\mathcal{X} = N_m \oplus R_m$, where N is a nilpotent operator and S is boundedly invertible. Hence,

$$R_k = \operatorname{ran} T_\mu^k = \operatorname{ran} N^k \oplus \operatorname{ran} S^k = \operatorname{ran} N^k \oplus R_m$$

is closed if and only if $\operatorname{ran} N^k$ is closed. The question therefore is whether there exist nilpotent operators with nonclosed range. And in fact there are. As an example consider any bounded operator S with nonclosed range on a Banach space $\mathcal{Z}$. Then $N : \mathcal{Z}^2 \to \mathcal{Z}^2$,

$$N := \begin{bmatrix} 0 & S \\ 0 & 0 \end{bmatrix}$$

satisfies $N^2 = 0$, and $\operatorname{ran} N = \operatorname{ran} S \times \{0\}$ is not closed.

Recall that an operator $T \in L(\mathcal{X})$ is called *upper semi-Fredholm* if $\dim \ker T < \infty$ and $\operatorname{ran} T$ is closed in $\mathcal{X}$. The operator T is called *Fredholm* if it is upper semi-Fredholm and, in addition, $\dim(\mathcal{X}/\operatorname{ran} T) < \infty$. The *Fredhom index* of a Fredholm operator $T \in L(\mathcal{X})$ is defined by

$$\operatorname{ind}(T) = \dim \ker T - \dim(\mathcal{X}/\operatorname{ran} T).$$

Next, we present two sufficient conditions for the spaces R_k to be closed.

Lemma 3.11 *The Wong subspaces R_k are closed in the following cases:*

(a) *E is upper semi-Fredholm.*
(b) *$\mu E - A$ is compact for some $\mu \in \mathbb{C}$.*

Proof

(a) It is well known (see [10, Lemma IV.5.29]) that an upper semi-Fredholm operator maps closed subspaces to closed subspaces. Since A is continuous, the claim follows right from the definition of the R_k.
(b) Let $A_0 := \mu E - A$ be compact. By Lemma 3.5 the R_k do not change if we pass in their definition from (E, A) to (E, A_0). Hence, the same lemma gives

$$R_{k+1} = (sE - A_0)^{-1} E R_k = \left(I + (sE - A_0)^{-1} A_0 \right) R_k$$

for $s \in \rho(E, A_0) \setminus \{0\}$. Since the operator $I + (sE - A_0)^{-1} A_0$ is Fredholm (with Fredholm index zero), see [10, Theorem IV.5.26], all R_k are closed.

$\square$

Corollary 3.12 *Assume $\dim \ker E < \infty$. Then the pencil (E, A) has finite index at most m if and only if E is Fredholm with Fredholm index zero and $\ker E \cap R_m = \{0\}$.*

Proof Let $\mu \in \rho(E, A)$. Note that $\dim \ker E < \infty$ implies $\dim \ker T_\mu < \infty$ and hence also $\dim \ker T_\mu^m < \infty$.

Assume that (E, A) has finite index at most m. Then, by Theorem 3.7, $R_m = \operatorname{ran} T_\mu^m$ is closed, and we have $\mathcal{X} = \ker T_\mu^m \oplus \operatorname{ran} T_\mu^m$. That is, T_μ^m is Fredholm with $\operatorname{ind}(T_\mu^m) = 0$. By Müller [14, Theorem 16.6], T_μ is Fredholm, and the index theorem [14, Theorem 16.12] yields $\operatorname{ind}(T_\mu^m) = m \cdot \operatorname{ind}(T_\mu)$ so that $\operatorname{ind}(T_\mu) = 0$, and hence E is Fredholm with $\operatorname{ind}(E) = \operatorname{ind}((\mu E - A)^{-1}) + \operatorname{ind}(E) = \operatorname{ind}((\mu E - A)^{-1} E) = \operatorname{ind}(T_\mu) = 0$. Also, $\ker E \cap R_m = N_1 \cap R_m \subset N_m \cap R_m = \{0\}$.

Conversely, assume that E is Fredholm with $\operatorname{ind}(E) = 0$ and $\ker E \cap R_m = \{0\}$. The latter is equivalent to $\ker T_\mu \cap \operatorname{ran} T_\mu^m = \{0\}$ and means that $\ker T_\mu^m = \ker T_\mu^{m+1}$, i.e., $\alpha(T_\mu) \leq m$. Moreover, it is evident that also T_μ is Fredholm with $\operatorname{ind}(T_\mu) = 0$ so that from [9, Thm. 4.3] we infer that $\delta(T_\mu) = \alpha(T_\mu) \leq m$. Now, it follows from Proposition 3.4 that (E, A) has finite index at most m. $\qquad\square$

We now apply the results to DAEs

$$\tfrac{d}{dt} Ex = Ax, \qquad Ex(0) = Ex_0. \tag{3.4}$$

Theorem 3.13 *Assume that the regular pencil (E, A) has finite index m. Then the IVP (3.4) has a solution if and only if $x_0 \in R_m \oplus \ker E$, in which case the solution is unique.*

Proof Choose some $\mu \in \rho(E, A)$ and define P_0 as in (2.7). Then $P_0 \mathcal{X} = N_m = \ker T_\mu^m$ and $(I - P_0)\mathcal{X} = R_m = \operatorname{ran} T_\mu^m$. Hence, $S_0 = T_\mu|_{P_0 \mathcal{X}}$ is nilpotent and hence so is $T = (\mu S_0 - I)^{-1} S_0$. By Lemma 2.2, the IVP (3.4) has a solution if and only if (2.9) has a solution. In fact, the DAE in (2.9) has only the trivial solution. Namely, if z_0 is a solution, then $0 = \tfrac{d}{dt} T^m z_0 = T^{m-1} z_0$ and thus $z_0(t) \in \ker T^{m-1}$ for all $t \geq 0$. Further processing this argument leads to $z_0(t) \in \ker T$ for all $t \geq 0$, and hence finally $z_0 = \tfrac{d}{dt} T z_0 = 0$. Consequently, the initial value condition in (2.9) is feasible if and only if $T P_0 x_0 = 0$, which means that $x_0 \in R_m \oplus \ker E$ as $\ker T = \ker E$. $\qquad\square$

Corollary 3.14 *If the regular pencil (E, A) has index $m = 1$, then (3.4) has a unique solution for all $x_0 \in \mathcal{X}$.*

Proof By Theorem 3.13, the IVP (3.4) has a solution for any $x_0 \in R_1 \oplus \ker E$. But $\ker E = N_1$ and $R_1 \oplus N_1 = \mathcal{X}$. $\qquad\square$

4 DAEs with Bounded Spectrum and Infinite Index

In this section, we assume that the spectrum $\sigma(E, A)$ of the pencil (E, A) is bounded. By Lemma 2.2, the IVP (2.1) then has a solution if and only if (2.9) has a solution. Therefore, we shall now focus on DAEs as in (2.9) of the form

$$\tfrac{d}{dt} Tx = x, \tag{4.1}$$

where $\sigma(T) = \{0\}$. Such operators are called *quasi-nilpotent*. A quasi-nilpotent operator T is nilpotent if and only if zero is a pole of its resolvent (cf. Proposition 3.4). In this case, the DAE (4.1) only has the trivial solution $x = 0$ (see the proof of Theorem 3.13). Otherwise, T is not nilpotent and zero is an essential singularity of the resolvent of T. In this case, the question arises whether solutions of (4.1) are still necessarily trivial. The next proposition shows that this is not the case, in general.

Proposition 4.1 *Let A be a generator of a C_0-semigroup $(T_t)_{t\geq 0}$ of operators $T_t \in L(\mathcal{X})$ with $\sigma(A) = \varnothing$. Then $T = A^{-1}$ is quasi-nilpotent, and for any $x_0 \in \mathcal{X}$ the IVP $\frac{d}{dt}Tx = x$, $x(0) = x_0$, has the unique solution $T_t x_0$.*

Proof First of all, it is easy to see that T is quasi-nilpotent. Let $x_0 \in \mathcal{X}$ be arbitrary. In fact, the proposed solution $\bar{x}(t) := T_t x_0$ is evidently continuous as a function from $\mathbb{R}_0^+$ to $\mathcal{X}$, and since $y_0 := A^{-1}x_0 \in \operatorname{dom} A$ and $T_t y_0$ solves the Cauchy problem $\dot{y} = Ay$, $y(0) = y_0$, we have

$$\tfrac{d}{dt}T\bar{x}(t) = \tfrac{d}{dt}A^{-1}T_t x_0 = \tfrac{d}{dt}T_t A^{-1}x_0 = AT_t A^{-1}x_0 = T_t x_0 = \bar{x}(t).$$

Hence, $\bar{x}$ is a solution of the IVP. If $\tilde{x}$ is another solution, set $z := \bar{x} - \tilde{x}$ and $w = A^{-1}z$. Then $\frac{d}{dt}Tz = z$ and $z(0) = 0$. In particular, $w = Tz \in C^1(\mathbb{R}_0^+, \mathcal{X})$ and $\dot{w} = z = Aw$, $w(0) = 0$. Consequently, $w = 0$ and so $z = Aw = 0$, i.e., $\tilde{x} = \bar{x}$. $\square$

Hence, if T^{-1} is the generator of a C_0-semigroup, the DAE $\frac{d}{dt}Tx = x$ behaves like an infinite-dimensional well-posed ODE—in a sense even better, because it has a unique "classical" solution for every initial value $x_0 \in \mathcal{X}$.

Example 4.2 Consider the Volterra operator $V : L^2(0, 1) \to L^2(0, 1)$, defined by

$$(Vu)(s) = \int_0^s u(\sigma)\, d\sigma.$$

It is well known that V is quasi-nilpotent and compact:

(a) The operator $T = -V$ is the inverse of the differential operator $A : L^2(0, 1) \supset \operatorname{dom} A \to L^2(0, 1)$ with $Af = -f'$ for f in

$$\operatorname{dom} A = \{u \in W^{1,2}(0, 1) : u(0) = 0\}.$$

It is well known that the operator A is the generator of the C_0-semigroup of contractions $(T_t)_{t\geq 0}$, given by

$$T_t u(s) = \begin{cases} u(s - t) & \text{for } t \leq s \leq 1 \\ 0 & \text{otherwise} \end{cases}.$$

Hence, the DAE $\frac{d}{dt}Tx = x$ has nontrivial solutions by Proposition 4.1.

(b) The DAE $\frac{d}{dt}Tx = x$ with $T = V$ only has the trivial solution on $[0, \infty)$. Indeed, if x is a solution on $[0, \infty)$ and $\tau > 1$, then $z(t) = x(\tau - t)$ for $t \in [0, \tau]$, solves $-\frac{d}{dt}Tz = z$ on $[0, \tau]$, and is therefore given by $z(t) = T_t z(0)$. Hence, $x(t) = T_{\tau-t}x(\tau) = 0$ for $t < \tau - 1$. Since $\tau > 1$ was arbitrary, the claim follows. We further would like to point out that the function $x(t) = T_{\tau-t}x(\tau)$ solves the IVP $\frac{d}{dt}Tx = x$, $x(0) = 0$, on $[0, \tau]$, but does not vanish.

In view of Example 4.2, the question arises for which quasi-nilpotent operators the DAE (4.1) only has the trivial solution on $[0, \infty)$. We pose this as an open problem.

Open Problem 4.3 For which quasi-nilpotent operators $T \in L(\mathcal{X})$ does the DAE $\frac{d}{dt}Tx = x$ only have the trivial solution $x = 0$ on $[0, \infty)$?

Remark 4.4 Note that if x is a solution of (4.1), then $T^n x$, $n \in \mathbb{N}$, is a solution. Thus, for any function f analytic on a neighborhood of zero, we have that $f(T)x$ is a solution.

In the following three subsections, we collect some necessary and sufficient conditions for the existence of solutions of the DAE $\frac{d}{dt}Tx = x$.

4.1   *Essentially Bounded Solutions on the Half-Axis*

In this subsection, we study L^∞-solutions of the IVP

$$\frac{d}{dt}Tx = x, \quad Tx(0) = Tx_0, \tag{4.2}$$

on $[0, \infty)$. For this, we assume that the Banach space $\mathcal{X}$ has the Radon–Nikodym property. It is well known that reflexive Banach spaces have this property. W. Arendt proved in [1] that $\mathcal{X}$ has the Radon–Nikodym property if and only if each Lipschitz-continuous function $F : [0, 1] \to \mathcal{X}$ is a.e. differentiable.

Theorem 4.5 *Let $\mathcal{X}$ have the Radon–Nikodym property. Then the IVP (4.2) has a solution $x \in L^\infty([0, \infty), \mathcal{X})$ if and only if*

$$\sup\left\{\left\|[(T - s)^{-1}T]^{n+1}x_0\right\| : s > 0, \ n \in \mathbb{N}_0\right\} < \infty. \tag{4.3}$$

The solution then is unique.

Proof In what follows, we make use of the Laplace transform

$$\mathcal{L}z(s) = \int_0^\infty e^{-st}z(t)\,dt, \quad s > 0,$$

which is well-defined for measurable functions $z : [0, \infty) \to \mathcal{X}$ with polynomial growth, i.e., $\|z(t)\| \leq M(1 + t)^n$ for some $n \in \mathbb{N}$ and $M > 0$ and all $t \geq 0$, and defines an analytic $\mathcal{X}$-valued function on $\mathbb{C}^+ = \{\lambda \in \mathbb{C} : \operatorname{Re}\lambda > 0\}$. It is easy to see that the Laplace transform Z of a function $z \in L^\infty([0, \infty), \mathcal{X})$ satisfies

$$\tfrac{1}{n!}\left\| s^{n+1} Z^{(n)}(s) \right\| \le \|z\|_\infty, \quad n \in \mathbb{N}_0, \ s > 0. \tag{4.4}$$

Let $x \in L^\infty([0, \infty), \mathcal{X})$ be a solution of (4.2). Applying the Laplace transform to $\frac{d}{dt}Tx = x$ gives $-Tx_0 + sT\mathcal{L}x(s) = \mathcal{L}x(s)$ and thus $X(s) := \mathcal{L}x(s) = (sT - I)^{-1}Tx_0$. By (4.4) for $n \in \mathbb{N}_0$ and $s > 0$, we have

$$\left\| [(T - \tfrac{1}{s})^{-1}T]^{n+1} x_0 \right\| = \left\| s^{n+1}(sT - I)^{-(n+1)} T^{n+1} x_0 \right\|$$
$$= \tfrac{1}{n!}\left\| s^{n+1} X^{(n)}(s) \right\| \le \|x\|_\infty,$$

which implies (4.3).

Conversely, assume that (4.3) holds and define the function $X(s) = (sT - I)^{-1}Tx_0$, $s > 0$. Then (4.3) implies that

$$\sup\left\{ \tfrac{1}{n!}\left\| s^{n+1} X^{(n)}(s) \right\| : s > 0, \ n \in \mathbb{N}_0 \right\} < \infty.$$

By Arendt [1, Theorem 1.4] this implies that there exists a unique $x \in L^\infty([0, \infty), \mathcal{X})$ such that $X = \mathcal{L}x$. Hence, we have $(sT - I)^{-1}Tx_0 = \mathcal{L}x(s)$ and thus $Tx_0 = (sT - I)\mathcal{L}x(s)$, i.e.,

$$\mathcal{L}[Tx_0](s) = \tfrac{1}{s}Tx_0 = \mathcal{L}[Tx](s) - \tfrac{1}{s}\mathcal{L}x(s).$$

Define $y(t) := \int_0^t x(\tau)\, d\tau$. Then y has polynomial growth and $\mathcal{L}y(s) = \tfrac{1}{s}\mathcal{L}x(s)$, which gives $\mathcal{L}[Tx - Tx_0 - y] = 0$. By the uniqueness theorem for the Laplace transform (see, e.g., [17, §5 Corollary 7.2]), we obtain $Tx - Tx_0 = y$, which is equivalent to (4.2). $\qquad\square$

Remark 4.6

(a) As is easily proved by induction, we have

$$[(T - s)^{-1}T]^{n+1} = (-1)^{n+1} \sum_{k=n+1}^{\infty} \binom{k-1}{n} s^{-k} T^k.$$

Therefore, for all $s > 0$,

$$\sum_{n=0}^{\infty} \left\| [(T - s)^{-1}T]^{n+1} \right\| \le \sum_{n=0}^{\infty} \sum_{k=n+1}^{\infty} \binom{k-1}{n} s^{-k} \|T^k\|$$
$$= \sum_{k=1}^{\infty} \left(\sum_{n=0}^{k-1} \binom{k-1}{n} \right) s^{-k} \|T^k\|$$
$$= \sum_{k=1}^{\infty} 2^{k-1} s^{-k} \|T^k\|.$$

In particular, we conclude that $\|[(T-s)^{-1}T]^{n+1}\|$ is uniformly bounded for all $n \in \mathbb{N}_0$ and $s \geq s_0 > 0$. Hence, the term "$s > 0$" in condition (4.3) can be equivalently replaced by "$s \in (0, s_0)$" for any $s_0 > 0$.

(b) In the case where T is injective, the IVP (4.2) is equivalent to $\dot{z} = Az$, $z(0) = z_0$, where $A = T^{-1}$, $z = Tx$, and $z_0 = Tx_0$. Condition (4.3) can then be rephrased as $\sup\{\|\lambda^{n+1}(\lambda - A)^{-(n+1)}x_0\| : \lambda > 0, n \in \mathbb{N}_0\} < \infty$, which resembles the Hille–Yosida condition for generators of bounded semigroups.

(c) As is seen from the proof of Theorem 4.5, the necessity of (4.3) for the existence of an L^∞-solution of the IVP does not require $\mathcal{X}$ to have the Radon–Nikodym property.

4.2 Square Integrable Solutions on Compact Intervals

Next, we refrain from the requirement that solutions must be defined on the entire positive time axis $[0, \infty)$. Instead, we shall consider the existence of solutions to (4.1) on intervals $[0, \tau]$ for $0 < \tau < \infty$.

Recall that if $\mathcal{X}$ is a Hilbert space, then the Fourier series of a function $F \in L^2([0, \tau]; \mathcal{X})$ converges (unconditionally) to F in $L^2([0, \tau]; \mathcal{X})$ (see [2]). In the proof of the next theorem, we make use of the following lemma which easily follows from the scalar case by exploiting an orthonormal basis of $\mathcal{X}$.

Lemma 4.7 *Let $\mathcal{X}$ be a Hilbert space, $F \in L^2([0, \tau]; \mathcal{X})$, and let $F(t) = \sum_{k \in \mathbb{Z}} c_k e^{2\pi i k t/\tau}$ be its Fourier series. Then $F \in W^{1,2}([0, \tau]; \mathcal{X})$ with $F(\tau) = F(0)$ if and only if $(k c_k)_{k \in \mathbb{Z}} \in \ell^2(\mathbb{Z}; \mathcal{X})$. In this case, we have $\dot{F}(t) = \frac{1}{\tau} \sum_{k \neq 0} 2\pi i k c_k \cdot e^{2\pi i k t/\tau}$.*

As we saw in Example 4.2 (b), there may be distinct solutions of the DAE $\frac{d}{dt}Tx = x$ on an interval with the same initial value $Tx(0) = Tx_0$. The following proposition shows that the existence of an L^2-solution of $\frac{d}{dt}Tx = x$ on $[0, \tau]$ can be characterized in terms of the difference $Tx(\tau) - Tx(0)$ and that this difference determines the solution uniquely.

Theorem 4.8 *Let $\mathcal{X}$ be a Hilbert space and let $y \in \mathcal{X}$. Then there exists a function $x \in L^2([0, \tau]; \mathcal{X})$ such that $Tx \in W^{1,2}([0, \tau]; \mathcal{X})$ and*

$$\frac{d}{dt}Tx = x, \qquad Tx(\tau) - Tx(0) = y, \tag{4.5}$$

if and only if $(\|(I - \frac{2\pi i k}{\tau}T)^{-1}y\|)_{k \in \mathbb{Z}} \in \ell^2(\mathbb{Z})$. In this case, the solution x is unique and is given by

$$x(t) = \frac{1}{\tau}\sum_{k \in \mathbb{Z}}(I - \frac{2\pi i k}{\tau}T)^{-1}y \cdot e^{\frac{2\pi i k}{\tau}t}. \tag{4.6}$$

Proof Assume that $(\|(I - \frac{2\pi i k}{\tau}T)^{-1}y\|)_{k\in\mathbb{Z}} \in \ell^2(\mathbb{Z})$ and let x be as in (4.6). Then

$$
Tx(t) = \tfrac{1}{\tau}Ty + \sum_{k\neq 0} \tfrac{1}{2\pi i k}(I - \tfrac{2\pi i k}{\tau}T)^{-1}\tfrac{2\pi i k}{\tau}Ty \cdot e^{\frac{2\pi i k}{\tau}t}
$$

$$
= \tfrac{1}{\tau}Ty - \Big(\sum_{k\neq 0} \tfrac{1}{2\pi i k}e^{\frac{2\pi i k}{\tau}t}\Big)y + \sum_{k\neq 0}\tfrac{1}{2\pi i k}(I - \tfrac{2\pi i k}{\tau}T)^{-1}y \cdot e^{\frac{2\pi i k}{\tau}t}
$$

$$
= \tfrac{1}{\tau}Ty + (\tfrac{t}{\tau} - \tfrac{1}{2})y + \sum_{k\neq 0}\tfrac{1}{2\pi i k}(I - \tfrac{2\pi i k}{\tau}T)^{-1}y \cdot e^{\frac{2\pi i k}{\tau}t}.
$$

Note that the coefficients $c_k = \frac{1}{2\pi i k}(I - \frac{2\pi i k}{\tau}T)^{-1}y$ in the series above satisfy $(\|kc_k\|) \in \ell^2(\mathbb{Z}\backslash\{0\})$. Hence, the function F represented by the series is in $H^1([0,\tau]; \mathcal{X})$ and satisfies $F(\tau) = F(0)$ so that $Tx(\tau) - Tx(0) = y$. Moreover, the series can be differentiated term-by-term resulting in $\frac{d}{dt}F$. Hence,

$$
\tfrac{d}{dt}Tx(t) = \tfrac{1}{\tau}y + \sum_{k\neq 0}\tfrac{1}{\tau}(I - \tfrac{2\pi i k}{\tau}T)^{-1}y \cdot e^{\frac{2\pi i k}{\tau}t} = x(t).
$$

Conversely, let $x \in L^2([0,\tau]; \mathcal{X})$ with $Tx \in W^{1,2}([0,\tau]; \mathcal{X})$ such that (4.5) holds. Write x as $x(t) = \sum_{k\in\mathbb{Z}} c_k e^{2\pi i k t/\tau}$ with $c_k \in \ell^2(\mathbb{Z}; \mathcal{X})$ and define $z(t) := Tx(t) + (\tfrac{1}{2} - \tfrac{t}{\tau})y$. Then $z \in W^{1,2}([0,\tau]; \mathcal{X})$ and

$$
\dot{z}(t) = \tfrac{d}{dt}Tx(t) - \tfrac{1}{\tau}y = x(t) - \tfrac{1}{\tau}y = c_0 - \tfrac{1}{\tau}y + \sum_{k\neq 0} c_k e^{\frac{2\pi i k}{\tau}t}.
$$

On the other hand,

$$
z(t) = \sum_{k\in\mathbb{Z}} Tc_k e^{\frac{2\pi i k t}{\tau}} + \tfrac{1}{2}y - \Big(\tfrac{1}{2} - \sum_{k\neq 0}\tfrac{1}{2\pi i k}e^{\frac{2\pi i k}{\tau}t}\Big)y
$$

$$
= Tc_0 + \sum_{k\neq 0}(Tc_k + \tfrac{1}{2\pi i k}y)e^{\frac{2\pi i k}{\tau}t}
$$

and

$$
z(\tau) - z(0) = \big(Tx(\tau) - \tfrac{1}{2}y\big) - \big(Tx(0) + \tfrac{1}{2}y\big) = y - y = 0.
$$

Hence,

$$
\dot{z}(t) = \sum_{k\neq 0}\tfrac{2\pi i k}{\tau}(Tc_k + \tfrac{1}{2\pi i k}y)e^{\frac{2\pi i k}{\tau}t}.
$$

From the identity theorem for Fourier series, we conclude that

$$c_0 = \tfrac{1}{\tau} y \qquad \text{and} \qquad \tau c_k = 2\pi i k T c_k + y, \quad k \in \mathbb{Z}\setminus\{0\}$$

and thus $c_k = \tfrac{1}{\tau}(I - \tfrac{2\pi i k}{\tau} T)^{-1} y$ for all $k \in \mathbb{Z}$, so that $(\|(I - \tfrac{2\pi i k}{\tau} T)^{-1} y\|)_{k \in \mathbb{Z}} \in \ell^2(\mathbb{Z})$, and x coincides with the function given in (4.6). $\qquad\square$

The following two corollaries are immediate consequences of Theorem 4.8.

Corollary 4.9 *Let $\mathcal{X}$ be a Hilbert space. If x is an L^2-solution of the DAE $\frac{d}{dt} T x = x$ on $[0, \tau]$ such that $T x(\tau) = T x(0)$ (or, equivalently, $\int_0^\tau x(s)\, ds = 0$), then $x = 0$.*

Corollary 4.10 *Let $\mathcal{X}$ be a Hilbert space. Then there exists an L^2-solution of the IVP $\frac{d}{dt} T x = x$, $T x(0) = T x_0$, on $[0, \tau]$ if and only if there is some $y \in \operatorname{ran} T$ such that $(\|(I - \tfrac{2\pi i k}{\tau} T)^{-1} y\|)_{k \in \mathbb{Z}} \in \ell^2(\mathbb{Z})$ and*

$$T x_0 = \tfrac{1}{\tau} T y - \tfrac{1}{2} y + \sum_{k \neq 0} \tfrac{1}{2\pi i k}(I - \tfrac{2\pi i k}{\tau} T)^{-1} y.$$

Remark 4.11 Note that the boundary condition in (4.5) implies $y \in \operatorname{ran} T$. If T is injective and we set $A := T^{-1}$, then $(I - \tfrac{2\pi i k}{\tau} T)^{-1} y = (A - \tfrac{2\pi i k}{\tau})^{-1} x$, where $T x = y$. In the case $T = -V$, where V is the Volterra operator from Example 4.2, it can be shown that $x \mapsto ((A - \tfrac{2\pi i k}{\tau})^{-1} x)_{k \in \mathbb{Z}}$ is a bounded operator from $\mathcal{X} = L^2(0, 1)$ to $\ell^2(\mathbb{Z}, \mathcal{X})$. Therefore, the ℓ^2-condition in Theorem 4.8 is satisfied for any $y \in \operatorname{ran} T$.

4.3 Analytic Solutions

Finally, we consider solutions x which are (real-)analytic in a neighborhood of $t = 0$. Note that in this case we can replace $\frac{d}{dt} T x$ by $T \dot{x}$.

Proposition 4.12 *Let $T \in L(\mathcal{X})$ be an injective quasi-nilpotent operator. Then the IVP*

$$T \dot{x} = x, \quad x(0) = x_0, \tag{4.7}$$

has an analytic solution on some interval $[0, r)$, $r > 0$, if and only if

$$x_0 \in \bigcap_{n=1}^{\infty} \operatorname{ran} T^n \qquad \text{and} \qquad \limsup_{k \to \infty} \tfrac{1}{k} \| T^{-k} x_0 \|^{1/k} < \infty. \tag{4.8}$$

In this case, the solution of the IVP is unique.

Let x be an analytic solution on $[0, r)$. If $x_0 = 0$, then $x = 0$. If $x_0 \neq 0$, then $r \leq a/e$, where $a := \liminf_{n \to \infty} n \|T^n\|^{1/n}$. In particular, if $a = 0$, then the only analytic solution of $T\dot{x} = x$ is the zero function.

Proof Let x be an analytic solution of $T\dot{x} = x$ and assume that $x(t) = \sum_{k=0}^{\infty} x_k t^k$ has radius of convergence $r > 0$, where $x_k \in \mathcal{X}$, $k \in \mathbb{N}$. Then $x_k = \frac{x^{(k)}(0)}{k!}$, and since $Tx^{(k+1)} = x^{(k)}$, we have

$$Tx_{k+1} = \frac{Tx^{(k+1)}(0)}{(k+1)!} = \frac{x^{(k)}(0)}{(k+1)!} = \frac{x_k}{k+1}.$$

Hence, $x_0 = k! \cdot T^k x_k$, i.e., $x_0 \in \bigcap_{n=1}^{\infty} \operatorname{ran} T^n$ and $x_k = (1/k!)T^{-k}x_0$ for all $k \in \mathbb{N}$. This also shows that x is unique as an analytic solution and, in particular, that $x = 0$ if $x_0 = 0$. Moreover,

$$\frac{1}{r} \geq \limsup_{k \to \infty} \frac{1}{(k!)^{1/k}} \|T^{-k}x_0\|^{1/k}.$$

By Stirling's formula, we have $k! \sim \sqrt{2\pi k} \left(\frac{k}{e}\right)^k$, and thus $(k!)^{1/k} \sim \frac{k}{e}$, and (4.8) follows. If $x_0 \neq 0$, we may estimate $\|x_0\| \leq k! \|T^k\| \|x_k\|$. This implies

$$\|x_0\|^{1/k} \leq \frac{(k!)^{1/k}}{k/e} \cdot \frac{1}{e} \cdot k \|T^k\|^{1/k} \cdot \|x_k\|^{1/k}.$$

Choose a subsequence k_n such that $k_n \|T^{k_n}\|^{1/k_n} \to a$. Then the left-hand side converges to 1, while the limit superior of the right-hand side does not exceed $\frac{a}{er}$, and hence $r \leq \frac{a}{e}$.

Conversely, if (4.8) holds, let $x(t) = \sum_{k=0}^{\infty} x_k t^k$ with $x_k = (1/k!)T^{-k}x_0$, $k \in \mathbb{N}$. Then (4.8) implies that x is a well-defined analytic function on some interval $[0, r)$, $r > 0$, and $kTx_k = x_{k-1}$ implies $T\dot{x} = x$. $\qquad\square$

Remark 4.13

(a) If the IVP with $x_0 \neq 0$ even has a solution which is an entire function, then the proposition implies that $n \|T^n\|^{1/n} \to \infty$.

(b) The Volterra operator V from Example 4.2 satisfies $n! \|V^n\| \to \frac{1}{2}$ (see [11]). Therefore, $a = e$, and thus any power series solution of (4.7) with $x_0 \neq 0$ has radius of convergence $r \leq 1$.

Acknowledgments The author would like to thank the reviewer for their very helpful comments and careful review of the paper.

Appendix

Proof of Proposition 2.3 Let C_1 and C_2 be two sets of Jordan curves as above surrounding σ such that C_1 is completely contained in the interior of C_2. Then we have (see (2.6))

$$P_\sigma^2 = \frac{1}{(2\pi i)^2} \int_{C_1} \int_{C_2} T_z T_s \, dz \, ds = \frac{1}{(2\pi i)^2} \int_{C_1} \int_{C_2} \frac{T_s - T_z}{z - s} \, dz \, ds$$

$$= \frac{1}{(2\pi i)^2} \int_{C_1} T_s \int_{C_2} \frac{1}{z - s} \, dz \, ds + \frac{1}{(2\pi i)^2} \int_{C_2} T_z \int_{C_1} \frac{1}{s - z} \, ds \, dz.$$

The interior integrals equal $2\pi i$ and 0, respectively, which implies $P_\sigma^2 = P_\sigma$. Hence, P_σ is indeed a projection.

Since $\frac{d\tau_\mu}{dz}(z) = (z - \mu)^{-2} = \tau_\mu^2(z)$ and μ is in the exterior of C, we have

$$0 = \frac{1}{2\pi i} \int_C \frac{1}{z - \mu} \, dz = -\frac{1}{2\pi i} \int_{\tau_\mu(C)} \frac{1}{\lambda} \, d\lambda;$$

hence, zero is in the exterior of $\tau_\mu(C)$. Also if $w \in \tau_\mu(\sigma)$, $w = \tau_\mu(s)$ with $s \in \sigma$, then

$$2\pi i = \int_C \frac{1}{z - s} \, dz = \int_{\tau_\mu(C)} \frac{1}{\lambda((\mu - s)\lambda - 1)} \, d\lambda = \int_{\tau_\mu(C)} \left(\frac{1}{\lambda - w} - \frac{1}{\lambda} \right) d\lambda.$$

Since zero is in the exterior of $\tau_\mu(C)$, this shows that w (and thus σ) is in the interior of $\tau_\mu(C)$.

Since $\tau_\mu : \mathbb{C} \setminus \{\mu\} \to \mathbb{C} \setminus \{0\}$ is a homeomorphism, it follows that $\tau_\mu(\sigma)$ is a spectral set for T_μ, cf. (2.5). Now,

$$P_\sigma = -\frac{1}{2\pi i} \int_C \tau_\mu(z) \left(T_\mu - \tau_\mu(z) \right)^{-1} T_\mu \, dz$$

$$= -\frac{1}{2\pi i} \int_C \tau_\mu(z) \left[I + \tau_\mu(z) \left(T_\mu - \tau_\mu(z) \right)^{-1} \right] dz$$

$$= -\frac{1}{2\pi i} \int_C \tau_\mu(z)^2 \left(T_\mu - \tau_\mu(z) \right)^{-1} dz = -\frac{1}{2\pi i} \int_{\tau_\mu(C)} \left(T_\mu - \lambda \right)^{-1} d\lambda.$$

This proves the Eq. (2.11) and also that $\tau_\mu(C)$ is positively oriented since otherwise $-P_\sigma$ would be a projection. $\qquad\square$

References

1. W. Arendt, Vector-valued laplace transforms and cauchy problems, Israel J. Math. **59**, 327–352 (1987)
2. W. Arendt, S. Bu, Fourier series in Banach spaces and maximal regularity.Oper. Theory: Adv. Appl. **201**, 21–39 (2009)
3. T. Berger, S. Trenn, The Quasi-Kronecker form for matrix pencils, SIAM J. Matrix Anal. Appl. **33**, 336–368 (2012)
4. T. Berger, S. Trenn, Addition to "The Quasi-Kronecker form for matrix pencils", SIAM J. Matrix Anal. Appl. **34**, 94–101 (2013)
5. T. Berger, A. Ilchmann, S. Trenn, The quasi-Weierstraß form for regular matrix pencils. Linear Algebra Appl. **436**, 4052–4069 (2012)
6. N. Dunford, J.T. Schwartz, *Linear Operators, Part I: General Theory* (Wiley, Hoboken, 1988)
7. T. Faulwasser, B. Maschke, F. Philipp, M. Schaller, K. Worthmann, Optimal control of port-Hamiltonian descriptor systems with minimal energy supply. SIAM J. Control Optim. **60**, 2132–2158 (2022)
8. S. Grabiner, Ranges of quasi-nilpotent operators. Illinois J. Math. **15**, 150–152 (1971)
9. M.A. Kaashoek, Ascent, descent, nullity and defect, a note on a paper by A. E. Taylor. Math. Ann. **172**, 105–115 (1967)
10. T. Kato, *Perturbation Theory for Linear Operators* (Springer, Berlin, 1995)
11. D. Kershaw, Operator norms of powers of the Volterra operator. J. Int. Equ. Appl. **11**, 351–362 (1999)
12. P. Kunkel, V. Mehrmann, *Differential-Algebraic Equations: Analysis and Numerical Solution* (European Mathematical Society Publishing House, Zürich, 2006)
13. B. Messirdi, A. Gherbi, M. Amouch, A spectral analysis of linear operator pencils on Banach spaces with application to quotient of bounded operators. Int. J. Anal. Appl. **7**, 104–128 (2015)
14. V. Müller, *Spectral Theory of Linear Operators and Spectral Systems in Banach Algebras*, 2nd edn. (Birkhäuser, Basel, 2007)
15. W. Rudin, *Functional Analysis*, 2nd edn. (McGraw-Hill, New York, 1991)
16. S. Trostorff, M. Waurick, On higher index differential-algebraic equations in infinite dimensions, in *The Diversity and Beauty of Applied Operator Theory*, ed. by A. Böttcher, D. Potts, P. Stollmann, D. Wenzel (Springer International Publishing, Cham, 2018)
17. D.V. Widder, *An Introduction to Transform Theory* (Academic, New York, 1971)
18. K. Wong, The eigenvalue problem $\lambda T x + S x$, J. Differ. Equ. **16**, 270–280 (1974)

BIBO Stability for Funnel Control: Semilinear Internal Dynamics with Unbounded Input and Output Operators

Anthony Hastir, René Hosfeld, Felix L. Schwenninger,
and Alexander A. Wierzba

1 Introduction

In this chapter, we report recent progress in the adaptive control of certain classes of nonlinear, infinite-dimensional systems. This is done by studying the interplay of two relatively well-studied concepts from finite-dimensional theory: *funnel control* and *bounded-input-bounded-output stability (BIBO stability)*. Whereas the latter is classical in the history of systems theory, funnel control has only been established in the last 20 years starting from seminal work by Ilchmann–Sangwin–Ryan [12];

This research was partially conducted with the financial support of F.R.S-FNRS, Belgium. A. H. is a FNRS Research Fellow under the grant CR 40010909. He is now supported by the German Research Foundation (DFG) under the grant HA 10262/2-1. The second and third authors are supported by the German Research Foundation (DFG) via the joint grant JA 735/18-1/SCHW 2022/2-1.

A. Hastir (✉)
Department of Mathematics and Namur Institute for Complex Systems (naXys), University of Namur, Namur, Belgium

School of Mathematics and Natural Sciences, IMACM, University of Wuppertal, Wuppertal, Germany
e-mail: anthony.hastir@unamur.be; hastir@uni-wuppertal.de

R. Hosfeld
School of Mathematics and Natural Sciences, IMACM, University of Wuppertal, Wuppertal, Germany
e-mail: hosfeld@uni-wuppertal.de

F. L. Schwenninger · A. A. Wierzba
Department of Applied Mathematics, University of Twente, Enschede, The Netherlands
e-mail: f.l.schwenninger@utwente.nl; a.a.wierzba@utwente.nl

© The Author(s), under exclusive license to Springer Nature Switzerland AG 2024
F. L. Schwenninger, M. Waurick (eds.), *Systems Theory and PDEs*,
Trends in Mathematics, https://doi.org/10.1007/978-3-031-64991-2_8

see the survey [7] and the references therein. For both infinite-dimensional theory exists only partially, in particular in the nonlinear case.

Funnel control has shown to be intimately connected to BIBO stability of the internal dynamics if the system allows for a relative degree; see [5, 6, 12] for the evolution of such "funnel theorems" and their applications to both finite- and infinite-dimensional systems. From an analytic point of view, this relative degree result means that the input–output behaviour is given by an ordinary differential equation coupled with a possibly infinite-dimensional system. In our setting, the latter *internal dynamics* are represented by a semilinear partial differential equation with various assumptions on the nonlinearity. The power of the above-mentioned "funnel theorems" lies in the fact that BIBO stability of the internal dynamics is essentially sufficient for guaranteeing that funnel control works.

The funnel controller is an adaptive model–free output error feedback whose objective is to let the output of a dynamical system follow a predetermined (time-varying) reference signal. In contrast to classical tracking, this objective is fulfilled in the sense that the output error has to remain within prescribed funnel boundaries. The funnel controller is adaptive in the sense that the time-varying gain function adapts to the current value of the output error. This field of adaptive control has attracted a lot of attention in the last few years. For systems with relative degree one, funnel control is extensively developed in [12] and [13]. Few years later, funnel control has been applied to Multiple Input Multiple Output (MIMO) systems with known strict relative degree; see [2]. The different types of dynamical systems for which funnel control is conducive are listed in [4].

BIBO stability of the internal dynamics is the crucial property when proving that funnel control indeed can be applied successfully, provided the system under investigation is of relative degree form. However, checking for BIBO stability for infinite-dimensional systems turns out to be subtle: in [7], a funnel controller was developed to regulate the position of a moving water tank, with internal dynamics being approximated by a linear wave equation. The main ingredient of that proof rests on carefully investigating the transfer function and showing that its inverse Laplace transform exists as a measure of bounded total variation; see also [6] and also the Callier–Desoer class, [8]. Indeed, for linear systems—including boundary control and observation—with finite-dimensional input and output spaces, this property characterizes BIBO stability [21]. A class of systems with semilinear internal dynamics was considered in [11], under additional assumptions on the linear part and global Lipschitz continuity of the nonlinearity. This extended earlier works [14] for linear systems involving a Byrnes–Isidori normal form. However, a drawback of this approach is that boundary control and observation seem to be excluded, which was one of the starting points of the present chapter.

We point out that funnel control has found application in a much wider context, but we focus our presentation on the theory relevant to our results. In particular, funnel control was recently applied to general systems given by partial differential equations, which are not of the form allowing for a relative degree; see [2] and [20]. Also, note that funnel control has been lately coupled to model-predictive-control (funnel MPC) for nonlinear systems with relative degree one; see [3].

In this chapter, we use L^∞-BIBO stability to treat semilinear internal dynamics allowing for unbounded input and output operators. This result is then used to conclude the feasibility of funnel control. As the latter is a direct application of known funnel theorems, the main contribution of the chapter lies in carefully exploring BIBO stability for such systems. To do so, we follow the idea of writing the nonlinear state space system as the interconnection of an extended linear system and a nonlinear feedback. The main result consists in showing that, provided that the nonlinear operator satisfies some global Lipschitz condition, the whole nonlinear system is L^∞-BIBO stable. To apply our results, we highlight the fact that the linear part of the dynamics has to satisfy some regularity assumptions, which covers a class of applications driven by parabolic partial differential equations.

The chapter is organized as follows: Sect. 2 aims at defining the notion of L^∞-BIBO stability for nonlinear infinite-dimensional systems with general input and output operators. A way of deriving BIBO stability of an extended linear system constructed as the original linear system with augmented input and output operators is detailed in Sect. 3. BIBO stability of the nonlinear system is shown in Sect. 4 in the case where the nonlinear operator is globally Lipschitz, viewed as an operator defined on a more regular space than the state space, constructed as the domain of the fractional powers of the opposite of the linear operator dynamics. An example of a heat equation with a locally Lipschitz nonlinearity and internal control is tackled in Sect. 5, which requires a different approach than in Sect. 4. Funnel control is applied in Sect. 6 to a nonlinear convection-reaction-diffusion system coupled to a linear ordinary differential equation. The needed assumption of BIBO stability of the internal dynamics is shown thanks to our main results. Conclusions and perspectives are given in Sect. 7.

2 BIBO Stability of Semilinear State Space Systems

In the following, let U, X, Y refer to Banach spaces and by $\mathcal{L}(X, Y)$, we denote the bounded linear operators from X to Y. Recall the notion of a system node going back to [23]. A quadruple $\Sigma(A, B, C, \mathbf{G})$ is called a *system node* for (U, X, Y) if A is the generator of a strongly continuous semigroup $(T(t))_{t\geq 0}$ of linear operators on X, $B \in \mathcal{L}(U, X_{-1})$, $C \in \mathcal{L}(X_1, Y)$ and $\mathbf{G} : \mathbb{C}_{\omega_0} \to \mathcal{L}(U, Y)$ is a transfer function, i.e. $\mathbf{G}$ is a holomorphic function satisfying the equation

$$\mathbf{G}(s) - \mathbf{G}(t) = C\left[(sI - A)^{-1} - (tI - A)^{-1}\right] B, \qquad s, t \in \mathbb{C}_{\omega_0},$$

where $\mathbb{C}_{\omega_0} = \{z \in \mathbb{C} : \Re(z) > \omega_0\}$ and ω_0 equals the growth bound of the semigroup. The space X_1 refers to the domain $D(A)$ of A equipped with the graph norm, whereas X_{-1} denotes the completion of X with respect to the norm $\|(\beta I - A)^{-1} \cdot \|_X$ for some $\beta \in \rho(A)$. For details on this concept, we refer to [23]. We use the following notion of a semilinear state space system and its mild solutions.

Definition 2.1 (Semilinear State Space System) Let $\Sigma(A, B, C, \mathbf{G})$ be a linear system node with state space X, semigroup generator $A : D(A) \subset X \to X$, input space U, input operator $B : U \to X_{-1}$, output space Y and output operator $C : X_1 \to Y$. Let $C\&D : D(C\&D) \to Y$ be the associated combined output/feedthrough operator [23, 24] given by

$$D(C\&D) = \left\{ \begin{bmatrix} x \\ u \end{bmatrix} \in X \times U : Ax + Bu \in X \right\}$$

$$C\&D \begin{bmatrix} x \\ u \end{bmatrix} = C \left[x - (sI - A)^{-1} Bu \right] + \mathbf{G}(s)u,$$

for some fixed $s \in \mathbb{C}_{\omega_0}$. Furthermore, let $f : Z \to X$ be a (possibly nonlinear) function with $Z \subset X$ a continuously embedded subspace. Then the pair (Σ, f) formally representing the equations

$$\begin{cases} \dot{x}(t) = Ax(t) + Bu(t) + f(x(t)), & x(0) = x_0, \\ y(t) = C\&D \begin{bmatrix} x(t) \\ u(t) \end{bmatrix}, \end{cases} \tag{1}$$

where $t > 0$, is called a *semilinear state space system*.

Remark 2.2 Here, the space Z will usually be some interpolation space $X_1 \subset X_\alpha \subset X$ with $0 \leq \alpha \leq 1$.

Definition 2.3 (Mild Solution of a Semilinear State Space System) Let $x_0 \in X$, $T > 0$ and $u \in L^1_{\mathrm{loc}}([0, T]; U)$. A triple (u, x, y) is called a *mild solution* to the semilinear state space system (1) on $[0, T]$ with initial value x_0 if

1. $x : [0, T] \to X_{-1}$ and $x(t) \in Z$ for almost all $t \in [0, T]$ and $f(x(\cdot)) \in L^1_{\mathrm{loc}}([0, T]; X)$
2. x solves

$$x(t) = T(t)x_0 + \int_0^t T(t - s) \left[f(x(s)) + Bu(s) \right] ds$$

in X_{-1} for all $t \in [0, T]$;
3. y is a Y-valued distribution given by

$$y(t) = \frac{d^2}{dt^2} \left((C\&D) \int_0^t (t - s) \begin{bmatrix} x(s) \\ u(s) \end{bmatrix} ds \right), \tag{2}$$

meaning that it acts on test functions $\varphi \in C_c^\infty(\mathbb{R}_{\geq 0}; Y')$ as

$$y[\varphi] = \int_0^\infty \left\langle \varphi''(t), (C\&D) \int_0^t (t - s) \begin{bmatrix} x(s) \\ u(s) \end{bmatrix} ds \right\rangle_{Y',Y} dt.$$

A triple (u, x, y) is called a *global mild solution* to the semilinear state space system, if $(u\mid_{[0,T]}, x\mid_{[0,T]}, y\mid_{[0,T]})$ is a mild solution on $[0, T]$ for every $T > 0$. If u and y are clear from the context, e.g. if $y = x$, we simply refer to x as the mild solution (u, x, y).

Remark 2.4 Part 3 of Theorem 2.3 is equivalent to the statement that $\left(\left[\begin{smallmatrix} u \\ f(x) \end{smallmatrix} \right], x, \left[\begin{smallmatrix} y \\ x \end{smallmatrix} \right] \right)$ is a generalized solution of the extended system node

$$\Sigma \left(A, [\, B \ I\,], \left[\begin{smallmatrix} C \\ I \end{smallmatrix} \right], \left[\begin{matrix} \mathbf{G} & C(\cdot I - A)^{-1} \\ (\cdot I - A)^{-1}B & (\cdot I - A)^{-1} \end{matrix} \right] \right)$$

as defined in [23, Definition 4.7.10]. Note that this also guarantees that the integral appearing in Eq. (2) indeed always lies in $D(C\&D)$, and thus the application of $C\&D$ is well-defined [23, Lemma 4.7.9].

With this solution concept, we can define BIBO stability for the considered semilinear state space systems.

Definition 2.5 (L^∞-BIBO Stability) Let (Σ, f) be a semilinear state space system. Then it is called L^∞-*BIBO stable* if the following two conditions are satisfied:

1. For any $u \in L^\infty_{\mathrm{loc}}(\mathbb{R}_{\geq 0}; U)$, there exists a global mild solution (u, x, y) with $x_0 = 0$ and $y \in L^\infty_{\mathrm{loc}}(\mathbb{R}_{\geq 0}; Y)$
2. For any $c_U > 0$, there exists a constant $c_Y > 0$ such that for any $t > 0$ and any global mild solution (u, x, y) of (Σ, f) with $x_0 = 0$, $u \in L^\infty_{\mathrm{loc}}(\mathbb{R}_{\geq 0}; U)$, the following implication holds:

$$\|u\|_{L^\infty([0,t];U)} < c_U \quad \Longrightarrow \quad \|y\|_{L^\infty([0,t];Y)} < c_Y.$$

We recall the following definition of admissible control operators [25].

Definition 2.6 (Admissible Control Operators) Let $A : D(A) \subset X \to X$ be the generator of a strongly continuous semigroup $(T(t))_{t\geq 0}$ and $1 \leq p \leq \infty$. Then $B \in \mathcal{L}(U, X_{-1})$ is called a L^p-*admissible control operator* (or just L^p-*admissible*) if for some (and hence for all) $t > 0$ there exists a (minimal) constant $C_t > 0$ such that

$$\left\| \int_0^t T(t - s)Bu(s)\, \mathrm{d}s \right\|_X \leq C_t \|u\|_{L^p([0,t];U)}$$

holds for all $u \in L^p([0, t]; U)$. If $C_\infty := \sup_{t>0} C_t < \infty$, then B is called *infinite-time L^p-admissible*.

Remark 2.7 In the above definition, we implicitly assumed that for some (and hence for all) $t > 0$ and all $u \in L^p(\mathbb{R}_{\geq 0}; U)$, it holds that

$$\int_0^t T(t - s)Bu(s)\, \mathrm{d}s \in X,$$

which is even sufficient for L^p-admissibility of B; see, for instance [25, Proposition 4.2]. Moreover, it can be concluded from the proof of [25, Proposition 2.5] that, if $(T(t))_{t \geq 0}$ is exponentially stable, i.e. there exists $M \geq 1$ and $\omega > 0$ such that $\|T(t)\| \leq M e^{-\omega t}$ for every $t \geq 0$, then L^p-admissibility and infinite-time L^p-admissibility of B are equivalent.

3 Linear Systems with Nonlinear Feedback and BIBO Stability

A way of approaching the question of BIBO stability for systems written as in (1) is explained in this section together with some preliminary result. One possible approach to study semilinear systems of the form of Eq. (1) that was, for instance, used in [10] is to rewrite the system in the form as depicted in Fig. 1 and consider the nonlinearity as a nonlinear feedback loop attached to an extended linear system.

This way it is possible to employ properties of the linear system to derive properties of the semilinear one. Here the most relevant property of the linear system for our discussions is naturally its L^∞-BIBO stability.

The following proposition provides a sufficient condition for this property.

Proposition 3.1 *Let* $\Sigma(A, B, C, \mathbf{G})$ *be a system node with* A *the generator of an exponentially stable semigroup. Then the extended system node* $\Sigma\left(A, [\,B\ I\,], \left[\begin{smallmatrix} C \\ I \end{smallmatrix}\right], \left[\begin{smallmatrix} \mathbf{G} & C(\cdot I - A)^{-1} \\ (\cdot I - A)^{-1} B & (\cdot I - A)^{-1} \end{smallmatrix}\right]\right)$ *is* L^∞-*BIBO stable if all of the following hold:*

- $\Sigma(A, B, C, \mathbf{G})$ *is* L^∞-*BIBO stable.*
- B *is an* L^∞-*admissible control operator.*
- $\Sigma(A, I, C, C(\cdot I - A)^{-1})$ *is* L^∞-*BIBO stable.*

Proof By [21, Corollary 6.2], from the L^∞-admissibility of B together with the exponential stability, it follows that the system node $\Sigma(A, B, I, (\cdot I - A)^{-1} B)$ is L^∞-BIBO stable. Analogously, the same follows for the system node $\Sigma(A, I, I, (\cdot I - A)^{-1})$.

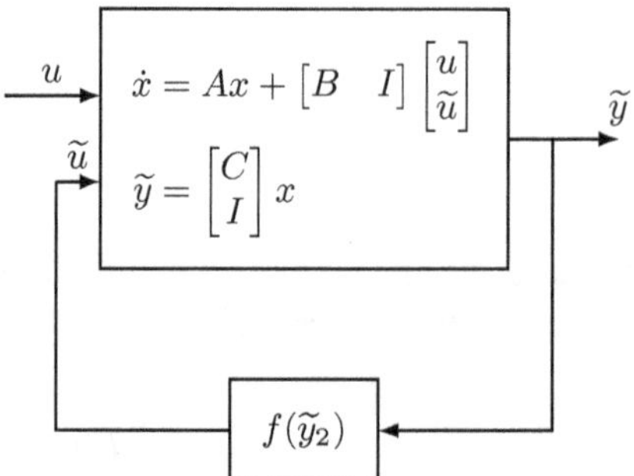

Fig. 1 Nonlinearity as feedback loop

By the L^∞-BIBO stability of the respective system nodes, it follows that there are constants $c, c_C, c_B, c_I > 0$ such that for any $\begin{bmatrix} u \\ \tilde{u} \end{bmatrix} \in L^\infty_{\mathrm{loc}}(\mathbb{R}_{\geq 0}; U \times X)$ and $x_0 = 0$, there are the following solutions which then, for any $t > 0$, satisfy the corresponding inequalities:

- (u, x, y) of $\Sigma(A, B, C, \mathbf{G})$ with $y \in L^\infty_{\mathrm{loc}}(\mathbb{R}_{\geq 0}; Y)$ and $\|y\|_{L^\infty([0,t];Y)} \leq c\|u\|_{L^\infty([0,t];U)}$
- (u, x_B, x_B) of $\Sigma(A, B, I, (\cdot I - A)^{-1}B)$ with $x_B \in L^\infty_{\mathrm{loc}}(\mathbb{R}_{\geq 0}; X)$ and $\|x_B\|_{L^\infty([0,t];X)} \leq c_B\|u\|_{L^\infty([0,t];U)}$
- $(\tilde{u}, x_C, y_C)$ of $\Sigma(A, I, C, C(\cdot I - A)^{-1})$ with $y_C \in L^\infty_{\mathrm{loc}}(\mathbb{R}_{\geq 0}; Y)$ and $\|y_C\|_{L^\infty([0,t];Y)} \leq c_C\|\tilde{u}\|_{L^\infty([0,t];X)}$
- $(\tilde{u}, x_I, x_I)$ of $\Sigma(A, I, I, (\cdot I - A)^{-1})$ with $x_I \in L^\infty_{\mathrm{loc}}(\mathbb{R}_{\geq 0}; Y)$ and $\|x_I\|_{L^\infty([0,t];X)} \leq c_I\|\tilde{u}\|_{L^\infty([0,t];X)}$

Clearly, we also have $x_B = x$ and $x_C = x_I$.

Then we find for the extended system that the state corresponding to the input $\begin{bmatrix} u \\ \tilde{u} \end{bmatrix}$ takes the form

$$\tilde{x}(t) = \int_0^t T(t-s)\begin{bmatrix} B & I \end{bmatrix}\begin{bmatrix} u(s) \\ \tilde{u}(s) \end{bmatrix} ds = x(t) + x_C(t).$$

We furthermore observe that the combined output/feedthrough operator $\widetilde{C\&D}$ for this extended system node acts as

$$\widetilde{C\&D}\begin{bmatrix} \tilde{x} \\ u \\ \tilde{u} \end{bmatrix} = \begin{bmatrix} C \\ I \end{bmatrix}\left(\tilde{x} - (\beta I - A)^{-1}\begin{bmatrix} B & I \end{bmatrix}\begin{bmatrix} u \\ \tilde{u} \end{bmatrix}\right)$$

$$+ \begin{bmatrix} G(\beta) & C(\beta I - A)^{-1} \\ (\beta I - A)^{-1}B & (\beta I - A)^{-1} \end{bmatrix}\begin{bmatrix} u \\ \tilde{u} \end{bmatrix}$$

$$= \begin{bmatrix} C \\ I \end{bmatrix}\left(x - (\beta I - A)^{-1}Bu\right) + \begin{bmatrix} C \\ I \end{bmatrix}\left(x_I - (\beta I - A)^{-1}\tilde{u}\right)$$

$$+ \begin{bmatrix} G(\beta)u + C(\beta I - A)^{-1}\tilde{u} \\ (\beta I - A)^{-1}Bu + (\beta I - A)^{-1}\tilde{u} \end{bmatrix}$$

$$= \begin{bmatrix} y + y_C \\ x + x_C \end{bmatrix}$$

and we thus observe that the output $\tilde{y}$ of the extended system node is given by

$$\tilde{y} = \begin{bmatrix} y + y_C \\ x_B + x_I \end{bmatrix}$$

a priori in a distributional sense, but thus also as $\widetilde{y} \in L^\infty_{\text{loc}}(\mathbb{R}_{\geq 0}; Y \times X)$. This shows the existence of a solution $\left(\left[\begin{smallmatrix} u \\ \widetilde{u} \end{smallmatrix}\right], \widetilde{x}, \widetilde{y}\right)$ with $\widetilde{y} \in L^\infty_{\text{loc}}(\mathbb{R}_{\geq 0}; Y \times X)$ of the extended system node.

Furthermore, we then find for all $t > 0$

$$\|\widetilde{y}\|_{L^\infty([0,t];Y \times X)} \lesssim \|y\|_{L^\infty([0,t];Y)} + \|y_C\|_{L^\infty([0,t];Y)} + \|x_B\|_{L^\infty([0,t];X)}$$

$$+ \|x_I\|_{L^\infty([0,t];X)}$$

$$\leq c\|u\|_{L^\infty([0,t];U)} + c_C\|u\|_{L^\infty([0,t];U)} + c_B\|\widetilde{u}\|_{L^\infty([0,t];X)}$$

$$+ c_I\|\widetilde{u}\|_{L^\infty([0,t];X)}$$

$$\lesssim \left\|\begin{bmatrix} u \\ \widetilde{u} \end{bmatrix}\right\|_{L^\infty([0,t];U \times X)}$$

which finishes the proof. $\qquad\square$

Remark 3.2

1. In the following, we will use the notation $\Sigma(A, B, C)$ to refer to a system node $\Sigma(A, B, C, \mathbf{G})$ if it is clear from the context which transfer function $\mathbf{G}$ is to be used.
2. One can straightforwardly extend Proposition 3.1 to the case of the extended linear system

$$\Sigma\left(A, [B \; \widetilde{B}], \begin{bmatrix} C \\ \widetilde{C} \end{bmatrix}, \begin{bmatrix} \mathbf{G} & C(\cdot I - A)^{-1}\widetilde{B} \\ \widetilde{C}(\cdot I - A)^{-1}B & \widetilde{C}(\cdot I - A)^{-1}\widetilde{B} \end{bmatrix}\right)$$

where $\widetilde{B} \in \mathcal{L}(U, X)$ and $\widetilde{C} \in \mathcal{L}(X, Y)$.
3. We note that the assumption that $\Sigma(A, I, C)$ is L^∞-BIBO stable excludes boundary observation if A even generates a strongly continuous group. Indeed, under this assumption, it is shown in [21, Proposition 6.6] that L^∞-BIBO stability of $\Sigma(A, I, C)$ implies that C must be a bounded operator.
4. The exponential stability assumed in the proposition cannot be dropped, as the subsystem $\Sigma(A, I, I)$ is obviously not BIBO if e.g. $A = 0$.

4 Different Types of Uniformly Lipschitz Nonlinearities

In this section, we consider global Lipschitz nonlinearities f defined on interpolation spaces between $D(A)$ and X, if A generates an analytic semigroup, and otherwise on X itself.

The standing assumption of this section is that the (unbounded) linear operator $A : D(A) \subset X \to X$ is the infinitesimal generator of an exponentially stable C_0-semigroup $(T(t))_{t \geq 0}$, i.e. there exist $M \geq 1$ and $\omega > 0$ such that for all $t \geq 0$ it holds that

$$\|T(t)\| \le M e^{-\omega t}. \tag{3}$$

Additionally, if A generates an analytic semigroup, then according to [19, Chapter 6], the fractional powers of $-A$, $(-A)^\alpha$, $0 \le \alpha \le 1$ are well-defined and the operator $(-A)^\alpha$ is closed, linear and invertible. Moreover, $D((-A)^\alpha)$ is dense in X. As an additional property, the analyticity entails that

$$\|(-A)^\alpha T(t)\| \le M_\alpha t^{-\alpha} e^{-\delta t} \tag{4}$$

holds for all $t > 0$, where the positive constant δ is such that the operator $A + \delta I$ is still the infinitesimal generator of an analytic and exponentially stable semigroup and $M_\alpha > 0$ depends only on α, see e.g. [19, Chapter 2, Theorem 6.13]. Note that (4) still holds with $\alpha = 0$ if A does not generate an analytic semigroup, where we denote $(-A)^0 = I$. Indeed, we can choose $\delta = \omega$ and $M_\alpha = M$ with constants M and ω from (3).

If A generates an analytic semigroup, we denote by X_α, $0 \le \alpha \le 1$ the space $D((-A)^\alpha)$ equipped with the norm $\|x\|_\alpha := \|(-A)^\alpha x\|_X$, and otherwise we set $\alpha = 0$ and write $(-A)^0 = I$, $X_0 = X$ and $\|\cdot\|_0 := \|\cdot\|_X$.

Remark 4.1 Assume that A generates an analytic semigroup.

1. Every operator $\tilde{B} \in \mathcal{L}(U, X_{-\beta})$ with $0 \le \beta < 1$ is L^p-admissible for $p > \frac{1}{1-\beta}$, see e.g. [22, Propositon 2.13].
2. It follows from point 1 that for $\alpha \in [0, 1)$ the extension of $(-A)^\alpha$ to an operator in $\mathcal{L}(X, X_{-\alpha})$ is L^∞-admissible. Using (4), we can give an upper bound of the admissibility constant. It holds for $\tilde{u} \in L^\infty(\mathbb{R}_{\ge 0}; X)$ that

$$\left\| \int_0^t T(t-s)(-A)^\alpha \tilde{u}(s)\, ds \right\|_X = \left\| \int_0^t (-A)^\alpha T(t-s)\tilde{u}(s)\, ds \right\|_X$$

$$\le M_\alpha \|\tilde{u}\|_{L^\infty([0,t];X)} \int_0^t (t-s)^{-\alpha} e^{-\delta(t-s)}\, ds$$

$$\le \frac{M_\alpha \Gamma(1-\alpha)}{\delta^{1-\alpha}} \|\tilde{u}\|_{L^\infty([0,t];X)}. \tag{5}$$

Estimate (5) is still valid for $\alpha = 0$ without the analyticity of the semigroup generated by A.

Assuming just bounded senalyticity of the semigroup, i.e. the semigroup is bounded on some sector, instead of exponential stability, (4) holds with $\delta = 0$, and thus, we obtain

$$\left\| \int_0^t T(t-s)(-A)^\alpha \tilde{u}(s)\, ds \right\|_X \le M \|\tilde{u}\|_{L^\infty([0,t];X)} \int_0^t (t-s)^{-\alpha}\, ds$$

which is also valid for not necessarily analytic semigroups if $\alpha = 0$. Note that exponential stability together with analyticity implies bounded analyticity.

Before discussing L^∞-BIBO stability of (1), we give a result on the existence of unique mild solutions of this system (without considering an output). The used methods are well known, and similar results for slightly different situations are available (see e.g. [19, Section 6.3] or more recently [18]). For the sake of completeness, we nevertheless give the details.

Lemma 4.2 *Let A be the generator of a bounded C_0-semigroup. If the semigroup is bounded analytic, let $\alpha \in [0, 1)$; else, set $\alpha = 0$. Moreover, let $B \in \mathcal{L}(U, X_{-(1-\alpha)})$ be such that $(-A)^\alpha B$ is L^∞-admissible and $f : \mathbb{R}_{\geq 0} \times X_\alpha \to X$ is locally Lipschitz in the following sense: there exists a measurable function $g : \mathbb{R}_{\geq 0}^2 \to \mathbb{R}_{\geq 0}$ such that the following properties hold:*

- $g(\cdot, 0) \in L^\infty_{\text{loc}}(\mathbb{R}_{\geq 0}; \mathbb{R}_{\geq 0})$
- $g(s, s) = 0$ *for all* $s \geq 0$
- *For every bounded set* $V \subseteq \mathbb{R}_{\geq 0} \times X_\alpha$, *there exists a constant* $L > 0$ *and* $0 < v \leq 1$ *such that for every* $(t_1, x_1), (t_2, x_2) \in V$, *it holds that*

$$\|f(t_1, x_1) - f(t_2, x_2)\|_X \leq L(g(t_1, t_2) + \|x_1 - x_2\|_\alpha) \tag{6}$$

Then for every $t_0 \geq 0$, $x_0 \in X_\alpha$ *and* $u \in L^\infty([t_0, \infty); U)$, *the system*

$$\begin{cases} \dot{x}(t) = Ax(t) + Bu(t) + f(t, x(t)), & t > t_0, \\ x(t_0) = x_0 \end{cases} \tag{7}$$

admits a unique mild solution $x \in L^\infty([t_0, t_1]; X_\alpha)$ *for some* $t_1 > t_0$, *i.e. x satisfies the implicit equation*

$$x(t) = T(t-t_0)x_0 + \int_{t_0}^t T(t-s)Bu(s)\,\mathrm{d}s + \int_{t_0}^t T(t-s)f(s, x(s))\,\mathrm{d}s, \qquad t \in [t_0, t_1].$$

Furthermore, if $t_{\max} > t_0$ *denotes the supremum over all t_1 such that (7) admits a solution on $[t_0, t_1]$ in the above sense, then we have the finite blow-up property, i.e.*

$$t_{\max} < \infty \quad \Longrightarrow \quad \lim_{t \nearrow t_{\max}} \|x(t)\|_\alpha = \infty.$$

Additionally, if there exists a nondecreasing function $k \in \mathcal{C}([t_0, \infty); \mathbb{R})$ *such that for every* $t \geq t_0$ *and* $x \in X_\alpha$,

$$\|f(t, x)\|_X \leq k(t)(1 + \|x\|_\alpha), \tag{8}$$

then the solution x lies in $L^\infty_{\text{loc}}([t_0, \infty); X_\alpha)$.

Proof First, we prove that for every $t_0 \geq 0$, $x_0 \in X_\alpha$ and $u \in L^\infty(\mathbb{R}_{\geq 0}; U)$, there exists $t_1 > t_0$ such that (7) admits a solution $x \in L^\infty([t_0, t_1]; X_\alpha)$. Moreover we prove that $t_1 = t_0 + \delta$ can be chosen such that $\delta > 0$ does not depend on the initial time t_0 and a set of bounded initial values x_0 in X_α. By $C_{1,t}$ and $C_{2,t}$ we denote the L^∞-admissibility constants of $(-A)^\alpha B$ and $(-A)^\alpha$, respectively. Inequality (5) yields that $C_{2,t} \to 0$ as $t \to 0^+$.

We denote by $M \geq 1$ the constant such that $\|T(t)\| \leq M$ for every $t \geq 0$. Let $t_1' > t_0$. For $r > 0$, $x_0 \in X_\alpha$ with $\|x_0\|_\alpha \leq r$ and $u \in L^\infty(\mathbb{R}_{\geq 0}; U)$, we choose

$$m := 2Mr + C_{1,\infty}\|u\|_{L^\infty([0,t_1'];U)} > 0.$$

Further, let $L > 0$ be a constant satisfying (6) for $V = ([t_0, t_1'] \times \{x \in X_\alpha \mid \|x\|_\alpha \leq m\}) \cup \{(0, 0)\}$. Choose $0 < \delta < t_1'$ such that

$$C_{2,\delta} \leq \min\left\{\frac{Mr}{L(\|g(\cdot, 0)\|_{L^\infty([0,t_1'];\mathbb{R}_{\geq 0})} + m) + \|f(0, 0)\|_X}, \frac{1}{2L}\right\}.$$

Note that δ depends on r, $(T(t))_{t \geq 0}$, α, u, f and $t_1' > t_0$ but not on $t_0 > 0$ and x_0 with $\|x_0\|_\alpha \leq r$.

For $y \in S := \{y \in L^\infty([t_0, t_0 + \delta]; X) \mid \|y\|_{L^\infty([t_0,t_0+\delta];X)} \leq m\}$ and $t \in [t_0, t_0 + \delta]$, we define

$$(Fy)(t) := T(t - t_0)(-A)^\alpha x_0 + \int_{t_0}^t T(t - s)(-A)^\alpha Bu(s)\,ds$$

$$+ \int_{t_0}^t T(t - s)(-A)^\alpha f(s, (-A)^{-\alpha} y(s))\,ds.$$

We will prove that F is a contraction on S. First note that F is well-defined since

$$\|(Fy)(t)\|_X \leq M\|x_0\|_\alpha + C_{1,\delta}\|u\|_{L^\infty([t_0,t];U)} + C_{2,\delta}\|f(\cdot, (-A)^{-\alpha} y(\cdot))\|_{L^\infty([t_0,t];X)}$$

$$\leq Mr + C_{1,\infty}\|u\|_{L^\infty(\mathbb{R}_{\geq 0};U)}$$

$$+ C_{2,\delta}\left(L(\|g(\cdot, 0)\|_{L^\infty([t_0,t];\mathbb{R}_{\geq 0})} + \|(-A)^{-\alpha} y\|_{L^\infty([t_0,t];X_\alpha)}) + \|f(0, 0)\|_X\right)$$

$$\leq Mr + C_{1,\infty}\|u\|_{L^\infty(\mathbb{R}_{\geq 0};U)}$$

$$+ C_{2,\delta}\left(L(\|g(\cdot, 0)\|_{L^\infty([0,t_1'];\mathbb{R}_{\geq 0})} + m) + \|f(0, 0)\|_X\right)$$

$$\leq m,$$

$$(9)$$

where we used (6) in the second last step. The contractivity of F follows from

$$\|(Fy_1)(t) - (Fy_2)(t)\|_X = \left\|\int_{t_0}^t T(t-s)(-A)^\alpha [f(s, (-A)^{-\alpha} y_1(s))\right.$$
$$\left. - f(s, (-A)^{-\alpha} y_2(s))] \, ds\right\|_X$$
$$\leq C_{2,\delta} L \|y_1 - y_2\|_{L^\infty([t_0,t];X)}$$

for every $y_1, y_2 \in S$. Thus, there exists a unique $y \in S$ such that

$$y(t) = T(t-t_0)(-A)^\alpha x_0 + \int_{t_0}^t T(t-s)(-A)^\alpha Bu(s) \, ds$$
$$+ \int_{t_0}^t T(t-s)(-A)^\alpha f(s, (-A)^{-\alpha} y(s)) \, ds$$

for almost every $t \in [t_0, t_0 + \delta]$. We obtain from (6) that $f(\cdot, (-A)^{-\alpha} y(\cdot)) \in L^\infty([t_0, t_0 + \delta]; X)$, and hence, the linear system

$$\begin{cases} \dot{x}(t) = Ax(t) + Bu(t) + f(t, (-A)^{-\alpha} y(t)), & t > t_0 \\ x(t_0) = x_0 \end{cases} \tag{10}$$

admits a unique mild solution $x \in L^\infty([t_0, t_0 + \delta]; X)$ given by

$$x(t) = T(t)x_0 + \int_{t_0}^t T(t-s)Bu(s) \, ds + \int_{t_0}^t T(t-s)f(s, (-A)^{-\alpha} y(s)) \, ds,$$

where each term on the right hand-side lies in $D((-A)^\alpha)$ for almost every $t \in [t_0, t_0 + \delta]$ by analyticity of the semigroup.

It follows that $(-A)^\alpha x(t) = (Fy)(t) = y(t)$ for almost every $t \in [t_0, t_0 + \delta]$, and thus, $x \in L^\infty([t_0, t_0 + \delta]; X_\alpha)$.

For given $t_0 \geq 0$, $x_0 \in X_\alpha$ and $u \in L^\infty(\mathbb{R}_{\geq 0}; U)$, we denote by $t_{\max}$ the supremum over all $t_1 > 0$ such that (7) admits a solution $x \in L^\infty([t_0, t_1]; X_\alpha)$. If $t_{\max} < \infty$ and $\lim_{t \nearrow t_{\max}} \|x(t)\|_\alpha < \infty$, then it follows that

$$r := \|x\|_{L^\infty([t_0,t_{\max}];X_\alpha)} < \infty.$$

Let $(t_n)_{n \in \mathbb{N}}$ be an increasing sequence in $[t_0, t_{\max})$ converging to $t_{\max}$. From the previous argumentation, we can find $\delta > 0$ independent of $n \in \mathbb{N}$ such that the system

$$\begin{cases} \dot{x}_n(t) = Ax_n(t) + Bu(t) + f(t, x_n(t)), & t > t_n, \\ x_n(t_n) = x(t_n) \end{cases}$$

admits unique solutions $x_n \in L^\infty([t_n, t_n + \delta]; X_\alpha)$. But then, we can extend the solution x by x_n to a solution on $[t_0, t_n + \delta)$. Since $t_n + \delta > t_{\max}$ for large n, this contradicts the maximality of $t_{\max}$.

The last statement follows as in [19, Theorem 6.3.3]. $\qquad\square$

Remark 4.3 In the situation of Theorem 4.2, if A generates an analytic semigroup and $\alpha \in (0, 1)$, then the solution x lies also in $C([t_0, t_1]; X)$ or $C([t_0, \infty); X)$, respectively, since x is the mild solution of the linear system (10) with L^p-admissible operators B and I for some $p \in [1, \infty)$ (c.f. Remark 4.1) as can be seen from [25, Proposition 2.3].

Remark 4.4

1. If A generates an exponentially stable and analytic semigroup, let $\alpha \in [0, 1)$; else, set $\alpha = 0$. If $B \in \mathcal{L}(U, X_{-(1-\alpha)})$ such that $(-A)^\alpha B$ is L^∞-admissible, then the mild solution of $\Sigma(A, [\,B\ I\,], [\begin{smallmatrix}C\\I\end{smallmatrix}])$ for $x_0 \in X_\alpha$, $u \in L^\infty(\mathbb{R}_{\geq 0}; U)$, $\tilde{u} \in L^\infty(\mathbb{R}_{\geq 0}; X)$ satisfies

$$\|x(t)\|_\alpha \leq M e^{-\omega t} \|x_0\|_\alpha + C_{1,\infty} \|u\|_{L^\infty([0,t];U)} + C_{2,\infty} \|\tilde{u}\|_{L^\infty([0,t];X)},$$

where $C_{i,\infty}, i = 1, 2$ are the infinite-time L^∞-admissibility constants of $(-A)^\alpha B$ and $(-A)^\alpha$, respectively, and constants M, ω from (3). From Remark 4.1, we know that $C_{2,\infty} \leq \frac{M_\alpha \Gamma(1-\alpha)}{\delta^{1-\alpha}}$ with constants from (4).

2. From the considerations in point 1 and the fact that the transfer function of $\Sigma(A, [\,B\ I\,], [\begin{smallmatrix}C\\I\end{smallmatrix}])$ is not only mapping into $\mathcal{L}(U, Y \times X)$ but also into $\mathcal{L}(U, Y \times X_\alpha)$, we obtain that $\Sigma(A, [\,B\ I\,], [\begin{smallmatrix}C\\I\end{smallmatrix}])$ is L^∞-BIBO stable with respect to the spaces $(U \times X, X, Y \times X)$ if and only if it is L^∞-BIBO stable with respect to the spaces $(U \times X, X_\alpha, Y \times X_\alpha)$. Hence, if one of the above system nodes is L^∞-BIBO stable, by linearity, there exist constants $K_1, K_2 > 0$ such that for $x_0 = 0$ and every $u \in L^\infty(\mathbb{R}_{\geq 0}; U)$, $\tilde{u} \in L^\infty(\mathbb{R}_{\geq 0}; X)$, the output $\tilde{y}$ satisfies

$$\|\tilde{y}\|_{L^\infty([0,t];Y \times X_\alpha)} \leq K_1 \|u\|_{L^\infty([0,t];U)} + K_2 \|\tilde{u}\|_{L^\infty([0,t];X)}. \tag{11}$$

Next, we present our main theorem on L^∞-BIBO stability of the semilinear state space system (1) for globally Lipschitz continuous functions $f : X_\alpha \to X$, i.e. there exists a constant $L > 0$ such that for every $x_1, x_2 \in X_\alpha$, it holds that

$$\|f(x_1) - f(x_2)\|_X \leq L \|x_1 - x_2\|_\alpha. \tag{12}$$

Theorem 4.5 *Let A generate an exponentially stable C_0-semigroup. If the semigroup is analytic, let $\alpha \in [0, 1)$; else, set $\alpha = 0$. Let $B \in \mathcal{L}(U, X_{-(1-\alpha)})$ be such that $(-A)^\alpha B$ is L^∞-admissible, f satisfies (12) with constant $L > 0$ and $\Sigma(A, [\,B\ I\,], [\begin{smallmatrix}C\\I\end{smallmatrix}])$ is L^∞-BIBO stable. If $C_{2,\infty} L < 1$, where $C_{2,\infty}$ is the infinite-time L^∞-admissibility constant of $(-A)^\alpha$, then the output y of (1) with initial value $x_0 = 0$ and input $u \in L^\infty(\mathbb{R}_{\geq 0}; U)$ satisfies the following inequality for some $K, \mathfrak{K} \geq 0$ and every $t \geq 0$*

$$\|y\|_{L^\infty([0,t];Y)} \le K \|u\|_{L^\infty([0,t];U)} + \Re, \tag{13}$$

This particularly means that the semilinear state space system (1) *is* L^∞-*BIBO stable.*

Proof First note that for any $u \in L^\infty(\mathbb{R}_{\ge 0}; U)$ and any initial value in X, the mild solution $x \in \mathcal{C}(\mathbb{R}_{\ge 0}; X)$ of (7) exists due to Theorem 4.2 by our assumptions on B and f. Further, note that x is also the state trajectory of the linear system node $\Sigma(A, [\,B\ I\,], [\,{}^C_I\,])$ with input $[\,{}^u_{f(x(\cdot))}\,] \in L^\infty_{\mathrm{loc}}(\mathbb{R}_{\ge 0}; U \times X)$ and that the corresponding output is given by $\tilde{y} = [\,{}^y_x\,]$, where y is given by (2). Hence, by Theorem 2.3, (u, x, y) is a mild solution of the semilinear system (Σ, f). Furthermore, because the linear system node $\Sigma(A, [\,B\ I\,], [\,{}^C_I\,])$ is L^∞-BIBO stable, we have $\tilde{y} \in L^\infty(\mathbb{R}_{\ge 0}; Y \times X_\alpha)$, and thus $y \in L^\infty(\mathbb{R}_{\ge 0}; Y)$.

Let $x_0 = 0$ and $u \in L^\infty(\mathbb{R}_{\ge 0}; U)$. We deduce from Theorem 4.4 and (12) that

$$\|x\|_{L^\infty([0,t];X_\alpha)} \le C_{1,\infty}\|u\|_{L^\infty([0,t];U)} + C_{2,\infty}L\|x\|_{L^\infty([0,t];X_\alpha)} + C_{2,\infty}\|f(0)\|_X$$

and thus, since $C_{2,\infty}L < 1$,

$$\|x\|_{L^\infty([0,t];X_\alpha)} \le \frac{C_{1,\infty}}{1 - C_{2,\infty}L}\|u\|_{L^\infty([0,t];U)} + \frac{C_{2,\infty}}{1 - LC_{2,\infty}}\|f(0)\|_X.$$

Combining this with (11) for $\tilde{u} = f(x)$ and applying (12), once more yield that

$$\|y\|_{L^\infty([0,t];Y)} \le \|\tilde{y}\|_{L^\infty([0,t];Y\times X_\alpha)} \le \left(K_1 + \frac{K_2LC_{1,\infty}}{1 - C_{2,\infty}L}\right)\|u\|_{L^\infty([0,t];U)} + \Re$$

with $\Re = \left(K_2 + \frac{K_2 C_{2,\infty}L}{1 - C_{2,\infty}L}\right)\|f(0)\|_X$. $\qquad\square$

Corollary 4.6 *Let the assumptions of Theorem 4.5 hold. By M_α, δ and K_2, we denote the constants from* (4) *and* (11), *respectively. If either $\frac{LM_\alpha\Gamma(1-\alpha)}{\delta^{1-\alpha}} < 1$ or $K_2L < 1$, then* (13) *holds, and hence,* (1) *is* L^∞-*BIBO.*

Proof By the definition of the admissibility constant, $C_{2,\infty}$ is the smallest, time-independent constant such that the mild solution x of $\Sigma(A, [\,B\ I\,], [\,{}^C_I\,])$ for $x_0 = 0$, $u = 0$ and $\tilde{u} \in L^\infty(\mathbb{R}_{\ge 0}; X)$ satisfies for every $t \ge 0$ that

$$\|x(t)\|_\alpha \le C_{2,\infty}\|\tilde{u}\|_{L^\infty([0,t];X)}.$$

Hence, it holds that $C_{2,\infty} \le K_2$ and by Remark 4.1 also $C_{2,\infty} \le \frac{M_\alpha\Gamma(1-\alpha)}{\delta^{1-\alpha}}$. The assertion is now a consequence of Theorem 4.5. $\qquad\square$

Remark 4.7 In the situation of Corollary 4.6, it is possible to improve the constants in (13) by replacing $C_{2,\infty}$ by K_2 or $\frac{M_\alpha\Gamma(1-\alpha)}{\delta^{1-\alpha}}$ suitably in the proof of Theorem 4.5.

Remark 4.8 Theorem 4.5 and Corollary 4.6 can be easily generalized to nonlinearities f depending also on time $t \ge 0$ and satisfying (6) and (8) for a bounded function $k \in \mathcal{C}(\mathbb{R}_{\geq 0}; \mathbb{R}_{\geq 0})$. Indeed, on has to replace the Lipschitz constant L in the smallness conditions by $\|k\|_{L^\infty(\mathbb{R}_{\geq 0}; \mathbb{R}_{\geq 0})}$ in the statement and the proof.

5 Locally Lipschitz Nonlinearities with Internal Control

We consider the following heat equation with Neumann boundary conditions, internal friction and which is subject to internal control and cubic nonlinearity on an open and bounded domain $\Omega \subset \mathbb{R}^d$, $d \leq 3$, with Lipschitz boundary $\partial \Omega$.

$$\begin{cases} \dfrac{\partial x}{\partial t}(\zeta, t) = \Delta x(\zeta, t) - x^3(\zeta, t) + Bu(t), & \zeta \in \Omega, t \in \mathbb{R}_{\geq 0}, \\[2mm] x(\zeta, 0) = x_0(\zeta), & \zeta \in \Omega, \\[2mm] \dfrac{\partial x}{\partial \nu}(\zeta, t) = 0, & \zeta \in \partial \Omega, t \in \mathbb{R}_{\geq 0}, \end{cases} \tag{14}$$

where ν is the outward pointing unit normal vector at the boundary and $\frac{\partial}{\partial \nu}$ is the Neumann trace operator, which coincides with the normal derivative on smooth functions.

In an abstract formulation, (14) may be written as $\dot{x}(t) = Ax(t) + f(x(t)) + Bu(t)$, $x(0) = x_0$, where:

- The state space X is the space of square integrable functions on Ω, i.e. $X = L^2(\Omega; \mathbb{R})$ equipped with the inner product $\langle f, g \rangle_X := \int_\Omega fg \, d\zeta$, $f, g \in X$.
- Thanks to the bilinear form $\mathfrak{a} : H^1(\Omega; \mathbb{R}) \times H^1(\Omega; \mathbb{R}) \to \mathbb{R}$ expressed as $\mathfrak{a}(\hat{x}, x) = \langle \nabla \hat{x}, \nabla x \rangle_X$, $\hat{x}, x \in H^1(\Omega; \mathbb{R})$, the operator $A : D(A) \subset X \to X$ is characterized as follows, see e.g. [16, Chapter 1]:

  - $D(A) = \{x \in H^1(\Omega; \mathbb{R}) : \exists \tilde{x} \in X, \forall \hat{x} \in H^1(\Omega; \mathbb{R}), \mathfrak{a}(\hat{x}, x) = -\langle \hat{x}, \tilde{x} \rangle_X \}$.
  - $\forall x \in D(A)$, $Ax = \tilde{x}$ with $\tilde{x}$ such that $\mathfrak{a}(\hat{x}, x) = -\langle \hat{x}, \tilde{x} \rangle_X$ for all $\hat{x} \in H^1(\Omega; \mathbb{R})$.

  As the bilinear form $\mathfrak{a}$ is symmetric, the operator A is self-adjoint. Moreover, it is a Riesz-spectral operator, and it generates a bounded analytic C_0-semigroup.
- $f : X_{\frac{1}{2}} \to X$ is defined by $f(x) = -x^3$.
- The input function u is assumed to take values in an arbitrary Banach space U and the control operator $B : U \to X$ is such that $B \in \mathcal{L}(U, X)$.

Observe that, thanks to this setting, the Hilbert space $X_{\frac{1}{2}}$ is given by $X_{\frac{1}{2}} = H^1(\Omega; \mathbb{R})$ with the norm $\|x\|_{\frac{1}{2}}^2 = \|x\|_{L^2(\Omega; \mathbb{R})}^2 + \|\nabla x\|_{L^2(\Omega; \mathbb{R})}^2$, which shows that f is well-defined (even continuous), since $H^1(\Omega; \mathbb{R})$ is continuously embedded in $L^6(\Omega; \mathbb{R})$, see e.g. [1, Chapter 4]. The following theorem gives an upper bound on the $X_{\frac{1}{2}}$-norm of the state trajectory of (14). The following result establishes well-posedness of the pde (14). Similar results are well known; e.g. for $u = 0$ see [8, Chapter 11]. Our focus lies in an estimate of the solution in the $X_{\frac{1}{2}}$-norm.

Theorem 5.1 *Let $X = L^2(\Omega; \mathbb{R})$ and $B \in \mathcal{L}(U, X)$. For any initial condition $x_0 \in X_{\frac{1}{2}}$ and control input $u \in L^\infty(\mathbb{R}_{\geq 0}; U)$, the heat equation (14) admits a unique mild solution $x \in H^1_{\text{loc}}(\mathbb{R}_{\geq 0}; X) \cap \mathcal{C}(\mathbb{R}_{\geq 0}; X_{\frac{1}{2}}) \cap L^2_{\text{loc}}(\mathbb{R}_{\geq 0}; X_1)$, which satisfies the estimate*

$$
\|(-A)^{\frac{1}{2}}x(t)\|_X^2 \leq \frac{1}{2}\left[\|x_0\|_X^2 + 2\|(-A)^{\frac{1}{2}}x_0\|_X^2 + \int_\Omega x_0^4(\zeta)d\zeta\right]e^{-\rho t}
$$

$$
+ \frac{\lambda}{2}\int_0^t e^{-\rho(t-s)}\|u(s)\|_U^2\,ds,
$$

for all $t \geq 0$ and some $\lambda, \rho > 0$ independent of t.

Proof Let $x_0 \in X_{\frac{1}{2}}$ and $u \in L^\infty(\mathbb{R}_{\geq 0}; U)$. Since $f : X_{\frac{1}{2}} \to X$ is locally Lipschitz, we deduce from Theorem 4.2 the existence of a unique mild solution $x \in \mathcal{C}([0, t_1]; X) \cap L^\infty([0, t_1]; X_{\frac{1}{2}})$ for some $t_1 > 0$. Consequently, $\tilde{u} := f(x(\cdot)) \in L^\infty([0, t_1]; X_{\frac{1}{2}}) \subseteq L^2([0, t_1]; X_{\frac{1}{2}})$. Since x is also the mild solution of the linear system

$$
\begin{cases}
\dot{x}(t) = Ax(t) + Bu(t) + \tilde{u}(t), \\
x(0) = x_0,
\end{cases}
$$

where the control operators B and I are bounded as operators into X and therefore also into $X_{\frac{1}{2}}$. The maximal regularity property of the analytic semigroup and [24, Proposition 6.5] yield that $x \in H^1([0, t_1]; X_{-\frac{1}{2}}) \cap \mathcal{C}([0, t_1]; X) \cap L^2([0, t_1]; X_{\frac{1}{2}})$ and

$$
\|x(t)\|_X^2 + 2\int_0^t \|x(s)\|_{\frac{1}{2}}^2\,ds = \|x_0\|_X^2 + 2\int_0^t \langle x(s), Bu(s)\rangle_X\,ds
$$

$$
+ 2\int_0^t \langle x(s), f(x(s))\rangle_X\,ds
$$

for every $t \in [0, t_1]$. Similarly, since $z = (-A)^{\frac{1}{2}}x$ is the mild solution of the linear system

$$
\begin{cases}
\dot{z}(t) = Az(t) + (-A)^{\frac{1}{2}}Bu(t) + (-A)^{\frac{1}{2}}\tilde{u}(t), \\
z(0) = (-A)^{\frac{1}{2}}x_0,
\end{cases}
$$

we obtain $z \in H^1([0, t_1]; X_{-\frac{1}{2}}) \cap C([0, t_1]; X) \cap L^2([0, t_1]; X_{\frac{1}{2}})$, which translates to $x \in H^1([0, t_1]; X) \cap C([0, t_1]; X_{\frac{1}{2}}) \cap L^2([0, t_1]; X_1)$. Similar as before, it holds that

$$\|x(t)\|_{\frac{1}{2}}^2 + 2\int_0^t \|x(s)\|_1^2 \, ds = \|x_0\|_{\frac{1}{2}}^2 - 2\int_0^t \langle Ax(s), Bu(s)\rangle_X \, ds$$

$$- 2\int_0^t \langle Ax(s), f(x(s))\rangle_X \, ds$$

for every $t \in [0, t_1]$. In order to get an ISS estimate for Eq. (14) on the space $X_{\frac{1}{2}}$, let us consider the following Lyapunov functional candidate:

$$V(x(t)) := \|x(t)\|_X^2 + 2\|(-A)^{\frac{1}{2}}x(t)\|_X^2 + \int_\Omega x^4(\zeta, t) \, d\zeta,$$

which is almost everywhere differentiable on $(0, t_1)$ by the regularity of the solution x and the representations of the squared X-norm and $X_{\frac{1}{2}}$-norm of $x(t)$. We obtain for the derivative

$$\dot{V}(x(t)) = -2\|x(t)\|_{\frac{1}{2}}^2 + 2\langle x(t), Bu(t)\rangle_X + 2\langle x(t), f(x(t))\rangle_X$$

$$- 4\|x(t)\|_1^2 - 4\langle Ax(t), Bu(t)\rangle_X$$

$$- 4\langle Ax(t), f(x(t))\rangle_X - 4\langle f(x(t)), Ax(t) + Bu(t) + f(x(t))\rangle_X$$

$$= -2\|x(t)\|_{\frac{1}{2}}^2 + 2\langle x(t), Bu(t)\rangle_X + 2\langle x(t), f(x(t))\rangle_X$$

$$- 4\|Ax(t) + f(x(t))\|_X^2 - 4\langle Ax(t) + f(x(t)), Bu(t)\rangle_X.$$

for every $t \in (0, t_1)$. According to a generalization of Young's inequality and the boundedness of the operator B from U into X, we have

$$\dot{V}(x(t)) \leq -2\|x(t)\|_{\frac{1}{2}}^2 - 2\int_\Omega x^4(\zeta, t) \, d\zeta + \epsilon\|x(t)\|_X^2 + \frac{1}{\epsilon}\|B\|_{\mathcal{L}(U,X)}^2 \|u(t)\|_U^2$$

$$- 4\|Ax(t) + f(x(t))\|_X^2$$

$$+ 2\eta\|Ax(t) + f(x(t))\|_X^2 + \frac{2}{\eta}\|B\|_{\mathcal{L}(U,X)}^2 \|u(t)\|_U^2,$$

for arbitrary $\epsilon, \eta > 0$. In order to get a satisfactory ISS estimate in $X_{\frac{1}{2}}$-norm, one shall choose ϵ and η such that $\epsilon < 2$ and $\eta < 2$. Moreover, by noting that $\|x(t)\|_X \leq \|x(t)\|_{\frac{1}{2}}$, the following inequalities are satisfied:

$$\dot{V}(x(t)) \leq (-2 + \epsilon)\|x(t)\|_{\frac{1}{2}}^2 - 2\int_\Omega x^4(\zeta, t)\,d\zeta + (-4 + 2\eta)\|Ax(t) + f(x(t))\|_X^2$$

$$+ \left(\frac{1}{\epsilon} + \frac{2}{\eta}\right)\|B\|_{\mathcal{L}(U,X)}\|u(t)\|_U^2$$

$$\leq -\kappa\|(-A)^{\frac{1}{2}}x(t)\|_X^2 - 2\int_\Omega x^4(\zeta, t)\,d\zeta + \lambda\|u(t)\|_U^2, \tag{15}$$

where $\kappa := -2 + \epsilon$ and $\lambda := \left(\frac{1}{\epsilon} + \frac{2}{\eta}\right)\|B\|_{\mathcal{L}(U,X)}$. Besides, observe that

$$V(x(t)) \geq \|(-A)^{\frac{1}{2}}x(t)\|_X^2 + \int_\Omega x^4(\zeta, t)\,d\zeta =: W(x(t)) \tag{16}$$

and that

$$V(x(t)) \leq 3W(x(t)). \tag{17}$$

Combining (15), (16) and (17) implies

$$\dot{V}(x(t)) \leq \frac{\max\{-\kappa, -2\}}{3}V(x(t)) + \lambda\|u(t)\|_U^2,$$

which, by using Gronwall's inequality, has the consequence that

$$V(x(t)) \leq V(x(0))e^{-\rho t} + \lambda\int_0^t e^{-\rho(t-s)}\|u(s)\|_U^2\,ds,$$

where $\rho := -\frac{\max\{-\kappa, -2\}}{3}$. This entails that the $X_{\frac{1}{2}}$-norm of $x(t)$ can be upper bounded as

$$\|(-A)^{\frac{1}{2}}x(t)\|_X^2 \leq \frac{1}{2}\left[\|x_0\|_X^2 + 2\|(-A)^{\frac{1}{2}}x_0\|_X^2 + \int_\Omega x_0^4(\zeta)\,d\zeta\right]e^{-\rho t}$$

$$+ \frac{\lambda}{2}\int_0^t e^{-\rho(t-s)}\|u(s)\|_U^2\,ds, \tag{18}$$

which completes the proof. $\qquad\square$

Corollary 5.2 *The heat equation with output*

$$\begin{cases} \dfrac{\partial x}{\partial t}(\zeta, t) = \Delta x(\zeta, t) - x^3(\zeta, t) + Bu(t), & \zeta \in \Omega, t \in \mathbb{R}_{\geq 0}, \\[2mm] x(\zeta, 0) = x_0(\zeta), & \zeta \in \Omega, \\[2mm] \dfrac{\partial x}{\partial \nu}(\zeta, t) = 0, & \zeta \in \partial\Omega, t \in \mathbb{R}_{\geq 0}, \\[2mm] y(t) = Cx(\cdot, t), & t \in \mathbb{R}_{\geq 0}, \end{cases}$$

with state space $X = L^2(\Omega; \mathbb{R})$, *input space* U, *control operator* $B \in \mathcal{L}(U, X)$, *output space* Y *and ouput operator* $C \in \mathcal{L}(X_{\frac{1}{2}}, Y)$ *is a* L^∞*-BIBO stable semilinear state space system as defined in Theorem 2.1.*

Proof Theorem 5.1 implies that (14) admits for every $u \in L^\infty(\mathbb{R}_{\geq 0}; U)$ and $x_0 = 0$, a mild solution $x \in H^1_{\mathrm{loc}}(\mathbb{R}_{\geq 0}; X) \cap C(\mathbb{R}_{\geq 0}; X_{\frac{1}{2}}) \cap L^2_{\mathrm{loc}}(\mathbb{R}_{\geq 0}; X_1)$ satisfying

$$\|x(t)\|^2_{\frac{1}{2}} \leq \frac{\lambda}{2} \int_0^t e^{-\rho(t-s)} \|u(s)\|^2_U \, ds \leq \frac{\lambda}{2\rho} \|u\|^2_{L^\infty([0,t];U)}.$$

Consequently, x is also the mild solution to the extended system node $\Sigma(A, [\,B\ I\,], [\begin{smallmatrix} C \\ I \end{smallmatrix}])$ with input $[\begin{smallmatrix} u \\ -x^3 \end{smallmatrix}]$ whose (distributional) output is $[\begin{smallmatrix} y \\ x \end{smallmatrix}]$ with $y(t) = (C\&D)\begin{bmatrix} x(t) \\ u(t) \end{bmatrix}$ for almost every $t \geq 0$ since $\begin{bmatrix} x(t) \\ u(t) \end{bmatrix} \in D(C\&D)$ for almost every $t \geq 0$. Now, as x takes only values in $X_{\frac{1}{2}} = D(C)$, it suffices to show that for all $t > 0$, we have that $\|Cx(\cdot)\|_{L^\infty([0,t];Y)} \lesssim \|u\|_{L^\infty([0,t];U)}$. But this bound directly follows from the estimate of $x(t)$ in the $X_{\frac{1}{2}}$-norm and the boundedness of C as operator from $X_{\frac{1}{2}}$ to Y. $\qquad\square$

In view of the machinery that has been considered to construct the ISS estimate (18), we have the feeling that such an approach should work for nonlinearities expressed as the opposite of odd monomials as well, provided that the subscript α for the space X_α is appropriately chosen. Indeed, such nonlinear operators satisfy the sectorial condition $\langle x, f(x)\rangle_X \leq 0$, which may be viewed as a condition for the energy to be nonincreasing. For instance, such a sectorial condition has already been used in [10] in order to prove the well-posedness of nonlinear infinite-dimensional systems like (1).

6 An Application to Funnel Control

The applicability of funnel control for a coupled ordinary differential equation–partial differential equation (ODE–PDE) system describing the evolution of chemical components in chemical reactors is studied in this section. Therefore, some notions related to funnel control are recalled in a first time. In the second part of this section, the different assumptions needed to apply funnel control are tested on the ODE–PDE system. Numerical simulations are then depicted to support the theoretical results. We emphasize the fact that the results that we present in this section extend the ones of [11] for the following reasons:

- The input and output operators are allowed to be unbounded here, while their boundedness with some additional regularity assumptions are needed in [11]. In particular, by denoting by $B : \mathbb{R} \to X$, $Bu = bu$ and $C : X \to \mathbb{R}$, $Cx = \langle c, x \rangle_X$, respectively, the conditions $b \in D(A)$ and $c \in D(A^*)$ are needed in [11], which may be viewed as restrictive.
- The assumption $\langle b, c \rangle_X \neq 0$ is required in [11] in such a way that the state space X can be split into two linear subspaces in order to deal with the Byrnes–Isidori forms.
- The nonlinearity is only allowed to be defined from the state space X into itself in [11]. Moreover, it is required there that the latter is globally Lipschitz continuous. More general classes of nonlinear operators are treated here.

We also note that already in the early days of funnel control, chemical reactors were studied [15].

6.1 General Considerations on Funnel Control for Systems with Relative Degree One

We recall the following framework yielding funnel control, which was already present in the early works of the field [12]; see also [4] and the references therein.

For the following input–output differential relation,

$$\dot{y}(t) = N(d(t), S(y)(t)) + M(d(t), S(y)(t))u(t), \quad y(0) = y_0, \tag{19}$$

it is supposed that the following conditions hold:

Assumption 6.1 *The disturbance $d \in L^\infty(\mathbb{R}_{\geq 0}; \mathbb{R})$, the nonlinear function N is in $C(\mathbb{R}^2; \mathbb{R})$ and the gain function $M \in C(\mathbb{R}^2; \mathbb{R})$ is positive in the sense that $M(d, \varrho) > 0$ for all $(d, \varrho) \in \mathbb{R}^2$.*

Assumption 6.2 *The map $S : C(\mathbb{R}_{\geq 0}; \mathbb{R}) \to L^\infty_{loc}(\mathbb{R}_{\geq 0}; \mathbb{R})$ is a (possibly nonlinear) operator which satisfies the following conditions:*

1. *Bounded trajectories are mapped into bounded trajectories (BIBO property) i.e. for all $k_1 > 0$, there exists $k_2 > 0$ such that for all $y \in C(\mathbb{R}_{\geq 0}; \mathbb{R})$,*

$$\|y\|_{L^\infty(\mathbb{R}_{\geq 0}; \mathbb{R})} \leq k_1 \quad \Longrightarrow \quad \|S(y)\|_{L^\infty([0,t]; \mathbb{R})} \leq k_2, \tag{20}$$

2. *The operator S is causal, i.e. for any $t \in \mathbb{R}_{\geq 0}$ and any $y, \hat{y} \in C(\mathbb{R}_{\geq 0}; \mathbb{R})$,*

$$y_{|[0,t)} = \hat{y}_{|[0,t)} \implies S(y)_{|[0,t)} = S(\hat{y})_{|[0,t)},$$

where $f_{|I}$ denotes the restriction of a function f to the interval I.

3. *S is locally Lipschitz in the following sense:*
   for all $t \in \mathbb{R}_{\geq 0}$ and all $y \in C([0, t]; \mathbb{R})$, there exist positive constants τ, δ and ρ such that for any $y_1, y_2 \in C(\mathbb{R}_{\geq 0}; \mathbb{R})$ with $y_{i|[0,t]} = y, i = 1, 2$ and $|y_i(s) - y(t)| < \delta$ for all $s \in [t, t + \tau]$ and $i = 1, 2$ it holds that

$$\|S(y_1) - S(y_2)\|_{L^\infty([t,t+\tau];\mathbb{R})} \leq \rho \|y_1 - y_2\|_{L^\infty([t,t+\tau];\mathbb{R})}. \tag{21}$$

The class of systems governed by (19) with Assumptions 6.1–6.2 is presented in [6, Section 1] for systems with (possible) memory and relative degree $r \in \mathbb{N}$. This class is quite general and encompasses systems with infinite-dimensional internal dynamics as shown in [6] and [14], for instance.

For systems written like in (19), a funnel controller is an adaptive model–free control method whose objective is to maintain the difference between the output and an a priori fixed reference signal within the following prescribed funnel $\mathcal{F}_\phi :=$ $\left\{(t, e(t)) \in \mathbb{R}_{\geq 0} \times \mathbb{R}, \ \phi(t)|e(t)| < 1\right\}$, where the function ϕ is assumed to belong to

$$\Phi := \left\{\phi \in C(\mathbb{R}_{\geq 0}; \mathbb{R}) : \phi, \dot{\phi} \in L^\infty(\mathbb{R}_{\geq 0}; \mathbb{R}), \phi(t) > 0, \forall t \in \mathbb{R}_{\geq 0} \text{ and } \liminf_{t \to \infty} \phi(t) > 0\right\}.$$

As described in [6, 14] or [5], a controller that achieves the described output tracking performance is given by

$$u(t) = \frac{-e(t)}{1 - \phi^2(t)e^2(t)}, \tag{22}$$

with $\phi \in \Phi$ and $\phi(0)|e(0)| < 1$. The following theorem, coming from [12], see also [5] with $r = 1$, characterizes the effectiveness of the controller (22) in terms of existence and uniqueness of solutions of the closed-loop systems and in terms of output tracking performance.

Theorem 6.3 ([5, 12]) *Consider System (19) with Assumptions 6.1–6.2. Let $y_{ref} \in W^{1,\infty}(\mathbb{R}_{\geq 0}; \mathbb{R}), \phi \in \Phi$ and $y_0 \in \mathbb{R}$ such that the condition $\phi(0)|e(0)| < 1$ holds. Then the funnel controller (22) applied to (19) results in a closed-loop system whose solution $y : [0, \omega) \to \mathbb{R}, \omega \in (0, \infty]$ has the following properties:*

1. *The solution is global, i.e. $\omega = \infty$.*
2. *The input $u : \mathbb{R}_{\geq 0} \to \mathbb{R}$, the gain function $k : \mathbb{R}_{\geq 0} \to \mathbb{R}$ and the output $y : \mathbb{R}_{\geq 0} \to \mathbb{R}$ are bounded.*
3. *The tracking error $e : \mathbb{R}_{\geq 0} \to \mathbb{R}$ evolves in the funnel $\mathcal{F}_\phi$ and is bounded away from the funnel boundaries in the sense that there exists $\epsilon > 0$ such that for all $t \geq 0, |e(t)| \leq \frac{1}{\phi(t)} - \epsilon$.*

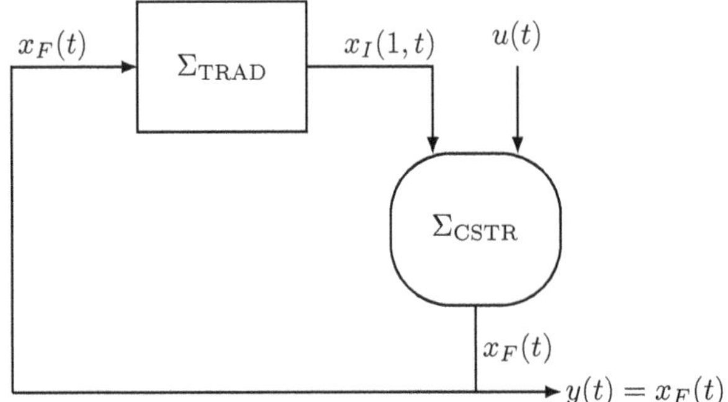

Fig. 2 Coupled CSTR-tubular reactor system

6.2 Dynamical Analysis of a Coupled ODE–PDE System for Funnel Control

Consider the system depicted in Fig. 2 comprised of a continuous stirred-tank reactor (CSTR) and a tubular reactor with axial dispersion (TRAD) that is similar to the one discussed in [17], dropping the terms describing disturbances for simplicity. Furthermore, in order for the system to be able to serve as an example for the type of systems discussed in this contribution, two changes have been made. First, the output has been chosen to be taken in the ODE part of the system in order for the system to be of relative degree one. Second, the Dirichlet boundary control has been supplanted with a Neumann boundary control ensuring that the original linear distributed system will be BIBO stable. Using the same notation as [17] this input–output system is then described by the coupled PDE-ODE system

$$
\Sigma_{\text{TRAD}} \begin{cases}
\dfrac{\partial x_I}{\partial t}(\zeta, t) = D\dfrac{\partial^2 x_I}{\partial \zeta^2}(\zeta, t) - v\dfrac{\partial x_I}{\partial \zeta}(\zeta, t) - \psi x_I(\zeta, t) + f(x_I(\zeta, t)), \\[2mm]
x_I(\zeta, 0) = 1, \\[2mm]
\dfrac{\partial x_I}{\partial \zeta}(0, t) = x_F(t), \quad \dfrac{\partial x_I}{\partial \zeta}(1, t) = 0,
\end{cases}
$$

$$
\Sigma_{\text{CSTR}} \begin{cases}
\dot{x}_F(t) = a_1 x_F(t) + a_2 u(t) + R x_I(1, t), \\[2mm]
x_F(0) = 1, \\[2mm]
y(t) = x_F(t),
\end{cases}
$$

where the temporal and the spatial variables satisfy $t \geq 0$ and $\zeta \in [0, 1]$, respectively, with $x_F(t) \in \mathbb{R}$ and $x_I(\cdot, t) \in L^2([0, 1]; \mathbb{R})$. The constants $v > 0$ and $D > 0$ are the transport and diffusion velocities in the tubular reactor, $R > 0$ describes the recycling within the system and a_1, a_2 and $\psi > 0$ are constants describing the chemical reactions within the two reactors. Furthermore, f is a nonlinear mapping from $L^2([0, 1]; \mathbb{R})$ to $L^2([0, 1]; \mathbb{R})$, such as e.g. the Lipschitz continuous function $f(x) = \frac{|x|}{|x|+1}$ from [9].

We can straightforwardly bring this system into the principle form of the system class in [7] as

$$\dot{y}(t) = S(y)(t) + a_2 u(t), \quad y(0) = 1$$

with the operator $S : C(\mathbb{R}_{\geq 0}; \mathbb{R}) \to L^\infty_{\mathrm{loc}}(\mathbb{R}_{\geq 0}; \mathbb{R})$ given by

$$S(\eta(\cdot)) = a_1 \eta(\cdot) + Rx(1, \cdot) \tag{23}$$

where x is the solution of the system

$$\begin{cases} \dfrac{\partial x}{\partial t}(\zeta, t) = D\dfrac{\partial^2 x}{\partial \zeta^2}(\zeta, t) - v\dfrac{\partial x}{\partial \zeta}(\zeta, t) - \psi x(\zeta, t) + f(x)(\zeta, t), \\[2mm] x(\zeta, 0) = 1, \\[2mm] \dfrac{\partial x}{\partial \zeta}(0, t) = \eta(t), \ \dfrac{\partial x}{\partial \zeta}(1, t) = 0. \end{cases} \tag{24}$$

While these internal dynamics are given in terms of a boundary control system, one could—using the methods laid out in [8, Chapter 10] and [22]—rewrite this system to arrive at one in the form of Equation (1).

The state space $X = L^2([0, 1]; \mathbb{R})$ is equipped with the following weighted inner product:

$$\langle f, g \rangle_\rho := \int_0^1 \rho(\zeta) f(\zeta) g(\zeta) \, d\zeta,$$

where $\rho(\zeta) := e^{-\frac{v}{D}\zeta}$. The operator A is defined as $Ax = D\frac{d^2 x}{d\zeta^2} - v\frac{dx}{d\zeta} + \psi x$, where x is in $D(A)$ expressed as

$$D(A) := \left\{ x \in H^2([0, 1]; \mathbb{R}) : \frac{dx}{d\zeta}(0) = 0 = \frac{dx}{d\zeta}(1) \right\}.$$

The boundary operator $B : \mathbb{R} \to X_{-1}$ is given as $Bu = -D\delta_0 u$, where δ_0 denotes the Dirac delta distribution at $\zeta = 0$. The observation operator $C : X_1 \to \mathbb{R}$ is the point measurement at $\zeta = 1$, i.e. $Cx = x(1)$. With this framework, it is easy to see that the operator A is self-adjoint by considering $\langle \cdot, \cdot \rangle_\rho$ as inner product. Moreover, A is a Riesz-spectral operator whose eigenvalues and normalized eigenfunctions are given by

$$\lambda_0 = -\psi, \ \lambda_n = -\frac{v^2 + 4D^2 n^2 \pi^2}{4D} - \psi, n \in \mathbb{N}_0$$

and

$$\phi_0(\zeta) = \sqrt{\frac{D}{v(1 - e^{\frac{-v}{D}})}} \, 1_{[0,1]}(\zeta), \quad \phi_n(\zeta) = \frac{\sqrt{2}v}{\sqrt{4n^2\pi^2 D^2 + v^2}}$$

$$\times \left[e^{\frac{v}{2D}\zeta} \left(\sin(n\pi\zeta) - \frac{2n\pi D}{v} \cos(n\pi\zeta) \right) \right],$$

respectively.

In order to apply Corollary 4.6 to deduce BIBO stability of the semilinear system, one shall first check that the extended linear system $(A, [\, B \; I \,], \left[\begin{smallmatrix} C \\ I \end{smallmatrix} \right])$ is L^∞-BIBO stable.

Proposition 6.4 *The extended linear system* $(A, [\, B \; I \,], \left[\begin{smallmatrix} C \\ I \end{smallmatrix} \right])$ *is* L^∞-*BIBO stable.*

Proof We use Proposition 3.1 to show this.

Condition 1: L^∞-control-admissibility of B Observe that any of the spaces $X_{-\eta}, 0 \leq \eta \leq 1$ can be expressed as

$$X_{-\eta} = \left\{ z \in X_{-1} : \sum_{n=0}^{\infty} \frac{|\langle z, \phi_n \rangle_X|^2}{|1 - \lambda_n|^{2\eta}} < \infty \right\}.$$

Now for $z = Bu = -D\delta_0 u$, we have that $|\langle z, \phi_n \rangle_X|^2 \sim n^0$ so that

$$\sum_{n=0}^{\infty} \frac{|\langle z, \phi_n \rangle_X|^2}{|1 - \lambda_n|^{2\eta}} \sim \sum_{n=0}^{\infty} \frac{1}{n^{4\eta}},$$

which converges for all $\eta > \frac{1}{4}$. Hence, $-D\delta_0 u \in X_{-\eta}$ for any $\eta \in (\frac{1}{4}, 1]$, and thus in particular, $B \in \mathcal{L}(U, X_{-\frac{1}{2}})$ which entails that B is L^∞-admissible; see Remark 4.1.

Condition 2: L^∞-BIBO stability of $\Sigma(A, B, C, \mathbf{G})$ This follows directly from [21, Proposition 4.3] after finding that (using the nomenclature of [21]) we have for $n > 0$ $b_n \sim n^0$, $c_n \sim n^0$ and $\mathrm{Re}(\lambda_n) \sim n^2$ so that

$$\sum_{n>0} \frac{|b_n c_b|}{|\mathrm{Re}(\lambda_n)|} \sim \sum_{n>0} \frac{1}{n^2} < \infty.$$

Condition 3: L^∞-BIBO stability of $\Sigma(A, I, C, C(\cdot I - A)^{-1})$ The boundedness of the operator C from $X_{\frac{1}{2}}{}^1$ into $\mathbb{R}$ directly implies this by [21, Proposition 6.4].

$\square$

---

[1] The space $X_{\frac{1}{2}}$ is the first order Solobev space, namely $H^1([0, 1]; \mathbb{R})$ with the following inner product:

Let us now fix $\alpha = \frac{1}{2}$. According to Corollary 4.6, the next assumption that is needed is $B \in \mathcal{L}(U, X_{-(1-\alpha)})$ with $(-A)^\alpha B$ being L^∞-admissible. Since $\alpha = \frac{1}{2}$, first observe that $-D\delta_0 \in X_{-(1-\alpha)}$. Second, note that $(-A)^\alpha B \in \mathcal{L}(U, X_{-\beta})$ for any $\beta \in (\frac{3}{4}, \frac{3}{2}]$. In particular, $(-A)^{\frac{1}{2}} B \in \mathcal{L}(U, X_{-\beta})$ for $\beta \in (\frac{3}{4}, 1)$. As a consequence of Remark 4.1, i., the operator $(-A)^\alpha B$ is L^∞-admissible.

The last condition we will check in order to have L^∞-BIBO stability of the nonlinear system (24) is $\delta^{-\frac{1}{2}} L M_{\frac{1}{2}} \Gamma(\frac{1}{2}) < 1$. First, observe that the nonlinear operator $f : X \to X$, $f(x) = \frac{|x|}{|x|+1}$ is globally Lipschitz continuous from X into X. Consequently, it preserves that property when defined from $X_{\frac{1}{2}}$ into X. One of its Lipschitz constant, computed as the supremum of the derivatives of the scalar function $f : \mathbb{R} \to [0, 1]$, $f(t) = \frac{|t|}{|t|+1}$, is given by $L = 1$.

We shall now give an estimate on the constant $M_{\frac{1}{2}}$, coming from the inequality (4). For this, let us assume that the constant δ satisfies $\delta \in [\varepsilon, \psi - \varepsilon]$ for some $\varepsilon > 0$ sufficiently small in such a way that the semigroup generated by the operator $A + \delta I$ is still analytic and exponentially stable. It can be shown that the operator norm of $(-A)^{\frac{1}{2}} T(t)$ is upper bounded as follows:

$$\|(-A)^{\frac{1}{2}} T(t)\| \le \frac{e^{-\delta t}}{t^{-\frac{1}{2}}} \sqrt{\max\left\{\frac{\psi}{\psi - \delta}, \frac{-\lambda_1}{-\lambda_1 - \delta}, e\right\}} \frac{e^{-\frac{1}{2}}}{\sqrt{2}} \le \frac{e^{-\delta t}}{t^{-\frac{1}{2}}} \sqrt{\max\left\{\frac{\psi}{\varepsilon}, e\right\}} \frac{e^{-\frac{1}{2}}}{\sqrt{2}},$$

where the relations $\delta \in (\varepsilon, \psi - \varepsilon)$ and $-\lambda_1 = \psi + c$ for some $c > 0$ have been used. This implies that a possible constant $M_{\frac{1}{2}}$ is $M_{\frac{1}{2}} = \sqrt{\max\left\{\psi\varepsilon^{-1}, e\right\}}(2e)^{-\frac{1}{2}}$. As a consequence, since $L = 1$, $\delta^{-\frac{1}{2}} L M_{\frac{1}{2}} \Gamma(\frac{1}{2}) < 1$ is equivalent to

$$\max\left\{\psi\varepsilon^{-1}, e\right\} < 2e\delta\pi^{-1}, \tag{25}$$

where $\Gamma(\frac{1}{2}) = \sqrt{\pi}$ has been used.

By using Corollary 4.6, one may conclude that a sufficient condition for the nonlinear system (24) to be L^∞-BIBO stable is that (25) holds. This result shows that the map S, mapping the input η to the output $x(1, \cdot)$, is L^∞-BIBO stable, which shows that the relation (20) of Assumption 6.2 is satisfied.

It remains to show that the map S is locally Lipschitz continuous in the sense of (21) to be able to apply funnel control. For this, let us consider the following proposition:

$$\langle f, g \rangle_{\frac{1}{2}} = D \int_0^1 e^{-\frac{v}{D}\zeta} \frac{\mathrm{d}f}{\mathrm{d}\zeta} \frac{\mathrm{d}g}{\mathrm{d}\zeta} \, \mathrm{d}\zeta + \psi \int_0^1 e^{-\frac{v}{D}\zeta} fg \, \mathrm{d}\zeta,$$

which is equivalent to the standard inner product whose $X_{\frac{1}{2}}$ is equipped with.

Proposition 6.5 *Under assumption (25), the operator S defined in (23) satisfies the inequality (21) of Assumption 6.2.*

Proof Let us fix $t \in \mathbb{R}_{\geq 0}$ and let us consider an arbitrary $\tau \in \mathbb{R}_{\geq 0}$. Moreover, let us pick any $\eta_1, \eta_2 \in \mathcal{C}(\mathbb{R}_{\geq 0}; \mathbb{R})$ that satisfy $\eta_1|_{[0,t]} = \eta = \eta_2|_{[0,t]}$ for some fixed $\eta \in \mathcal{C}([0, t]; \mathbb{R})$. As a consequence, the mild solutions of (24) with η_1 and η_2 as inputs, denoted by x_1 and x_2, respectively, are equal up to time t. For $\tilde{t} \in [t, t + \tau]$, these mild solutions are expressed as

$$(-A)^{\frac{1}{2}} x_i(\tilde{t}) = (-A)^{\frac{1}{2}} T(\tilde{t} - t) x_i(t) + \int_t^{\tilde{t}} T(\tilde{t} - s)(-A)^{\frac{1}{2}} f(x_i(s)) \, ds$$

$$+ \int_t^{\tilde{t}} T(\tilde{t} - s)(-A)^{\frac{1}{2}} B \eta_i(s) \, ds,$$

$i = 1, 2$. Making the difference between x_1 and x_2 and taking the X-norm on both sides implies that

$$\|x_1 - x_2\|_{L^\infty([t,t+\tau]; X_{\frac{1}{2}})} \leq M_{\frac{1}{2}} L \Gamma(\tfrac{1}{2}) \delta^{-\frac{1}{2}} \|x_1 - x_2\|_{L^\infty([t,t+\tau]; X_{\frac{1}{2}})}$$

$$+ C_{2,\infty} \|\eta_1 - \eta_2\|_{L^\infty([t,t+\tau]; \mathbb{R})},$$

where $C_{2,\infty}$ and $M_{\frac{1}{2}} \Gamma(\tfrac{1}{2}) \delta^{-\frac{1}{2}}$ are the infinite-time L^∞-admissibility constants of the operators, $(-A)^{\frac{1}{2}} B$ and $(-A)^{\frac{1}{2}}$, respectively. Assumption (25) together with the boundedness of the trace operator from $X_{\frac{1}{2}}$ to $\mathbb{R}$ concludes the proof. $\qquad\square$

According to Theorem 6.3, funnel control is conducive for (24) provided that the initial error between the output and the tracked reference is in the prescribed funnel. Some numerical simulations are reported in the following section.

6.3 Numerical simulations

As parameters for the PDE and the ODE in (24), we consider the following values $D = 0.1, v = 0.4, \psi = 2.8, a_1 = -1, a_2 = 2, R = 3$. The reference signal that the output $x_F(t)$ is supposed to track is set as $y_{ref}(t) = \frac{1}{2} \cos(t)$, while the prescribed funnel boundaries in which the output error evolves are fixed to $\frac{-1}{\phi(t)}$ and $\frac{1}{\phi(t)}, \phi(t) = (2e^{-2t} + 0.2)^{-1}$. The spatial interval $[0, 1]$ is discretized into n equal pieces, $n = 100$. Then the PDE–ODE system (24) in closed-loop with the funnel controller (22) is discretized by using finite differences, and it is integrated afterward with the ODE solver `ode23s` of Matlab©. The resulting state of the PDE (24), namely $x_I(\zeta, t)$, is depicted in Fig. 3. The error between the output $x_F(t)$ and the

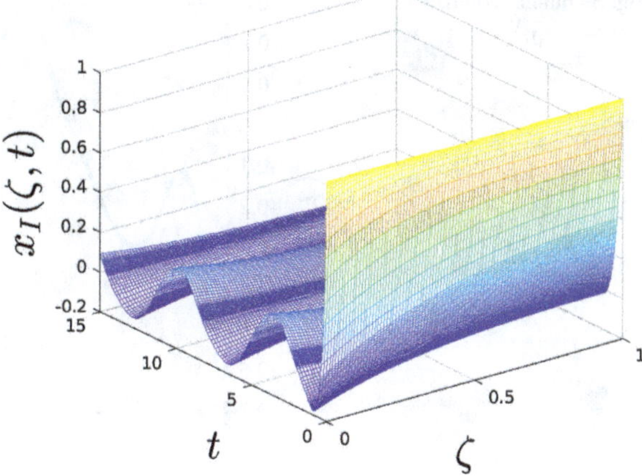

Fig. 3 State variable $x_I(\zeta, t)$ of the PDE (24)

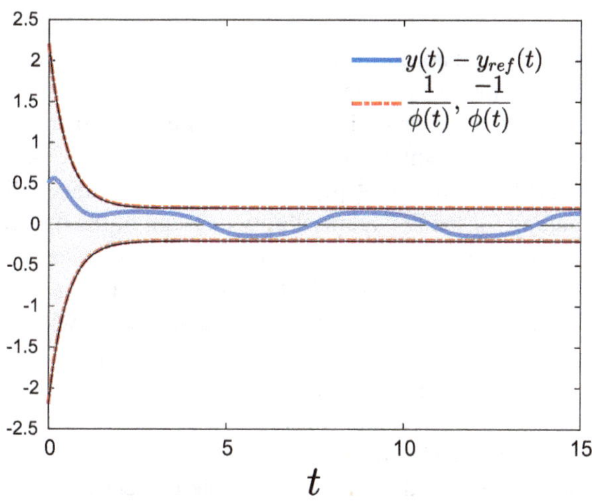

Fig. 4 Output error tracking $e(t) = y(t) - y_{ref}(t)$ with the funnel whose boundaries are the functions $-\frac{1}{\phi(t)}$ and $\frac{1}{\phi(t)}$

reference signal $y_{ref}(t)$ together with the prescribed funnel is given in Fig. 4. The funnel controller is depicted in Fig. 5.

7 Conclusion and Perspectives

BIBO stability for nonlinear infinite-dimensional systems with general input and output operators has been considered in this chapter. By expressing the nonlinear system as the interconnection between an extended linear system and a nonlinear feedback, it has been shown how BIBO stability of the original linear system was linked to BIBO stability of the extended linear system. Then, by considering

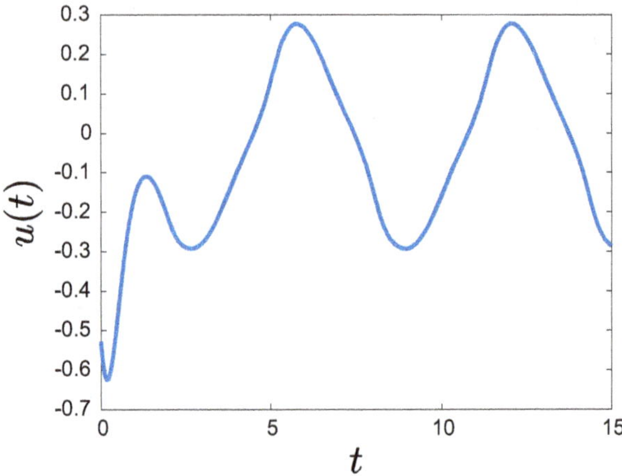

Fig. 5 Funnel controller $u(t) = \frac{-e(t)}{1-\phi^2(t)e^2(t)}$

different types of globally Lipschitz nonlinear operators, L^∞-BIBO stability of the original nonlinear system has been studied. In the case where the linear operator A generates an analytic semigroup, we showed that sufficient conditions for the nonlinear system to be BIBO stable are the L^∞-admissibility of the operator $(-A)^\alpha B$ for some $\alpha \in [0, 1)$ together with the inequality $C_{2,\infty}L < 1$, $C_{2,\infty}$ and L standing for the L^∞-admissibility constant of the operator $(-A)^\alpha$ and the Lipschitz constant of the nonlinear operator. This is the main contribution of our chapter. We applied our results to build a funnel controller for a nonlinear tubular reactor model coupled with a continuous stirred tank reactor with unbounded input and output operators.

As future works, relaxing the global Lipschitz continuity assumption should be investigated. One could, for instance, allow for dissipative nonlinear operators and take advantage of this feature to prove existence of solutions as well as BIBO stability, as it is done in the Example of Sect. 5. This would considerably enlarge the applicability of the results. Moreover, investigating BIBO stability in the case where the nonlinearity enters into the dynamics via an unbounded operator and in the case where an unbounded operator acts on the second component of the output $\tilde{y}$ is a challenge for further research. We finally mention that the assumption that the extended linear system is BIBO requires that the original linear system is exponentially stable; see Theorem 3.2. This can be an unnecessary strong assumption, as the nonlinearity may in fact cause the stability rather than the linear dynamics; see Sect. 5.

Acknowledgments The authors are thankful to Prof. H.J. Zwart (UTwente) for valuable discussions on the subject.

References

1. R. Adams, J. Fournier, *Sobolev Spaces*. Volume 140 of Pure and Applied Mathematics, 2nd ed. (Elsevier Science/Academic Press, Amsterdam, 2003)
2. T. Berger, T. Breiten, M. Puche, T. Reis, Funnel control for the monodomain equations with the FitzHugh-Nagumo model. J. Differ. Equ. **286**, 164–214 (2021)
3. T. Berger, D. Dennstädt, A. Ilchmann, K. Worthmann, Funnel model predictive control for nonlinear systems with relative degree one. SIAM J. Control Optim. **60**(6), 3358–3383 (2022)
4. T. Berger, A. Ilchmann, E. Ryan, Funnel control of nonlinear systems. Math. Control Signals Syst **33**, 151–194 (2021)
5. T. Berger, H. H. Lê, T. Reis, Funnel control for nonlinear systems with known strict relative degree. Automatica **87**, 345–357 (2018)
6. T. Berger, M. Puche, F.L. Schwenninger, Funnel control in the presence of infinite-dimensional internal dynamics. Syst. Control Lett. **139**, 104678 (2020)
7. T. Berger, M. Puche, F.L. Schwenninger, Funnel control for a moving water tank. Automatica **135**, 109999 (2022)
8. R. Curtain, H. Zwart, *Introduction to Infinite-Dimensional Systems Theory: A State-Space Approach*, volume 71 of Texts in Applied Mathematics Book Series (Springer, New York, 2020)
9. C. Delattre, D. Dochain, J. Winkin, Observability analysis of nonlinear tubular (bio)reactor models: a case study. J. Process Control **14**(6), 661–669 (2004)
10. A. Hastir, F. Califano, H. Zwart, Well-posedness of infinite-dimensional linear systems with nonlinear feedback. Syst. Control Lett. **128**, 19–25 (2019)
11. A. Hastir, J.J. Winkin, D. Dochain, Funnel control for a class of nonlinear infinite-dimensional systems. Automatica **152**, 110964 (2023)
12. A. Ilchmann, E. Ryan, C. Sangwin, Tracking with prescribed transient behaviour. ESAIM - Control Optim. Cal. Var. **7**, 471–493 (2002)
13. A. Ilchmann, E.P. Ryan, S. Trenn, Tracking control: performance funnels and prescribed transient behaviour. Syst. Control Lett. **54**(7), 655–670 (2005)
14. A. Ilchmann, T. Selig, C. Trunk, The Byrnes–Isidori form for infinite-dimensional systems. SIAM J. Control Optim. **54**(3), 1504–1534 (2016)
15. A. Ilchmann, S. Trenn, Input constrained funnel control with applications to chemical reactor models. Syst. Control Lett. **53**(5), 361–375 (2004)
16. T. Kato, *Perturbation Theory for Linear Operators*. Classics in Mathematics (Springer Berlin Heidelberg, 1995)
17. S. Khatibi, G. Cassol, S. Dubljevic, Linear model predictive control for a coupled CSTR and axial dispersion tubular reactor with recycle. Mathematics **8**, 711 (2020)
18. A. Mironchenko, Well-posedness and properties of the flow for semilinear boundary control systems (2022)
19. A. Pazy, *Semigroups of Linear Operators and Applications to Partial Differential Equations*. Applied Mathematical Sciences (Springer, New York, 1983)
20. M. Puche, F. Schwenninger, T. Reis, Funnel control for boundary control systems. Evol. Equ. Control Theory **10**(3), 519–544 (2021)
21. F. Schwenninger, A. Wierzba, H. Zwart, On BIBO stability of infinite-dimensional linear state-space systems. SIAM J. Control Optim. **62**(1), 22–41 (2024)
22. F.L. Schwenninger, Input-to-state stability for parabolic boundary control: linear and semilinear systems, in *Control Theory of Infinite-Dimensional Systems*, ed. by J. Kerner, H. Laasri, D. Mugnolo (Springer, Cham, 2020), pp. 83–116
23. O. Staffans, *Well-Posed Linear Systems*. Encyclopedia of Mathematics and its Applications (Cambridge University Press, Cambridge, 2005)
24. M. Tucsnak, G. Weiss, Well-posed systems—the LTI case and beyond. Automatica **50**(7), 1757–1779 (2014)
25. G. Weiss, Admissibility of unbounded control operators. SIAM J. Control Optim. **27**(3), 527–545 (1989)

On Checking L^p-Admissibility for Parabolic Control Systems

Philip Preußler and Felix L. Schwenninger

1 Introduction

1.1 State Space Systems and Admissibility

We consider abstract linear systems of the form

$$\dot{x}(t) = Ax(t) + Bu(t), \qquad t > 0, \tag{1}$$

on an infinite-dimensional Banach space X which we call the *state space*. Here we require that the operator $A\colon X \supset \mathcal{D}(A) \to X$ generates a strongly continuous semigroup of linear operators, or C_0-semigroup, on X, denoted by $T = (T(t))_{t \geq 0}$. The *state trajectory*, defined for $t \geq 0$, is denoted by x, and the *input function* (or *control function*), also defined for $t \geq 0$, is denoted by u. We require u to be U-valued, where U is a Banach space that we call the *input space*. Moreover, inputs u enter the system through the *control operator B* only.

In the field of infinite-dimensional systems theory, there is a wide array of literature dealing with bounded control operators, that is to say, systems with $B \in \mathcal{L}(U, X)$. However, many systems arising from controlled partial differential equations lead to control operators that fail to be bounded as maps from U to X, for example, in the case of control acting on the boundary. This leads us to the study of unbounded control operators B.

Here B is called an *unbounded* control operator if B is a linear operator which is bounded as a map from U to X_{-1}, but not in $\mathcal{L}(U, X)$. The extrapolation space

P. Preußler (✉) · F. L. Schwenninger
Department of Applied Mathematics, University of Twente, Enschede, The Netherlands
e-mail: p.n.preusler@utwente.nl; f.l.schwenninger@utwente.nl

© The Author(s), under exclusive license to Springer Nature Switzerland AG 2024
F. L. Schwenninger, M. Waurick (eds.), *Systems Theory and PDEs*,
Trends in Mathematics, https://doi.org/10.1007/978-3-031-64991-2_9

$X_{-1} \supset X$ is defined as the completion of X with the weaker norm

$$\|z\|_{X_{-1}} := \|(sI - A)^{-1}z\|_X$$

for some complex number s in the resolvent set $\rho(A)$.

This allows us to consider (1) in the ambient space X_{-1}. To that end, consider a fixed initial value $x(0) \in X$ and an unbounded control operator B. The mild solution of the controlled system (1) is given by

$$x(t) = T(t)x(0) + \int_0^t T(t-s)Bu(s)\,ds, \qquad t \geq 0, \tag{2}$$

and takes values which a priori only lie in X_{-1}. As we are interested in the case with continuous state trajectories taking values in X—such as in the case of bounded B— we want to find conditions on B such that (2) lies in X for all $t \geq 0$, bringing us to the property of admissibility [28, 50, 51].

Definition 1.1 (L^p**-Admissible Control Operators**) For $1 \leq p \leq \infty$, the control operator $B \in \mathcal{L}(U, X_{-1})$ is called *finite-time* L^p*-admissible for* T (*or for* A) if the convolution type Bochner integral

$$\int_0^t T(t-s)Bu(s)\,ds$$

is an element of X for any $t \geq 0$, and there exists $C > 0$ such that

$$\left\| \int_0^t T(t-s)Bu(s)\,ds \right\|_X \leq C\|u\|_{L^p([0,t],U)}$$

holds for all input functions $u \in L^p([0, t], U)$.

The operator B is called *infinite-time* L^p*-admissible for* T (*or for* A) if the constant $C > 0$ from above can be chosen independently of t.

Remark 1.2

1. If the context is clear, one can also drop the reference to the semigroup T or the generator A when referring to admissibility. Moreover, one can also drop the letter L and speak of p-admissibility of control operators. Lastly, when the distinction between finite-time and infinite-time admissibility is not made, one generally speaks about infinite-time admissibility.
2. We also note that—due to translation invariance of the set of input functions— one can replace $T(t-s)$ by $T(s)$ in the integrals above. Moreover, infinite-time

p-admissibility is equivalent to the existence of $C > 0$ such that

$$\left\| \int_0^\infty T(s) Bu(s)\, ds \right\|_X \leq C \|u\|_{L^p([0,\infty),U)}$$

holds for all $u \in L^p([0, \infty), U)$.

3. The property that $B \in \mathcal{L}(U, X_{-1})$ is p-admissible can be rephrased by saying that the mapping $\Phi_t : u \mapsto \int_0^t T(t-s) Bu(s)\, ds$ is bounded from L^p to X— rather than only mapping to X_{-1}—for some and hence all $t > 0$. Dualizing this statement, we arrive at an admissibility notion for *observation operators* $C : \mathcal{D}(A) \to Y$, where $\mathcal{D}(A)$ is equipped with the graph norm of A and the Banach space Y is called the *output space*.

    More precisely, we fix $p \in (1, \infty)$ and denote the Hölder conjugate of p by p'. Then, provided that the dual semigroup $T' = (T(t)')_{t \geq 0}$ is strongly continuous on the Banach space dual X', $B : U \to X_{-1}$ is p-admissible if and only if the operator $C = B' : \mathcal{D}(A') \to U'$ satisfies

$$\int_0^t \| B' T(t)' x \|_Y^{p'}\, dt \leq K \|x\|_{X'}^p, \qquad x \in \mathcal{D}(A')$$

for some $K > 0$, see, e.g., [51, Theorem 6.9].

1.2 Boundary Control Systems

The most natural form of many systems of partial differential equations with control on the boundary of the domain is not the state space form as in (1) above. Rather, in these systems the control enters via the boundary condition; for example, in the Dirichlet trace sense, $x|_{\partial\Omega} = u$. This then leads to the type of systems known as *boundary control systems*, see, e.g., Salamon [39] and Tucsnak and Weiss [46]. Boundary control systems take the form

$$\begin{cases} \dot{x}(t) = \mathfrak{A}x(t) & \text{on } (0, \infty), \\[1mm] \mathfrak{B}x(t) = u(t) & \text{on } (0, \infty), \\[1mm] x(0) = x_0. \end{cases} \qquad (3)$$

Typically, $\mathfrak{A}$ is a differential operator and $\mathfrak{B}$ is a boundary trace operator, such that the system corresponds to a (partial) differential equation with the control input acting via the boundary of the spatial domain. To relate this formulation to the state space form, we need some well-posedness assumptions.

Definition 1.3 (Boundary Control Systems) Given Banach spaces X and U and closed operators

$$\mathfrak{A}\colon X \supset \mathcal{D}(\mathfrak{A}) \to X, \qquad \mathfrak{B}\colon X \supset \mathcal{D}(\mathfrak{B}) = \mathcal{D}(\mathfrak{A}) \to U,$$

where we have normed the space $\mathcal{D}(\mathfrak{A})$ with the graph norm. If now the restriction of $\mathfrak{A}$ to $\ker \mathfrak{B}$ is the generator of a C_0-semigroup on X and $\mathfrak{B}$ has a right inverse $B_0 \in \mathcal{L}(U, \mathcal{D}(\mathfrak{A}))$, then we call the pair of operators $(\mathfrak{A}, \mathfrak{B})$ and the corresponding equations (3) a *(linear) boundary control system.*

Remark 1.4 One can translate a boundary control system on X into a state space system on X by setting

$$A := \mathfrak{A}|_{\ker \mathfrak{B}}, \qquad B = (\mathfrak{A} - A_{-1})B_0, \tag{4}$$

where $A_{-1}\colon X_{-1} \supset \mathcal{D}(A_{-1}) \to X_{-1}$ is the unique isometric extension of A to the space X_{-1} with domain $\mathcal{D}(A_{-1}) = X$, see, e.g., [9, 42, 46]. More on the rewriting procedure for boundary control systems into state space form can, e.g., be found in [41] or [46, Chapter 10].

This correspondence allows us to transfer concepts formulated in the state space system framework—including admissibility—to the boundary control context. The intended meaning here is that the operator B that results from $\mathfrak{B}$ through the rewriting procedure is admissible for $A = \mathfrak{A}|_{\ker \mathfrak{B}}$. While the right inverse involved in the construction of B is generally not unique, the operator B is uniquely determined. Yet, computing B from the expression given in (4) above is usually not feasible. However, on Hilbert spaces, the relation

$$\langle \mathfrak{A}x, y \rangle - \langle x, A^*y \rangle = \langle \mathfrak{B}x, B^*y \rangle, \tag{5}$$

see, e.g., [41, Proposition 2.9], holds for all $x \in \mathcal{D}(\mathfrak{A})$ and $y \in \mathcal{D}(A^*)$. With this identity, we can identify the control operator B, simplifying the computation in contrast to (4).

1.3 Current State of L^p-Admissibility Theory

When verifying p-admissibility, there is a vast difference between the cases $p = 2$ and $p \neq 2$, a result of the missing Hilbert space structure in the latter case. This discrepancy is reflected in the amount of results available in the literature on 2-admissibility compared to p-admissibility, see also the survey article [28] by Jacob and Partington that deals mostly with the case $p = 2$. As an example, for $p = 2$ characterizations of admissibility using Lyapunov equations and Lyapunov inequalities are available, compare also [46, Chapter 5]. These methods do not generalize to $p \neq 2$ as they rely on the Hilbert space structure of the state space

X, of the input space U, and of the L^2 space in the domain of the input functions, illustrating the added difficulty of checking for p-admissibility.

This disparity could lead to the question of practical relevance of the more general concept of p-admissibility compared to 2-admissibility. In other words, one could ask if there are any common systems that are not 2-admissible but only p-admissible for some $p > 2$. Concretely, we are interested in finding examples of systems arising from partial differential equations—on a natural state space such as $L^2(\Omega)$ for some domain $\Omega \subset \mathbb{R}^n$—such that the corresponding control operator B is not 2-admissible. Examples of this type exist, for instance, the heat equation on the spatial domain $\Omega = [0, 1]$ with Dirichlet boundary control and scalar input space $U = \mathbb{C}$. Here the control operator B resulting from rewriting the boundary control system into state space form turns out to be $B = \delta_0'$, which is p-admissible for all $p > 4$ and not p-admissible for $p \leq 4$, see [34] and also Sect. 2.2 in the sequel. Moreover, p-admissibility (in a weighted variant) appears as a prerequisite to the abstract Kato method in the article [17] by Haak and Kunstmann to derive results for Navier–Stokes equations.

In the context of abstract linear systems, p-admissibility for $p \neq 2$ has been studied in the operator-theoretic framework by and since Weiss's [50, 51] seminal works on observation and control operators. Weiss also justifies the assumption $B \in \mathcal{L}(U, X_{-1})$ in the definition of an admissible operator by showing that each "admissible" operator B (which would be bounded only from U to some space V in which X lies dense) is equivalent to an operator $\hat{B}$ that is bounded $U \to X_{-1}$ in the sense that they lead to the same abstract linear system. Furthermore, tests for p-admissibility are given for some special cases, such as 1-admissibility on reflexive X, as Weiss [50] shows that B is admissible in this situation if and only if $B \in \mathcal{L}(U, X)$.

It should be mentioned that the concept of admissibility—with respect to L^2—had previously appeared in Salamon [39] as hypothesis (S2). Connections between state space systems and boundary control systems are also discussed in the aforementioned work by Salamon, and we also mention the work by Curtain and Pritchard [7] for more on this topic.

Among approaches to characterizing admissibility, important examples are Carleson measure criteria for diagonal systems, which are known as such due to being based on certain discrete measures with weights depending on the system in question being Carleson measures. A method based on these conditions for checking 2-admissibility with scalar input space $U = \mathbb{C}$ appears in Ho and Russell [24], was extended to an equivalent condition by Weiss [49], and was further expanded to control operators defined on $\ell^2(\mathbb{N})$ by Hansen and Weiss [21]. Generalizations of these results to the context of L^p-admissibility for diagonal semigroups on state spaces of ℓ^q-type using Laplace–Carleson embeddings can be found in works by Haak [15], Unteregge [47], Jacob, Partington, and Pott [31], and Jacob, Partington, Pott, Rydhe, and Schwenninger [32]. In this frame of reference we also mention that extensions to normal semigroups have appeared in Weiss [48], Hansen and Weiss [21], Unteregge [47], and Staffans [42].

Another property of note which is used to derive results for admissibility is the Weiss property and its L^p-version. For $p \in [1, \infty]$, the *p-Weiss property for control operators* for the generator A is the statement that for any Banach space U and any $B \in \mathcal{L}(U, X_{-1})$ we have the equivalence

$$B \text{ is infinite-time } p\text{-admissible} \quad \Leftrightarrow \quad \sup_{\text{Re}\,\lambda > 0} (\text{Re}\,\lambda)^{\frac{1}{p}} \|R(\lambda, A)B\| < \infty;$$

for the analogous formulation for observation operators, see Sect. 5 in the sequel.

The property originates from conjectures formulated by Weiss in [52], which postulated the equivalence of a resolvent condition and admissibility. In [52] the author remarked that the conjecture is not true on general state spaces X that are not Hilbert spaces. Later it was shown that the conjecture is false even in the Hilbert space setting. For counterexamples, see Jacob, Partington, and Pott [30] based on the right shift semigroup; compare with Zwart and Jacob [54] and Zwart, Jacob, and Staffans [55] with counterexamples where the semigroups are even analytic. However—in special cases—positive results on the Weiss property were also found, e.g. by Jacob and Partington [29] for contraction semigroups with $p = 2$ and finite-dimensional input spaces and by Le Merdy [35] in the case of 2-admissibility for bounded analytic semigroups satisfying a square function estimate and arbitrary input spaces.

In conjunction with Le Merdy's article [35], the connection with functional calculus methods and square function estimates must also be mentioned, which are used as a powerful tool to generate estimates in the admissibility context. With these methods, Le Merdy showed that if A has dense range and generates a bounded analytic semigroup, then admissibility of $(-A)^{1/2}$ in the observation operator sense—which is a square function estimate for A—is equivalent to A having the 2-Weiss property. That is to say that $C: X_1 \to Y$ is 2-observation-admissible for A if and only if

$$\sup_{\text{Re}\,\lambda > 0} \sqrt{\text{Re}\,\lambda} \, \|CR(\lambda, A)\| < \infty.$$

Note that for Hilbert spaces, the Weiss property is specifically satisfied if A is densely defined and generates an analytic contraction semigroup.

Further results in this direction appear in the work [20] by Haak and Le Merdy, where the H^∞ functional calculus is used to show the equivalence of certain square function estimates and weighted 2-admissibility for analytic semigroups. The functional calculus is also an integral part of the dissertation [14] by Haak, where extensions of Le Merdy's results to L^p-admissibility are treated with L^p estimates in lieu of the square function estimates that appear in the L^2 setting. In particular, p-admissibility is characterized under the assumption that $(-A)^{1/p}$ is p-admissible for A. The article [19] by Haak and Kunstmann features generalized results in this direction. Specifically, the authors investigate the more general property of weighted L^p-admissibility of type α. Here so-called L_*^p functional calculus estimates are used, generalizing the square function estimates appearing in the condition of admissibility of $(-A)^{1/2}$. With this assumption, the authors prove that the Weiss resolvent condition in the L^p version is equivalent to p-admissibility of B. They also show that $-A$ has L_*^p estimates if and only if the state space X embeds into the real interpolation space $(\dot{X}_{-1}, \dot{X}_1)_{1/2,p}$. This leads to results for concrete example systems if full expressions for these interpolation spaces are known, such as the Neumann controlled heat equation on L^q spaces or zero smoothness Besov spaces $B_{q,p}^0$. Furthermore, a test for the resolvent condition in the Weiss property via real interpolation spaces is given.

For more applications of functional calculus methods to proofs of admissibility, we refer to Haak [16], where the interplay of the Weiss condition and admissibility with respect to the weak-type space L$^{2,\infty}$ is discussed. Adding to the positive results on the Weiss conjecture, in [4] the authors Bounit, Driouich, and El-Mennaoui characterize the p-Weiss condition for bounded analytic semigroups by admissibility of $(-A)^{1/p}$ without resorting to the H$^\infty$ functional calculus.

Finally we remark that the above list is not exhaustive, as there are even more methods of showing L^p-admissibility such as semigroup generation criteria. These are methods based on block operator matrices generating C_0-semigroups; see Grabowski and Callier [11], Engel [8], and the recent article [25] by Hosfeld, Jacob, and Schwenninger, which treats the more general case of admissibility with respect to Orlicz spaces. We further stress that ∞-admissibility, which is formally very closely related to the notion of input-to-state stability (ISS), see, e.g., the survey article [37], is not the focus of the present chapter. However, as p-admissibility for any finite p implies ∞-admissibility, the discussed results serve as natural sufficient conditions for ISS.

1.4 Aims

The methods of verifying p-admissibility—for $p \neq 2$—that we set our focus on can be grouped into

(i) Sufficient conditions for analytic semigroups using mapping properties of boundary trace operators, see, e.g., [34, 36, 41]

(ii) Tests arising from Laplace–Carleson embeddings for diagonal semigroups with finite-dimensional input spaces, see, e.g., [15, 31, 32, 47]

(iii) Applying p-Weiss property type results in cases where the property is valid, see, e.g., [4, 14, 18, 19, 35, 52]

Our aim is to provide a comparison—with some extensions—of the aforementioned methods of verifying p-admissibility while reviewing applicability to infinite-dimensional input spaces and potential problems in applications. In this setting, we also comment on the case $p > 2$ posing even more difficulties than the case $p \leq 2$.

Elaborating on this, we discuss that (i) even leads to a sort of characterization of p-admissibility for values of p strictly greater than a certain threshold, which depends on the interpolation space that the control operator B maps into, see also [36].

For (ii) we note that L^p-admissibility of the input element $b \in \mathcal{L}(\mathbb{C}, X_{-1})$ is equivalent to boundedness of the Laplace transform operator $\mathscr{L}$, which is to be understood as a mapping

$$\mathscr{L} : L^p([0, \infty)) \to L^2(\mathbb{C}_+, \mu).$$

The measure μ is given by the discrete measure $\mu = \sum_k |b_k|^2 \delta_{-\lambda_k}$ in the diagonal case, i.e., in the presence of a Riesz basis of eigenvectors of the generator with eigenvalues $\{\lambda_k\}_{k \in \mathbb{N}}$, cf. results by Jacob, Partington, and Pott [31]. However, the conditions given in their work are not applicable if the generator does not have compact resolvents as the spectrum is then not discrete. For instance, if the generator A is a normal operator, the spectral theorem yields a spectral measure E (which is not necessarily discrete) associated with A, admitting the representation

$$A = \int_{\sigma(A)} z \, dE(z).$$

Using this fact, we are able to generalize the diagonal results. The extensions are established using a suitable generalization of the discrete measure μ related to the measure $\langle Eb, b \rangle$; see also similar considerations for the special case $p = 2$ in [21, 42, 47, 48]. We also discuss a related result for systems with semigroups generated by multiplication operators on L^q spaces, generalizing the setting where one has a q-Riesz basis. This is related to diagonal systems on ℓ^q sequence spaces as in [31].

As an outlook and relating to (iii), we provide an argument based on the p-Weiss property which allows the extension of results for scalar input spaces to infinite-dimensional input spaces. In certain special cases where the p-Weiss property does hold, such as for negative semidefinite self-adjoint generators when $p \le 2$, this provides a tool to assess p-admissibility for the critical value of p, that is to say, the value of p that defines the threshold between admissibility and non-admissibility.

2 Interpolation Spaces, Extrapolation Spaces, and Admissibility

2.1 Mapping Properties and Their Relation to Admissibility

Here and throughout the rest of the chapter, we denote the resolvent of an operator A by $R(z, A) := (zI - A)^{-1}$ and the range of A by the symbol $\mathcal{R}(A)$. Moreover, we denote the *growth bound* of the semigroup T by $\omega(T)$, and we recall that T is called *exponentially stable* if $\omega(T) < 0$. A C_0-semigroup T is called *analytic* if there exists $\varphi \in (0, \pi/2]$ such that T can be extended to an analytic map on the open sector of angle 2φ around the positive real line defined by $S_\varphi := \{z \in \mathbb{C} \setminus \{0\} : |\mathrm{Arg}\, z| < \varphi\}$. In the presence of an analytic semigroup T and a generator A such that $0 \in \rho(A)$, we define a scale of interpolation and extrapolation spaces X_β for $\beta \in (-1, 1)$. These spaces lie between the spaces $X_1 = \mathcal{D}(A)$, $X_0 := X$, and X_{-1}. Using this scale of spaces allows for the description of finer mapping properties of control operators. This is meant in the sense that if B maps into a space $X_{-\beta}$ for some $\beta \in (0, 1)$, then we can use this information to make statements about admissibility. In defining these spaces, we make use of fractional powers $(-A)^{-\beta} : X \to X$, which are defined by the operator-valued integral

$$(-A)^{-\beta} := \int_\gamma z^{-\beta} R(z, -A)\, \mathrm{d}z,$$

where the curve γ is the boundary of a sector of the form S_φ as above that contains $\sigma(-A)$, see also [9, Chapter II.5] and [41]. The corresponding positive power $(-A)^\beta$ is then defined as the inverse of $(-A)^{-\beta}$, i.e.,

$$(-A)^\beta := ((-A)^{-\beta})^{-1} : \mathcal{R}((-A)^{-\beta}) \to X.$$

Let A generate an analytic semigroup on X with growth bound less than 0. For any $\beta \in (0, 1)$, define the space $X_{-\beta}$ as the completion of X with respect to the norm given on X by $\|x\|_{-\beta} := \|(-A)^{-\beta}x\|_X$. Furthermore, we define the space X_β as the set $\mathcal{D}((-A)^\beta) = \mathcal{R}((-A)^{-\beta})$ with the norm $\|x\|_\beta := \|x\|_X + \|(-A)^\beta x\|_X$. These spaces are also known as *abstract Sobolev spaces*. For general analytic semigroups, the spaces X_β are defined by considering the shifted generator $A - \lambda I$ for sufficiently large λ, noting that this definition is independent of the choice of λ and hence also consistent.

In the following we present a characterization of p-admissibility that is derived from the mapping property that B be bounded into $X_{-\beta}$ for some $\beta \in (0, 1)$, compare also [36]. We also record that for $p \in [1, \infty]$ we use the notation p' for the Hölder conjugate (or conjugate exponent) satisfying $1/p + 1/p' = 1$ with the usual convention for the conjugated exponents of 1 and ∞. The following lemma is essentially well known in the context of admissibility for analytic semigroups. For the sake of the reader, we provide the argument.

Lemma 2.1 *Let A be the generator of an analytic semigroup T on X. Fix $0 < \beta < 1$. If B is bounded as an operator $U \to X_{-\alpha}$ for all $\alpha > \beta$, then B is finite-time p-admissible for all $p > \frac{1}{1-\beta} = (\beta^{-1})'$.*

Proof We fix $0 < \beta < 1$, and let $B \in \mathcal{L}(U, X_{-1})$. Without loss of generality, assume that T is exponentially stable. Assume $B \in \mathcal{L}(U, X_{-\alpha})$ for an arbitrary $\alpha > \beta$. Then $(-A)^{-\alpha}B \in \mathcal{L}(U, X)$. Note that for A generating an analytic semigroup, we have

$$\|(-A)^\alpha T(t)\|_{\mathcal{L}(X)} \le C_\alpha t^{-\alpha} e^{t\omega}$$

for $t > 0$ and $\omega > \omega(T)$ and some constant $C_\alpha > 0$, see [38, Theorem 6.13]. Using these properties, we can deduce that

$$
\begin{aligned}
\|T(t)B\|_{\mathcal{L}(U,X)} &= \|T(t)(-A)^\alpha(-A)^{-\alpha}B\|_{\mathcal{L}(U,X)} \\
&\leq \|T(t)(-A)^\alpha\|_{\mathcal{L}(X)}\|(-A)^{-\alpha}B\|_{\mathcal{L}(U,X)} \\
&\leq C_\alpha t^{-\alpha}e^{t\omega}\|(-A)^{-\alpha}B\|_{\mathcal{L}(U,X)} \\
&\leq C_\alpha t^{-\alpha}e^{t\omega}\|B\|_{\mathcal{L}(U,X_{-\alpha})}
\end{aligned}
$$

holds for $t > 0$ and $\omega > \omega(T)$. Applying this to the left-hand side of the defining inequality for L^p-admissibility, we get

$$
\begin{aligned}
\left\|\int_0^t T(t-s)Bu(s)\,\mathrm{d}s\right\|_X &\leq \int_0^t \|T(t-s)Bu(s)\|_X\,\mathrm{d}s \\
&\leq C_\alpha \int_0^t (t-s)^{-\alpha}e^{(t-s)\omega}\|B\|_{\mathcal{L}(U,X_{-\alpha})}\|u(s)\|_U\,\mathrm{d}s.
\end{aligned}
$$

We can now use Hölder's inequality to further estimate this quantity and find

$$
\int_0^t (t-s)^{-\alpha}e^{(t-s)\omega}\|u(s)\|_U\,\mathrm{d}s \leq \|u\|_{L^p([0,t],U)}\left(\int_0^t ((t-s)^{-\alpha}e^{(t-s)\omega})^{p'}\mathrm{d}s\right)^{1/p'}.
$$

The right-hand side is finite if and only if $p' < \frac{1}{\alpha}$. Combining the two above inequalities, we get

$$
\left\|\int_0^t T(t-s)Bu(s)\,\mathrm{d}s\right\|_X \leq K\|u\|_{L^p([0,t],U)}
$$

for $p' < \frac{1}{\alpha}$ and some constant $K > 0$. Thus B is p-admissible for all $p > \frac{1}{1-\beta}$. $\square$

We want to show that the converse of this statement also holds. To do this, we need to introduce the scale of Favard spaces, which are another useful class of intermediate spaces as there are embeddings that allow us to derive L^p estimates. We give a short summary of the needed results and properties of these spaces.

Definition 2.2 (Favard Spaces [9]) Let T be an exponentially stable C_0-semigroup. For $0 < \alpha \leq 1$, we define the α-*Favard norm* by

$$
\|x\|_{F_\alpha} := \sup_{t>0} t^{-\alpha}\|T(t)x - x\|_X
$$

and the *Favard space* F_α as the set of all $x \in X$ with finite α-Favard norm.

To extend the definition to arbitrary real indices α, we decompose the number $\alpha \in \mathbb{R}$ into an integer part and a fractional part; that is to say, we write $\alpha = k + \beta$

(uniquely) with $k \in \mathbb{Z}$ and $0 < \beta \leq 1$ and define F_α to be the Favard space of order β for the extended (respectively, restricted) semigroup T_k. Moreover, if T is not exponentially stable, we define F_α as the α-Favard space corresponding to the semigroup $(e^{-\omega t} T(t))_{t \geq 0}$ for some $\omega > \omega(T)$.

Note that this definition yields the same space for any such choice of ω; for more on this topic, we refer to Engel and Nagel [9, Chapter II.5].

We now quote results about Favard spaces needed in the sequel.

Remark 2.3 For A generating a C_0-semigroup T, the following hold:

1. If $\alpha > \beta$, we have

$$X_\alpha \subset F_\alpha \hookrightarrow X_\beta \subset F_\beta,$$

cf. [9, Proposition II.5.14, Proposition II.5.33]; note that X_α is defined alternatively there.

2. If T is exponentially stable, the α-Favard norm with $0 < \alpha \leq 1$ is equivalent to

$$\|x\|_{F_\alpha} := \sup_{\lambda > 0} \lambda^\alpha \|A R(\lambda, A) x\|_X,$$

cf. [9, Proposition II.5.12].

We give a characterization of the property that B maps into extrapolation spaces of the form $X_{-\beta}$ for $\beta \in (0, 1)$ by admissibility of B, providing a counterpart to Lemma 2.1 above. In this context we note that similar results have appeared in [36].

Lemma 2.4 *Let A generate an analytic semigroup T on X. Let $\beta \in (0, 1)$, and assume that for all $p > \frac{1}{1-\beta}$ the resolvent condition in the p-Weiss condition holds for A and a given $B \in \mathcal{L}(U, X_{-1})$, i.e.,*

$$\exists \alpha > \omega(T): \quad \sup_{\operatorname{Re}\lambda > \alpha} (\operatorname{Re}\lambda)^{1/p} \|R(\lambda, A) B\|_{\mathcal{L}(U,X)} < \infty. \tag{6}$$

Then B is bounded as a mapping $B: U \to X_{-\alpha}$ for all $\alpha > \beta$.

Proof Again we assume without loss of generality that T is exponentially stable. For a fixed $q > \frac{1}{1-\beta}$, we have by the resolvent condition (6) that

$$\|R(\lambda, A) B u_0\|_X \leq K (\operatorname{Re}\lambda)^{-1/q} \|u_0\|_U, \qquad \operatorname{Re}\lambda > 0.$$

Using the extension A_{-1} of the generator A, this implies

$$(\operatorname{Re}\lambda)^{1/q} \|A R(\lambda, A) B u_0\|_{X_{-1}} \leq K \|u_0\|_U \tag{7}$$

for all $\lambda \in \mathbb{C}$ with $\mathrm{Re}\,\lambda > 0$. By exponential stability of T and $0 < \gamma < 1$, Item 2 in Remark 2.3 implies that we can consider the norm

$$\|\|x\|\|_{F_{-\gamma}} = \sup_{\lambda > 0} \lambda^{1-\gamma} \|AR(\lambda, A)x\|_{X_{-1}}, \qquad x \in X_{-1},$$

which is equivalent to the usual norm on $F_{-\gamma} \subset X_{-1}$. Recall that for negative parameters the Favard space was defined in terms of the extended generator on X_{-1}. Using (7), we get

$$\|\|Bu_0\|\|_{F_{-(1-1/q)}} = \sup_{\lambda > 0} \lambda^{1-(1-1/q)} \|AR(\lambda, A)Bu_0\|_{X_{-1}} \leq K \|u_0\|_U.$$

Relating this to abstract Sobolev spaces, notice that for any choice of $\alpha \in (\beta, 1]$ there exists a $q > 1$ such that $\alpha > 1 - \frac{1}{q} > \beta$. In fact, a possible choice is $q = (1 - \frac{\alpha+\beta}{2})^{-1}$. Embedding relations between Favard and abstract Sobolev spaces from Remark 2.3 then imply $F_{-(1-1/q)} \hookrightarrow X_{-\alpha}$. As a consequence, we also have

$$\|Bu_0\|_{X_{-\alpha}} \leq K_1 \|Bu_0\|_{F_{-(1-1/q)}} \leq K_2 \|u_0\|_U$$

for all $\alpha > \beta$. Hence we have shown that $B \in \mathcal{L}(U, X_{-\alpha})$. $\qquad\square$

Note that Lemma 2.4 is in particular applicable if B is even infinite-time p-admissible for all $p > \frac{1}{1-\beta}$. Indeed, specifying the input function $u = e^{-\lambda(\cdot)}u_0$ for $u_0 \in U$ and using the representation of the resolvent as the Laplace transform of the semigroup, the definition of infinite-time p-admissibility yields

$$\|R(\lambda, A)Bu_0\|_X \leq K (\mathrm{Re}\,\lambda)^{-1/p} \|u_0\|_U, \qquad \lambda \in \rho(A).$$

The next theorem combines Lemmas 2.1 and 2.4 into a characterization of p-admissibility of B for p greater than a threshold in terms of boundedness of B into spaces of type $X_{-\alpha}$.

Theorem 2.5 *Let A be the generator of an analytic and exponentially stable semigroup T on X, and let $B \in \mathcal{L}(U, X_{-1})$ be a control operator. Then the following are equivalent for $\beta \in (0, 1)$:*

(a) B is finite-time L^p-admissible for all $p > \frac{1}{1-\beta}$.
(b) B is a bounded map into $X_{-\alpha}$ for all $\alpha > \beta$.

Proof If $B \in \mathcal{L}(U, X_{-1})$ is L^p-admissible for all $p > \frac{1}{1-\beta}$, then B is also infinite-time L^p-admissible for all $p > \frac{1}{1-\beta}$ due to the assumption of exponential stability. Since in this case the resolvent condition (6) holds for all $p > \frac{1}{1-\beta}$, we have $B \in \mathcal{L}(U, X_{-\alpha})$ for $\alpha > \beta$ by Lemma 2.4. The implication (b) $\Rightarrow$ (a) directly follows from an application of Lemma 2.1. $\qquad\square$

The statement in the previous theorem does not hold in the endpoint case $p = \frac{1}{1-\beta}$ and $\alpha = \beta$, as the following result shows.

Proposition 2.6 *For any $p \in (1, \infty)$, there exist an analytic semigroup generator on $X = \ell^2$ and a p-admissible $B : \mathbb{C} \to X_{-1}$ that is not in $\mathcal{L}(\mathbb{C}, X_{-1/p'})$.*

Proof We adapt an example given in [46, Example 5.3.11] to $p \neq 2$. In fact, we let $1 < p < \infty$ be arbitrary. As in [46, Example 5.3.11], we consider the semigroup generated by the diagonal operator A, that is, $Ae_k = \lambda_k e_k$, $k \in \mathbb{N}$, with $\{e_k\}_{k \in \mathbb{N}}$ referring to the canonical basis and with eigenvalue sequence defined by $\lambda_k = -2^k$ for $k \in \mathbb{N}$.

Due to the identification of $\mathcal{L}(\mathbb{C}, X_{-1})$ with the space X_{-1} itself, the claim amounts to the existence of a sequence $b \in X_{-1}$ such that b is p-admissible but fails to be an element of $X_{-1/p'}$. We consider the cases $1 < p \leq 2$ and $2 < p < \infty$ separately and use the appropriate Carleson measure criteria from [31, Theorem 3.2, Theorem 3.5]; see also Theorem 3.2 below.

Let $1 < p \leq 2$, and set $b = \{2^{k/p'}\}_{k \in \mathbb{N}}$. This choice of b represents an element of X_{-1} as

$$\sum_{k=1}^{\infty} \frac{|b_k|^2}{|\lambda_k|^2} = \sum_{k=1}^{\infty} 2^{2k/p'} 2^{-2k} = \sum_{k=1}^{\infty} 2^{-2k/p} < \infty.$$

To prove p-admissibility using the Carleson measure criterion, we need to show that there exists $K > 0$ such that $\mu(Q_I) \leq K|I|^{2/p'}$, where $\mu = \sum_k |b_k|^2 \delta_{-\lambda_k}$ and Q_I is the Carleson square defined for an interval I on the imaginary axis symmetric about 0 (an interval of the form $i[-a, a]$ for some $a > 0$) by

$$Q_I := \{z \in \mathbb{C} : i\,\mathrm{Im}\,z \in I, 0 < \mathrm{Re}\,z < |I|\},$$

see also Fig. 1.

We therefore proceed by showing that there exists $K > 0$ with

$$\sum_{k \in \mathbb{N}} 2^{2k/p'} \delta_{2^k}((0, |I|)) \leq K|I|^{2/p'}$$

for all choices of I as above. Since $|I| \geq 2^{\lfloor \log_2 |I| \rfloor}$, we set $m = \lfloor \log_2 |I| \rfloor$ and get

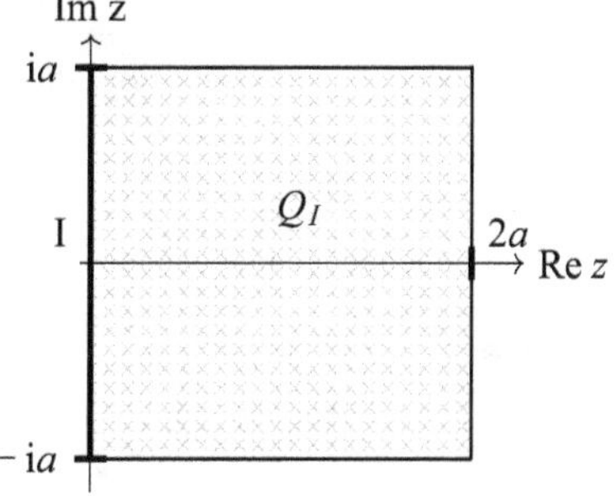

Fig. 1 The Carleson square Q_I for the interval $I = i[-a, a]$

$$\sum_{k\in\mathbb{N}} 2^{2k/p'} \delta_{2^k}((0,|I|)) = \sum_{k=1}^{m} (2^{2/p'})^k = \frac{2^{2/p'}}{2^{2/p'}-1}((2^m)^{2/p'}-1)$$

$$\leq K_{p'}|I|^{2/p'},$$

which implies p-admissibility of b.

Now, if b were in $X_{-1/p'}$, then we would have $(-A)^{-1/p'}b \in X$. But due to $((-A)^{-1/p'}b)_k \equiv 1$, this condition is violated, and it follows that $b \notin \mathcal{L}(\mathbb{C}, X_{-1/p'})$.

In the case $p > 2$, we consider the same semigroup, but now we let b be the control operator represented by $b_k = k^{-1/2}2^{k/p'}$, $k \in \mathbb{N}$. We have $b \in X_{-1}$ since

$$\sum_{k=1}^{\infty} \frac{|b_k|^2}{|\lambda_k|^2} = \sum_{k=1}^{\infty} k^{-1}\, 2^{2k/p'}\, 2^{-2k} \leq \sum_{k=1}^{\infty} 2^{-2k/p} < \infty.$$

In this case, the Carleson measure criterion reads

$$\{2^{-2n/p'}\mu(S_n)\}_{n\in\mathbb{N}} \in \ell^{p/(p-2)}(\mathbb{N}),$$

where $\mu = \sum_{k\in\mathbb{N}} |b_k|^2 \delta_{-\lambda_k}$ and $S_n = \{z \in \mathbb{C} : \operatorname{Re} z \in (2^{n-1}, 2^n]\}$.

For this choice of semigroup and control operator, the condition means that

$$\sum_{k=1}^{\infty} \left|k^{-1/2}\, 2^{k/p'}\right|^{\frac{2p}{p-2}} \cdot 2^{-\frac{2k}{p'}\frac{p}{p-2}} < \infty,$$

which holds true here as

$$\sum_{k=1}^{\infty} \left|k^{-1/2}\, 2^{k/p'}\right|^{\frac{2p}{p-2}} \cdot 2^{-\frac{2k}{p'}\frac{p}{p-2}} = \sum_{k=1}^{\infty} k^{-\frac{p}{p-2}} < \infty.$$

However,

$$\|(-A)^{-1/p'}b\|_{\ell^2} = \sum_{k=1}^{\infty} 2^{-2k/p'}\, |k^{-1/2}\, 2^{k/p'}|^2 = \sum_{k=1}^{\infty} k^{-1}$$

with a divergent sum, thereby showing that b is unbounded as an operator $\mathbb{C} \to X_{-1/p'}$. $\qquad\square$

2.2 Example: Admissibility for the Heat Equation via Interpolation

As indicated before, we are interested in L^p-admissibility for the heat equation on the spatial domain $[0, 1]$ with Dirichlet control at the boundary point 1, given through the system's equations

$$\begin{cases} x_t = x_{\xi\xi} & \text{in } (0, 1) \times (0, \infty), \\ x = 0 & \text{on } \{\xi = 0\} \times (0, \infty), \\ x = u & \text{on } \{\xi = 1\} \times (0, \infty), \\ x = x_0 & \text{on } (0, 1) \times \{t = 0\}, \end{cases}$$

with initial heat distribution given by x_0 and heat injection with magnitude $u(t)$ at the point 1.

In this simple example, all properties of the system can basically be derived from the solution formulae given through the spectral decomposition. Yet, it serves as the most natural and well known example where 2-admissibility is not satisfied if the state space is chosen to be $X = L^2([0, 1])$, see also [34]. We formulate this as a boundary control system by defining

$$\mathfrak{A}\colon X \supset \mathcal{D}(\mathfrak{A}) \to X, \quad f \mapsto f''$$

on the domain $\mathcal{D}(\mathfrak{A}) = \{f \in H^2([0, 1]) : f(0) = 0\}$. The abstract boundary operator is then given by

$$\mathfrak{B} = \delta_1\colon X \supset \mathcal{D}(\mathfrak{A}) \to U, \quad f \mapsto f(1).$$

The Sobolev embedding theorem (see, e.g., [1, Theorem 4.12]) implies that each H^2 function in spatial dimension $n = 1$ is continuous, ensuring that $\mathfrak{B}$ is well-defined. We want to translate this system into state space form to make statements about admissibility. Define the operator A by

$$A = \mathfrak{A}|_{\ker \mathfrak{B}}\colon \mathcal{D}(A) \to X, \quad f \mapsto f''$$

on the domain $\mathcal{D}(A) = \{f \in H^2([0, 1]) : f(0) = f(1) = 0\}$. Using the identity (5)—as both X and U are Hilbert spaces—one integrates by parts twice to find that $\mathfrak{B}x \cdot B^*y = -x(1)\, y'(1)$ for all $x \in \mathcal{D}(\mathfrak{A})$ and $y \in \mathcal{D}(A)$. Thus we can identify that B is given by

$$B = \delta_1'\colon \mathbb{C} \to X_{-1}, \quad z \mapsto z\delta_1',$$

where $\delta_1'(f) = -f'(1)$.

To proceed, we consider the orthonormal basis $\{e_n : n \in \mathbb{N}\}$ with $e_n(\xi) = \sqrt{2}\sin(n\pi\xi)$. Then the diagonal representation $Ae_n = \lambda_n e_n$ holds with $\lambda_n = -n^2\pi^2$. The control operator B can be represented by the sequence $\{b_n\}_{n\in\mathbb{N}}$ for $b_n = (-1)^n\sqrt{2}n\pi$ as in [27, Example 5.1].

Using the representation of the system in diagonal form, we show that $b = \{b_n\}_{n\in\mathbb{N}}$ is an element of the interpolation space $X_{-\beta}$ for $\beta > 3/4$. By Lemma 2.1, this then will imply that the control operator is p-admissible for all $p > \frac{1}{1-3/4} = 4$. It remains to show that

$$\sum_{n=1}^{\infty} \frac{|b_n|^2}{|\lambda_n|^{2\beta}} < \infty$$

for all $\beta > \frac{3}{4}$. We have

$$\sum_{n\in\mathbb{N}} \frac{|b_n|^2}{|\lambda_n|^{2\beta}} = \sum_{n\in\mathbb{N}} \frac{2n^2\pi^2}{\pi^{4\beta}n^{4\beta}} = \frac{2}{\pi^{4\beta-2}} \sum_{k\in\mathbb{N}} \frac{1}{n^{4\beta-2}},$$

which is finite for $\beta > \frac{3}{4}$. Invoking Lemma 2.1 then implies p-admissibility for all $p > 4$.

Remark 2.7 One can also use the interpolation space argument to show that the Dirichlet controlled heat equation is admissible for $p > 4$ in multiple spatial dimensions. To show this, we follow [41] and fix a dimension $n \geq 2$ and a domain $\Omega \subset \mathbb{R}^n$ with sufficiently smooth boundary $\partial\Omega$. The heat equation with Dirichlet boundary control is represented by the system

$$\begin{cases} \dot{x}(t) = \Delta x(t) \\[4pt] \gamma_0 x = u, \\[4pt] x(0) = x_0 \in X \end{cases}$$

on $X = L^2(\Omega)$ with input space $U = L^2(\partial\Omega)$, where γ_0 is the Dirichlet trace operator on $\partial\Omega$. Relating the above to abstract boundary control systems, we set

$$\mathfrak{A} = \Delta, \qquad \mathcal{D}(\mathfrak{A}) = H^2(\Omega) + DL^2(\partial\Omega), \qquad \mathfrak{B} = \gamma_0,$$

where $D \in \mathcal{L}(L^2(\partial\Omega), L^2(\Omega))$ denotes the Dirichlet map which maps $g \in L^2(\partial\Omega)$ to the unique solution $h \in L^2(\Omega)$ of

$$\Delta h = 0, \qquad \gamma_0 h = g;$$

see [46, Proposition 10.6.6]. Since by [46, Proposition 10.6.4] the composition $\gamma_0 D$ yields the identity on $L^2(\partial\Omega)$, we have $\ker\mathfrak{B} = H^2(\Omega) \cap H_0^1(\Omega)$. The general rewriting procedure for boundary control systems leads to the self-adjoint operator

$$A = \mathfrak{A}|_{\ker \mathfrak{B}} = \Delta, \qquad \mathcal{D}(A) = H^2(\Omega) \cap H^1_0(\Omega).$$

As X and U are Hilbert spaces, we use the formula (5) and Green's first identity to deduce that

$$\langle \mathfrak{A}x, y \rangle_X - \langle x, Ay \rangle_X = \langle \mathfrak{B}x, B^*y \rangle_U = -\langle \mathfrak{B}x, \tfrac{\partial y}{\partial v} \rangle_U,$$

whereupon we find that B^*y is the normal derivative operator $\frac{\partial y}{\partial v}$ on $\partial\Omega$. By [34, Lemma 3.1.1], B^* is bounded $H^\beta(\Omega) \cap H^1_0(\Omega) \to L^2(\partial\Omega)$ for all $\beta > \frac{3}{2}$. Hence, for all $\beta > 3/4$, this yields

$$B^* \in \mathcal{L}(X_\beta, U^*) = \mathcal{L}(H^{2\beta}(\Omega) \cap H^1_0(\Omega), L^2(\partial\Omega)), \qquad B^*: f \mapsto \frac{\partial f}{\partial v}\Big|_{\partial\Omega}.$$

By duality—as $X = L^2(\Omega)$ is reflexive—we conclude that $(-A)^{-\beta}B$ is bounded as an operator $U \to X$ for all $\beta > 3/4$. Lemma 2.1 then implies that B is p-admissible for all $p > 4$.

Remark 2.8 (Notes on Neumann Control) Although it does not appear using the notion of *admissibility* explicitly, we remark that the heat equation on the state space $X = L^2(\Omega)$ with Neumann control acting via the boundary—and inputs in $U = L^2(\partial\Omega)$—is featured in the book by Lasiecka and Triggiani [34, Remark 3.3.1.4] (and the result is folklore). There the control operator B is computed and shown to satisfy $(-A)^{-(1/4+\varepsilon)}B \in \mathcal{L}(U, X)$ for any $\varepsilon > 0$. This allows the application of the interpolation space argument as in Sect. 2.1.

Furthermore, p-admissibility of the Neumann controlled heat equation is characterized by an embedding argument due to Haak and Kunstmann, see [19, Proposition 2.4]. We also refer to Byrnes, Gilliam, Shubov, and Weiss [5] for the case $p = 2$. Denoting the Neumann Laplacian on $\Omega \subset \mathbb{R}^n$ by Δ_N, it follows from Haak and Kunstmann's results that the control operator $B = \gamma'_0$ with input space $L^2(\partial\Omega)$—resulting from the Neumann controlled heat equation on $X = L^2(\Omega)$— is p-admissible for $A = -\Delta_N + I$ if and only if $p \geq \frac{4}{3}$. This is a special case of a more general result they obtain in the context of weighted admissibility on state spaces such as $L^q(\Omega)$. In the unweighted case that we consider here, the result yields p-admissibility whenever U embeds into the Besov space $B^{2/p-1-1/q}_{q,\infty}(\partial\Omega)$. We employ the Besov space embedding

$$B^0_{2,2} \hookrightarrow B^{-s}_{2,\infty}, \qquad s \geq 0,$$

see, e.g., [45, Section 2.3.2, Remark 2] or [40, Section 2.3], such that in the case of the Neumann controlled heat equation on $X = L^2(\Omega)$ with $U = L^2(\partial\Omega)$ one arrives at the condition

$$U = \mathrm{B}^0_{2,2}(\partial\Omega) \hookrightarrow \mathrm{B}^{2/p-3/2}_{2,\infty}(\partial\Omega),$$

which is satisfied if and only if $p \geq \frac{4}{3}$, as claimed.

3 Characterizations for Multiplier-Like Systems

We are interested in testing p-admissibility also in cases where the system is not necessarily of diagonal form. In this section, we formulate extensions of Laplace–Carleson type results to normal semigroups and multiplication semigroups. For this, recall the following result due to Jacob, Partington, and Pott for diagonal semigroups.

Theorem 3.1 ([31, Theorem 2.1]) *Let A generate a diagonal semigroup on $X = \ell^q(\mathbb{Z})^1$ with eigenvalues given by $\{\lambda_k\}_{k\in\mathbb{Z}}$. Let $p \in [1, \infty)$. Then $B = \{b_k\}_{k\in\mathbb{Z}} \in \mathcal{L}(\mathbb{C}, X_{-1})$ is infinite-time L^p-admissible if and only if the Laplace transform*

$$\mathscr{L}: \mathrm{L}^p([0, \infty)) \to \mathrm{L}^q(\mathbb{C}_+, \mu), \quad f \mapsto \left(z \mapsto \int_0^\infty \mathrm{e}^{-zt} f(t)\, \mathrm{d}t\right)$$

is a bounded operator for the measure

$$\mu = \sum_{k\in\mathbb{Z}} |b_k|^q \delta_{-\lambda_k}.$$

Note that by using the special form of a diagonal semigroup the proof of the above result is rather direct and the difficulty in assessing admissibility only transfers to checking the boundedness of the Laplace–Carleson embedding. This is where arguments from harmonic analysis, and in particular from the theory of Carleson measures, take effect—at least under further assumptions on the range of p and the location of the eigenvalues (in sectors or strips). The latter situations reflect whether the diagonal semigroup is analytic or even a group. In those cases, admissibility indeed reduces to elementary conditions on certain sequences lying in corresponding ℓ^r-spaces, such as employed in Sects. 2.1 and 3.1.

For convenience, we recall the admissibility results for diagonal semigroups on ℓ^q spaces from [31].

Proposition 3.2 ([31, Theorem 3.2, Theorem 3.5]) *Fix $1 \leq q < \infty$, and assume that $A: \ell^q(\mathbb{Z}) \supset \mathcal{D}(A) \to \ell^q(\mathbb{Z})$ generates a diagonal semigroup on $X = \ell^q(\mathbb{Z})$*

---

[1] More generally, one can assume that the semigroup is diagonal with respect to a q-Riesz basis on an arbitrary Banach space X.

with eigenvalues $\{\lambda_k\}_{k\in\mathbb{Z}}$ lying in the left half plane such that $-\lambda_k \in S_\theta$ for some angle $\theta \in (0, \pi/2)$ and all $k \in \mathbb{Z}$. Consider the operator $B \in \mathcal{L}(\mathbb{C}, X_{-1})$ represented by the sequence $\{b_k\}_{k\in\mathbb{N}}$.

If $1 < p \leq q$, then the following statements are equivalent:

1. *B is p-admissible.*
2. *There exists $K > 0$ such that*

$$\mu(Q_I) \leq K|I|^{q/p'}$$

holds for all intervals $I \subset i\mathbb{R}$ symmetric about 0; that is, intervals of the form $i[-a, a]$ for some $a > 0$, where $Q_I := \{z \in \mathbb{C} : i\,\mathrm{Im}\,z \in I, 0 < \mathrm{Re}\,z < |I|\}$.
3. *There exists $K > 0$ such that $\|R(z, A)B\|_X \leq Kz^{-1/p}$ for all $z > 0$.*

Here $Q_I = \{z \in \mathbb{C} : i\,\mathrm{Im}\,z \in I, 0 < \mathrm{Re}\,z < |I|\}$.
If instead $q < p < \infty$, the following are equivalent:

1. *B is p-admissible.*
2. *$\{2^{-nq/p'}\mu(S_n)\}_{k\in\mathbb{Z}} \in \ell^{p/(p-q)}(\mathbb{Z})$.*
3. *$\{2^{n/p}\|R(2^n, A)B\|_{\ell^q}\}_{k\in\mathbb{Z}} \in \ell^{p/(p-q)}(\mathbb{Z})$.*

Here $\mu = \sum_{k\in\mathbb{N}} |b_k|^2\delta_{-\lambda_k}$ and $S_n = \{z \in \mathbb{C} : \mathrm{Re}\,z \in (2^{n-1}, 2^n]\}$.

Thus, in this section we are looking for (more general) measures μ such that the equivalence

$$\mathcal{L} \in \mathcal{L}(L^p([0, \infty)), L^q(\mathbb{C}_+, \mu)) \Leftrightarrow \left\|\int_0^\infty T(s)Bu(s)\,ds\right\|_X \leq C\|u\|_{L^p([0,\infty))}$$

holds for normal semigroups and multiplication semigroups.

3.1 Non-admissibility for a Dirichlet Controlled Heat Equation

As an application of the Laplace–Carleson method from [31], we want to show that the Dirichlet controlled heat equation in one spatial dimension as above is not L^p-admissible for any $p \leq 4$.

It is enough to show the claim for $p = 4$. To show that Dirichlet boundary control does not lead to a 4-admissible control operator for this system, we apply the necessary and sufficient condition given by the Laplace–Carleson method from [31, Theorem 3.5], as above in Proposition 2.6 and Proposition 3.2. Keeping the same notation, i.e., $\mu = \sum_{k\in\mathbb{N}} |b_k|^2\delta_{-\lambda_k}$ and $S_n = \{z \in \mathbb{C} : \mathrm{Re}\,z \in (2^{n-1}, 2^n]\}$, we insert $p = 4$ into the condition, which yields

$$B\ 4\text{-admissible} \Leftrightarrow s := \{2^{-3n/2}\mu(S_n)\}_{n\in\mathbb{Z}} \in \ell^2(\mathbb{Z}).$$

In showing that s is not square-summable, we use the following expression for $\mu(S_n)$: Recall that $\lambda_k = -k^2\pi^2$ [27, Example 5.1]. For $n \in \mathbb{Z}$, we have

$$2^{-2n/p'}\mu(S_n) = 2^{-2n/p'}\sum_{k\in\mathbb{N}}|\sqrt{2}k\pi|^2\delta_{k^2\pi^2}(\{z \in \mathbb{C} : \mathrm{Re}\, z \in (2^{n-1}, 2^n]\})$$

$$= 2^{1-2n/p'}\pi^2\sum_{k\in\mathbb{N}\cap\left(\frac{\sqrt{2^{n-1}}}{\pi},\frac{\sqrt{2^n}}{\pi}\right]}k^2. \tag{8}$$

Combining (8) with the fact that $k \mapsto k^2$ is an increasing function on $[0, \infty)$, we estimate

$$\mu(S_n) = 2\pi^2\sum_{k\in\mathbb{N}\cap\left(\frac{\sqrt{2^{n-1}}}{\pi},\frac{\sqrt{2^n}}{\pi}\right]}k^2 \geq 2\pi^2\int_{\lfloor 2^{(n-1)/2}/\pi\rfloor}^{\lfloor 2^{n/2}/\pi\rfloor}k^2\,dk = \frac{2\pi^2}{3}\left(\left\lfloor\frac{2^{n/2}}{\pi}\right\rfloor^3 - \left\lfloor\frac{2^{(n-1)/2}}{\pi}\right\rfloor^3\right).$$

Then we get

$$\mu(S_n) \geq \frac{2\pi^2}{3}\left(\left\lfloor\frac{2^{n/2}}{\pi}\right\rfloor^3 - \left\lfloor\frac{2^{(n-1)/2}}{\pi}\right\rfloor^3\right),$$

which, in combination with the estimate $\lfloor b\rfloor^3 - \lfloor a\rfloor^3 \geq (b-1)^3 - a^3$, gives

$$\mu(S_n) \geq \frac{2\pi^2}{3}\left(\frac{(1 - 2^{-3/2})}{\pi^3}2^{3n/2} - \frac{3}{\pi^2}2^{2n/2} + \frac{3}{\pi}2^{n/2} - 1\right).$$

Multiplying both sides with the prefactor $2^{-3n/2}$, we have

$$2^{-3n/2}\mu(S_n) \geq \frac{2\pi^2}{3}\left(\frac{(1 - 2^{-3/2})}{\pi^3} - \frac{3}{\pi^2}2^{-n/2} + \frac{3}{\pi}2^{-2n/2} - 2^{-3n/2}\right),$$

where the expression on the right-hand side does not converge to zero in the limit as $n \to \infty$. Therefore the sequence s is not square-summable, and consequently B is not 4-admissible.

3.2 Normal Semigroups

Proposition 3.3 (Laplace–Carleson Embeddings for Normal Semigroups) *Let A generate a strongly continuous semigroup of normal operators on a Hilbert space*

X. Denote the spectral measure of A by E and the spectral measure of the extension A_{-1} by E_{-1}. Consider a control operator $b \in X_{-1} \simeq \mathcal{L}(\mathbb{C}, X_{-1})$, and define the measure

$$\mu(S) := \int_S (1 + |z|^2)\, d\langle E_{-1}(z)b, b\rangle_{X_{-1}} = (1 + |z|^2)(E_{-1})_{b,b}(S) \tag{9}$$

on Borel subsets S of $\mathbb{C}$.

Then b is L^p-admissible for T if and only if the Laplace transform is a bounded operator when considered as a map

$$\mathcal{L} : L^p([0, \infty)) \to L^2(\sigma(-A), \mu).$$

Proof By the spectral theorem applied to the normal operator A_{-1}, we have the representations

$$(rI - A_{-1}) = \int_{\sigma(-A)} (r - z)\, dE_{-1}(z) \quad (r \in \mathbb{C}), \qquad T(t) = \int_{\sigma(-A)} e^{-tz}\, dE_{-1}(z).$$

Let u be a locally integrable complex-valued function on $[0, \infty)$. To prove the claim, we show that

$$\|\mathcal{L}u\|_{L^2(\sigma(-A),\mu)} \simeq \left\| \int_0^\infty T(s)bu(s)\, ds \right\|_X,$$

where $a \simeq b$ means there exist $C_1, C_2 > 0$ such that $a \le C_1 b$ and $b \le C_2 a$.

To begin the proof, we expand the expression on the right-hand side and get

$$\left\| \int_0^\infty T(s)bu(s)\, ds \right\|_X^2 = \left\| \int_{\sigma(-A)} \int_0^\infty e^{-sz} u(s)\, ds\, dE_{-1}(z)b \right\|_X^2$$

$$= \left\| (rI - A_{-1}) \int_{\sigma(-A)} \int_0^\infty e^{sz} u(s)\, ds\, dE_{-1}(z)b \right\|_{X_{-1}}^2$$

$$= \left\| \int_{\sigma(-A)} (r - z) \int_0^\infty e^{sz} u(s)\, ds\, dE_{-1}(z)b \right\|_{X_{-1}}^2$$

$$= \left\| \int_{\sigma(-A)} (r - z) \cdot (\mathcal{L}u)(z)\, dE_{-1}(z)b \right\|_{X_{-1}}^2$$

with $r \in \rho(A)$ being the element of $\rho(A)$ that was implicitly used to define the space X_{-1}.

Given a measurable function $f : \sigma(A) \to \mathbb{C}$, one has the following norm representation:

$$\left\|\left(\int_{\sigma(-A)} f \, dE_{-1}\right)b\right\|_{X_{-1}}^2 = \int_{\sigma(-A)} |f|^2 \, d(E_{-1})_{b,b},$$

see, e.g., [6, Theorem 4.7]. Specifying the function $z \mapsto (r - z) \cdot (\mathscr{L}u)(z)$ as the function f gives

$$\left\|\int_{\sigma(-A)} (r - z) \cdot (\mathscr{L}u)(z) \, dE_{-1}(z)b\right\|_{X_{-1}}^2$$

$$= \int_{\sigma(-A)} |(r - z) \cdot (\mathscr{L}u)(z)|^2 \, d\langle E_{-1}(z)b, b\rangle_{X_{-1}}$$

$$= \int_{\sigma(-A)} |(\mathscr{L}u)(z)|^2 |r - z|^2 \, d(E_{-1})_{b,b}(z)$$

$$\simeq \int_{\sigma(-A)} |(\mathscr{L}u)(z)|^2 |r - z|^2 \frac{1 + |z|^2}{|r - z|^2} \, d(E_{-1})_{b,b}(z)$$

$$= \|\mathscr{L}u\|_{L^2(\sigma(-A),\mu)}^2,$$

where we have used that $r \in \rho(A)$ implies that

$$\frac{1 + |z|^2}{|r - z|^2} \simeq C$$

on $\sigma(A)$ for some $C > 0$. Then we also have

$$\left\|\int_0^\infty T(s)bu(s) \, ds\right\|_X^2 \simeq \|\mathscr{L}u\|_{L^2(\sigma(-A),\mu)}^2.$$

Hence p-admissibility of b is equivalent to boundedness of the Laplace transform.
□

Remark 3.4 We note that in the literature the specific case of normal operators with discrete spectrum is treated in full detail, whereas the general case is often argued to follow similarly as for diagonal semigroups; cf., e.g., [21, 42, 52] with the exception of [48, Theorem 19], where an approximation argument is used.

More precisely, in these references the measure μ is chosen to be $\langle E(\cdot)b, b\rangle$, which is, in fact, only well-defined on bounded Borel sets for $b \in X_{-1}$ (see also the proof of Theorem 19 in [48]). In Proposition 3.3 we circumvent this technical subtlety by employing a slightly different definition of μ, which however coincides with the abovementioned choice for diagonal semigroups.

3.3 Conditions for Multiplication Semigroups

Comparable results to the ones in the previous subsection for normal semigroups on Hilbert spaces hold for semigroups generated by multiplication operators on L^q spaces.

Let (Ω, μ) be a σ-finite measure space. Given a measurable function $a \colon \Omega \to \mathbb{C}$, we let the multiplier M_a on $L^q(\Omega)$, $1 < q < \infty$, be defined by

$$g \mapsto M_a g = ag, \qquad g \in \mathcal{D}(M_f) = \{g \in L^q(\Omega) : ag \in L^q(\Omega)\}.$$

For multiplication semigroups, the extrapolation and interpolation spaces have convenient forms, see, e.g., [9, Example II.5.7]. Suppose that a is complex-valued with $\operatorname{ess\,sup}_\Omega \operatorname{Re} a < 0$. The semigroup generated by the multiplication operator M_a on X is given by $(T(t))_{t \geq 0}$, where $T(t)f = e^{ta}f$. Moreover, for $n \in \mathbb{Z}$, the abstract Sobolev spaces X_n are given by the weighted Lebesgue spaces $X_n = L^q(\Omega, |a|^{nq}\mu)$.

Lemma 3.5 *Let $1 < q < \infty$, and assume that the multiplier M_a generates a strongly continuous semigroup T on $L^q(\Omega, \mu)$ with $\operatorname{Re} a < 0$ μ-a.e. on Ω. Then we have that*

$$M_{|\operatorname{Re} a|^{-1/q}} b \in L^q(\Omega, \mu) \text{ implies that } b \text{ is } L^{q'}\text{-admissible for } T.$$

Proof By the functional calculus for multiplication operators, the semigroup T is given by $(T(t)f)(s) = e^{sa(t)}f(s)$. Therefore we need to show that there exists a constant $M > 0$ such that

$$\left\| \int_0^\infty e^{sa} b u(s)\, ds \right\|_{L^q(\Omega,\mu)} \leq M \|u\|_{L^{q'}([0,\infty))}$$

holds for all $u \in L^{q'}([0, \infty))$.

Since

$$\left\| \int_0^\infty e^{sa} b u(s)\, ds \right\|_{L^q(\Omega,\mu)}^q = \int_\Omega \left| \int_0^\infty e^{sa(t)} b(t) u(s)\, ds \right|^q d\mu(t),$$

we can use Hölder's inequality to get

$$\int_\Omega \left| \int_0^\infty e^{sa(t)} b(t) u(s)\, ds \right|^q d\mu(t)$$

$$\leq \int_\Omega \left| \left(\int_0^\infty |e^{sa(t)} b(t)|^q\, ds \right)^{1/q} \left(\int_0^\infty |u(s)|^{q'}\, ds \right)^{1/q'} \right|^q d\mu(t)$$

$$= \|u\|^{q}_{L^{q'}([0,\infty))} \int_{\Omega} |b(t)|^{q} \int_{0}^{\infty} |e^{sa(t)}|^{q} \, ds \, d\mu(t)$$

$$= \|u\|^{q}_{L^{q'}([0,\infty))} \int_{\Omega} |b(t)|^{q} \int_{0}^{\infty} e^{qs\,\mathrm{Re}(a(t))} \, ds \, d\mu(t)$$

$$= C(q)\|u\|^{q}_{L^{q'}([0,\infty))} \int_{\Omega} |b|^{q} \, \frac{1}{|\mathrm{Re}\,a|} \, d\mu.$$

This implies that b is q'-admissible for T if $b \in L^{q}(\Omega, |\mathrm{Re}\,a|^{-1}\mu)$. $\square$

Lemma 3.6 *Fix $1 < q < \infty$. Given a measurable complex-valued function a on some σ-finite measure space (Ω, μ) such that $\mathrm{Re}\,a$ is essentially bounded from above, let M_a be the corresponding multiplier on $L^{q}(\Omega, \mu)$. Then $b \in \mathcal{L}(\mathbb{C}, X_{-1}) \simeq X_{-1} = L^{q}(\Omega, |a|^{-1}\mu)$ is $L^{q'}$-admissible for T if and only if the Laplace transform induces a bounded map*

$$\mathcal{L} \colon L^{q'}([0,\infty)) \to L^{q}(\Omega, \nu), \quad u \mapsto \mathcal{L}u = \int_{0}^{\infty} e^{-t(\cdot)} u(t) \, dt,$$

where the measure ν is defined by $\nu := (-a)_{*}|b|^{q}\mu$.

Proof We have

$$\left\| \int_{0}^{\infty} T(s)bu(s) \, ds \right\|^{q}_{L^{q}(\Omega,\mu)} = \int_{\Omega} |b(t)|^{q} \left| \int_{0}^{\infty} e^{sa(t)}u(s) \, ds \right|^{q} d\mu(t)$$

$$= \int_{\Omega} |b(t)|^{q} |(\mathcal{L}u)(-a(t))|^{q} \, d\mu(t)$$

$$= \int_{\Omega} |(\mathcal{L}u) \circ (-a)|^{q} \, d(|b|^{q}\mu)$$

$$= \|(\mathcal{L}u)\|^{q}_{L^{q}(\Omega,(-a)_{*}|b|^{q}\mu)};$$

thus b is $L^{q'}$-admissible if and only if

$$\|(\mathcal{L}u)\|_{L^{q}(\Omega,\nu)} \leq \|u\|_{L^{q'}([0,\infty))},$$

which is what we wanted to show. $\square$

Due to invariance of admissibility under similarity transformations, this result extends to generators that are similar to a multiplier or unitarily equivalent to a multiplier and thus applies in particular to normal operators on a Hilbert space by a version of the spectral theorem for unbounded operators. Furthermore, the diagonal case follows as a special case:

Example 3.7 (Diagonal Semigroups) Let A generate a diagonal semigroup on the state space $X = \ell^{q}(\mathbb{N}) = L^{q}(\mathbb{N}, \xi)$, where $\xi = \sum_{k} \delta_{k}$ is the counting measure on $\mathbb{N}$.

Componentwise, this means that $(Ax)_k = \lambda_k x_k$ holds for all $k \in \mathbb{N}$ and all $x \in X$. Interpreting A as a multiplication operator on X with symbol $a \colon \mathbb{N} \to \mathbb{C}$, $k \mapsto \lambda_k$, we also identify the control operator $b \in X_{-1}$ with a function $b \colon \mathbb{N} \to \mathbb{C}$, $k \mapsto b_k$. By Lemma 3.6, the input element b is admissible for T if and only if the Laplace transform is a bounded map $\mathscr{L} \colon L^p([0, \infty)) \to L^q(\Omega, \nu)$ with measure ν given by

$$\nu = (-a)_* |b|^q \xi = |b|^q \sum_k \delta_{-\lambda_k} = \sum_k |b_k|^q \delta_{-\lambda_k},$$

which is the same measure as μ in Theorems 3.2 and 3.5 in [31].

4 Examples with Generators Having Non-discrete Spectra

By considering generators with non-discrete spectrum, we can find examples to illustrate the usage of the Laplace–Carleson type condition in the non-diagonal setting. Candidates are the Dirichlet or Neumann Laplacians on a suitable domain, since their spectra need not be discrete. In passing, we note that the Neumann Laplacian on Ω has purely discrete spectrum if and only if the Rellich–Kondrachov compactness theorem holds for the domain Ω. Hence a possible example would be the Neumann Laplacian on a sufficiently rough domain that fails the Rellich–Kondrachov embedding theorem, see, e.g., [23]. Another operator with non-discrete spectrum, which we will consider in the sequel, is the Laplacian on the whole space $\mathbb{R}^n$.

4.1 The Laplacian on the Full Space

Example 4.1 Let $n \leq 3$, and consider the state space $X = L^2(\mathbb{R}^n)$ and the Laplacian $A = \Delta$ on X with domain $\mathcal{D}(A) = H^2(\mathbb{R}^n)$, which generates a C_0-semigroup on X.

We choose the Dirac delta distribution δ_0 as the control operator b and claim that $b \in X_{-1}$ if and only if $n \leq 3$.

Proof of the Claim By self-adjointness of A, the domains $\mathcal{D}(A)$ and $\mathcal{D}(A^*)$ are identical. Therefore the extrapolation space X_{-1} is given by the dual of $\mathcal{D}(A) = H^2(\mathbb{R}^n)$ with respect to the pivot space $X = L^2(\mathbb{R}^n)$. As the spatial domain is the full space $\mathbb{R}^n$, we have $H^2(\mathbb{R}^n) = H_0^2(\mathbb{R}^n)$. Hence—by definition of Sobolev spaces with negative integer indices—it holds that $X_{-1} = H^{-2}(\mathbb{R}^n)$. In the Fourier domain the space $H^{-2}(\mathbb{R}^n)$ has the representation

$$H^{-2}(\mathbb{R}^n) = \{u \in \mathcal{S}'(\mathbb{R}^n) : (1 + |\cdot|^2)^{-1} \hat{u} \in L^2(\mathbb{R}^n)\},$$

where $\mathcal{S}'(\mathbb{R}^n)$ denotes the space of tempered distributions on $\mathbb{R}^n$. As the Fourier transform of δ_0 computes to $\hat{\delta}_0 \equiv 1$ in the sense of tempered distributions, we have $\delta_0 \in X_{-1}$ if and only if

$$\int_{\mathbb{R}^n} \frac{1}{(1 + |\xi|^2)^2}\, d\xi < \infty.$$

By transforming to spherical coordinates in $\mathbb{R}^n$, this is equivalent to finiteness of

$$\int_0^\infty \frac{r^{n-1}}{(1 + r^2)^2}\, dr,$$

which is the case if and only if $n \leq 3$. $\qquad\qquad\qquad\qquad\qquad\qquad\qquad\square$

The spectral measure $E_{g,g}$ corresponding to the operator A and some $g \in L^2(\mathbb{R}^n)$ is computed in, for example, [44, Equation (7.34)] using the representation of the Laplacian by its Fourier symbol $\xi \mapsto -|\xi|^2$ and is given by the Lebesgue measure λ^n weighted with the density

$$\rho_{g,g}(z) = \chi_{(-\infty,0]}(z)|z|^{\frac{n}{2}-1} \int_{\partial B_1^n(0)} \left|\hat{g}(|z|^{1/2} w)\right|^2 d\mathcal{H}^{n-1}(w);$$

in particular the spectrum of A is purely absolutely continuous and $\sigma(A) = (-\infty, 0]$.

In the above formula $\partial B_1^n(0) := \{y \in \mathbb{R}^n : |y| = 1\}$ denotes the unit sphere in $\mathbb{R}^n$, $\hat{g}$ denotes the Fourier transform of $g \in L^2(\mathbb{R}^n)$, and $\mathcal{H}^{n-1}$ is the $(n-1)$-dimensional Hausdorff measure on $\mathbb{R}^n$. Note that this representation of $\rho_{g,g}$ is also valid in dimension $n = 1$ using the zero-dimensional Hausdorff measure.

For our example of the control operator $b = \delta_0 \in X_{-1}$, we need to consider the extension of $E_{g,g}$ for $g \in H^{-2}(\mathbb{R}^n) = X_{-1}$. For $g \in X_{-1}$, the density $(\rho_{-1})_{g,g}$ is then given by

$$(\rho_{-1})_{g,g}(z)$$

$$= (1 + |z|^2)\chi_{(-\infty,0]}(z)|z|^{\frac{n}{2}-1} \int_{\partial B_1^n(0)} \left|(\mathcal{F}(R(1, A)g))(|z|^{1/2} w)\right|^2 d\mathcal{H}^{n-1}(w).$$

We use $\hat{\delta}_0 \equiv 1$ to get

$$(\rho_{-1})_{\delta_0,\delta_0}(z) = (1 + |z|^2)\chi_{(-\infty,0]}(z)|z|^{\frac{n}{2}-1} \int_{\partial B_1^n(0)} \left|\frac{1}{1 + ||z|^{1/2} w|^2}\right|^2 d\mathcal{H}^{n-1}(w)$$

$$= (1 + |z|^2)\chi_{(-\infty,0]}(z)|z|^{\frac{n}{2}-1} \int_{\partial B_1^n(0)} (1 + |z|)^{-2}\, d\mathcal{H}^{n-1}(w)$$

$$= \frac{1 + |z|^2}{(1 + |z|)^2}\chi_{(-\infty,0]}(z)|z|^{\frac{n}{2}-1} \cdot C(n).$$

Since $\frac{1}{2} \leq \frac{1+|z|^2}{(1+|z|)^2} \leq 1$ holds for $z \in \mathbb{C}$, we can reduce to the density

$$(\rho_{-1})_{\delta_0,\delta_0}(z) = \chi_{(-\infty,0]}(z)|z|^{\frac{n}{2}-1} \tag{10}$$

in applications of Proposition 3.3.

To make statements concerning finite-time p-admissibility of b, we shift the generator of the semigroup such that the resulting semigroup is exponentially stable and apply the test from Proposition 3.3 for infinite-time admissibility.

Concretely, since $0 \in \sigma(A)$, to apply Proposition 3.3 we consider the shifted semigroup generated by the operator $A_1 = A - I$, such that $\sigma(A_1) = \sigma(A) - 1 = (-\infty, -1]$. Then by Proposition 3.3 and using the expression for the density $(\rho_{-1})_{\delta_0,\delta_0}$ from (10), the proof of finite-time L^p-admissibility of b reduces to finding a Laplace transform bound of the form

$$\int_{-\infty}^{-1} \left| \int_0^\infty e^{ts} u(t)\, dt \right|^2 (-s)^{\frac{n}{2}-1}\, ds \leq C \|u\|_{L^p([0,\infty))}^2 \tag{11}$$

for some $C > 0$. Using the Hölder inequality, we have

$$\int_{-\infty}^{-1} \left| \int_0^\infty e^{ts} u(t)\, dt \right|^2 (-s)^{\frac{n}{2}-1}\, ds = \int_1^\infty \left| \int_0^\infty e^{-ty} u(t)\, dt \right|^2 y^{\frac{n}{2}-1}\, dy$$

$$\leq \int_1^\infty \|e^{-(\cdot)y}\|_{L^{p'}([0,\infty))}^2 \|u\|_{L^p([0,\infty))}^2\, y^{\frac{n}{2}-1}\, dy. \tag{12}$$

As $\|e^{-(\cdot)y}\|_{L^{p'}([0,\infty))}^2 = (yp')^{-2/p'}$, the integral above stays bounded if

$$-\frac{2}{p'} + \frac{n}{2} - 1 < -1 \Leftrightarrow p > \frac{4}{4-n}. \tag{13}$$

Since δ_0 only defines an element of X_{-1} for $n \leq 3$, one arrives at the condition $p > 4/3$ in the one-dimensional setting, $p > 2$ in two dimensions, and $p > 4$ for $n = 3$.

4.2 The Laplacian on a Half Line

Example 4.2 We consider the Dirichlet controlled heat equation on the half axis $[0, \infty)$. Set $X = L^2([0, \infty))$, and consider the Laplacian $A = \frac{d^2}{dt^2}$ on the domain $\mathcal{D}(A) = H_0^1([0, \infty)) \cap H^2([0, \infty))$. We select the scalar input space $U = \mathbb{C}$ and choose the control operator $b = \delta_0' \in X_{-1} \simeq \mathcal{L}(\mathbb{C}, X_{-1})$. Our aim is to investigate p-admissibility of b for this example system.

We want to compute the resolvent of A to aid us in finding the spectral measure of A. To that end, for $g \in L^1([0, \infty)) \cap L^2([0, \infty))$, define the odd extension $\tilde{g} \in L^1(\mathbb{R}) \cap L^2(\mathbb{R})$ by

$$\tilde{g}(x) = \begin{cases} g(x) & x \geq 0 \\ -g(-x) & x < 0. \end{cases}$$

Denoting the Fourier variable on the full real axis by η and employing the representation of the resolvent of the Laplacian by its symbol in the Fourier domain, we have

$$R(z, A)g = \mathcal{F}^{-1}\big((z + \eta^2)^{-1}(\mathcal{F}\tilde{g})(\eta)\big)\Big|_{[0,\infty)},$$

which allows for an explicit integral representation of the resolvent. In fact, we have

$$\begin{aligned} R(z, A)g(x) &= \frac{1}{\sqrt{2\pi}}\mathcal{F}^{-1}\left(\frac{1}{z + \eta^2}\int_{-\infty}^{\infty} e^{-iy\eta}\tilde{g}(y)\,dy\right)(x) \\ &= \frac{1}{\sqrt{2\pi}}\mathcal{F}^{-1}\left(\frac{1}{z + \eta^2}\int_{0}^{\infty} (e^{-iy\eta} - e^{iy\eta})g(y)\right)(x) \\ &= \frac{2}{i\sqrt{2\pi}}\mathcal{F}^{-1}\left(\frac{1}{z + \eta^2}\int_{0}^{\infty} \sin(y\eta)g(y)\,dy\right)(x) \\ &= \frac{1}{\pi i}\int_{-\infty}^{\infty} \frac{e^{ix\eta}}{z + \eta^2}\int_{0}^{\infty} \sin(y\eta)g(y)\,dy\,d\eta \\ &= \frac{1}{\pi i}\int_{0}^{\infty} \frac{e^{ix\eta} - e^{-ix\eta}}{z + \eta^2}\int_{0}^{\infty} \sin(y\eta)g(y)\,dy\,d\eta \\ &= \frac{2}{\pi}\int_{0}^{\infty} \frac{\sin(x\eta)}{z + \eta^2}\int_{0}^{\infty} \sin(y\eta)g(y)\,dy\,d\eta. \end{aligned}$$

For brevity, in the sequel we denote the one-sided sine transform by $\mathcal{S}$, i.e.,

$$(\mathcal{S}f)(x) := \int_{0}^{\infty} f(\xi)\sin(x\xi)\,d\xi, \qquad f \in L^1([0, \infty)) \cap L^2([0, \infty)).$$

Still considering $g \in L^1([0, \infty)) \cap L^2([0, \infty))$, we compute

$$\begin{aligned} \langle g, R(z, A)g\rangle_{L^2([0,\infty))} &= \int_{0}^{\infty} g(t)\overline{(R(z, A)g)(t)}\,dt \\ &= \frac{2}{\pi}\int_{0}^{\infty} b(t)\int_{0}^{\infty} \frac{\sin(\tau t)}{\bar{z} + \tau^2}\int_{0}^{\infty} \sin(\tau s)\overline{g(s)}\,ds\,d\tau\,dt \end{aligned}$$

$$= \frac{2}{\pi} \int_0^\infty g(t) \int_0^\infty \frac{\sin(\tau t)}{\overline{z} + \tau^2} (\mathcal{S}\,\overline{g})(\tau)\, d\tau\, dt$$

$$= \frac{2}{\pi} \int_0^\infty \frac{(\mathcal{S}\,\overline{g})(\tau)}{\overline{z} + \tau^2} \int_0^\infty \sin(\tau t) g(t)\, dt\, d\tau$$

$$= \frac{2}{\pi} \int_0^\infty \frac{(\mathcal{S}\,\overline{g})(\tau)}{\overline{z} + \tau^2} (\mathcal{S}g)(\tau)\, d\tau$$

$$= \frac{2}{\pi} \int_0^\infty \frac{|(\mathcal{S}g)(\tau)|^2}{\overline{z} + \tau^2}\, d\tau.$$

We set $\rho = \tau^2$ and get

$$\langle g, R(z, A)g\rangle_{L^2((0,\infty))} = \frac{1}{\pi} \int_0^\infty \frac{|(\mathcal{S}g)(\sqrt{\rho})|^2}{\overline{z} + \rho}\, \frac{d\rho}{\sqrt{\rho}}$$

$$= \frac{1}{\pi} \int_{\mathbb{C}} \chi_{[0,\infty)}(\rho)\, \frac{|(\mathcal{S}g)(\sqrt{|\rho|})|^2}{\overline{z} + \rho}\, \frac{d\rho}{\sqrt{|\rho|}}$$

$$= \frac{1}{\pi} \int_{\mathbb{C}} \chi_{(-\infty,0]}(\rho)\, \frac{|(\mathcal{S}g)(\sqrt{|\rho|})|^2}{\overline{z} - \rho}\, \frac{d\rho}{\sqrt{|\rho|}}$$

$$= \int_{\mathbb{C}} \frac{1}{\overline{z} - \rho}\, dE_{g,g}(\rho),$$

where $E_{g,g}$ is Lebesgue measure weighted with the function $\chi_{(-\infty,0]}(\cdot)|(\mathcal{S}g) (\sqrt{|\cdot|})|^2/(\pi\sqrt{|\cdot|})$. We note that—like the Fourier transform—the sine and cosine transforms extend to operators defined on L^2 functions by density; cf., e.g., [10].

Returning to the control operator $b = \delta_0'$, we note that

$$(\mathcal{S}\delta_0')(t) = \langle \delta_0', \sin((\cdot)t)\rangle = -\frac{d}{d\xi}\sin(t\xi)\big|_{\xi=0} = -t.$$

In the same manner as for the Laplacian on the full Euclidean space as in Sect. 4.1, we can determine the measure μ from Proposition 3.3. This measure μ is formally given by the expression "$\langle E(\cdot)\delta_0', \delta_0'\rangle_X$"—which is not surprising regarding the extension procedure used by Weiss in [48] mentioned in Remark 3.4—giving a measure that is equivalent (in the sense of mutual absolute continuity) to the measure defined by

$$E_{\delta_0',\delta_0'}(z) = \chi_{(-\infty,0]}(z)\, \frac{|(\mathcal{S}\delta_0')(\sqrt{|z|})|^2}{\sqrt{|z|}}\, \lambda^1(z) = \chi_{(-\infty,0]}(z)|z|^{1/2}\lambda^1(z);$$

note also that the associated Radon–Nikodým densities are bounded away from 0 and ∞.

In terms of p-admissibility of b, the computation in (11), (12), and (13) then yields finite-time admissibility of b for all $p > 4$.

Example 4.3 Proceeding in a similar manner as in the prior case of Dirichlet control, we can consider the Neumann controlled heat equation on the nonnegative real axis. On the same state space $X = L^2([0, \infty))$, we now consider the Laplace operator $A = \frac{d^2}{dt^2}$ on the domain $\mathcal{D}(A) = \{f \in H^2([0, \infty)) : f'(0) = 0\}$ and the control operator $b = \delta_0$ on the input space $U = \mathbb{C}$.

Analogously as in the Dirichlet setting, but using the even extension of functions defined on the half line, the resolvent of A computes to

$$(R(z, A)g)(x) = \frac{2}{\pi} \int_0^\infty \frac{\cos(x\eta)}{z + \eta^2} \int_0^\infty \cos(y\eta)g(y)\,dy\,d\eta$$

for $g \in L^1([0, \infty)) \cap L^2([0, \infty))$. For the spectral measure, an analogous calculation as in the Dirichlet case yields

$$E_{g,g}(z) = \chi_{(-\infty,0]}(z)\frac{|(\mathcal{C}x)(\sqrt{|z|})|^2}{\pi\sqrt{|z|}}\lambda^1(z), \qquad g \in L^1([0, \infty)) \cap L^2([0, \infty)).$$

Here $\mathcal{C}$ denotes the one-sided cosine transform defined by $(\mathcal{C}f)(x) := \int_0^\infty f(\xi)\cos(x\xi)\,d\xi$ for $f \in L^1([0, \infty)) \cap L^2([0, \infty))$, and we also denote by $\mathcal{C}$ the extension of the transform to $L^2([0, \infty))$ and to tempered distributions.

As $\mathcal{C}\delta_0 \equiv 1$, the same argument as in the Dirichlet case above lets us determine the measure μ from Proposition 3.3 to be equivalent to

$$E_{\delta_0,\delta_0}(z) = \chi_{(-\infty,0]}(z)|z|^{-1/2}\lambda^1(z).$$

Since this is the same density as in the full space case with $n = 1$ in Example 4.1, the computation in (11), (12), and (13) there yields finite-time L^p-admissibility of $b = \delta_0$ for A for all $p > \frac{4}{3}$.

5 Applications to Infinite-Dimensional Input Spaces

Laplace–Carleson embeddings have been proven to be useful in the context of admissibility with finite-dimensional input spaces. In this section we propose a partial extension to infinite-dimensional input spaces. We note that the question of a full characterization of p-admissibility in terms of these creuteria—without restrictive additional assumptions—seems to be open at this moment in time.

However, imposing additional assumptions on the system, we can extend admissibility results for finite-dimensional input spaces to infinite-dimensional ones. One of such conditions is that A *has the p-Weiss property*.

Definition 5.1 (p-Weiss Property) Let $p \in [1, \infty]$ and A generate a C$_0$-semigroup on the Banach space X. Then we say that the *p-Weiss property for control operators* holds for A if for any Banach space U and any $B \in \mathcal{L}(U, X_{-1})$, we have that

$$B \text{ is infinite-time } p\text{-admissible} \quad \Leftrightarrow \quad \sup_{\operatorname{Re}\lambda>0} (\operatorname{Re}\lambda)^{\frac{1}{p}} \|R(\lambda, A)B\| < \infty.$$

The *p-Weiss property for observation operators* holds if for all $C \in \mathcal{L}(X_1, Y)$, Y Banach space,

$$C \text{ is infinite-time } p\text{-admissible} \quad \Leftrightarrow \quad \sup_{\operatorname{Re}\lambda>0} (\operatorname{Re}\lambda)^{\frac{1}{p'}} \|CR(\lambda, A)\| < \infty.$$

For general semigroup generators A, the p-Weiss property does not hold, as indicated in the introduction; see, e.g., [30, 54, 55] for $p = 2$. In special cases—such as the case of self-adjoint generators A—it does hold; see Proposition 5.4.

Theorem 5.2 (Uniform Weak Admissibility and Weiss Property) *Let A be the generator of a C$_0$-semigroup on the Banach space X. Fix an exponent $q \in (1, \infty)$, and assume that the q-Weiss property holds for A. Then the following are equivalent for $B \in \mathcal{L}(U, X_{-1})$:*

(a) B is L^q-admissible for A.
(b) For all $u \in U$ with $\|u\|_U = 1$, the input element $Bu \in X_{-1}$ is L^q-admissible and the admissibility constants, i.e., the numbers

$$C(u) := \inf\left\{ C>0 : \forall v \in L^q([0, t]) \; \left\| \int_0^t T(t-s)Buv(s)\mathrm{d}s \right\| \leq C\|v\|_{L^q([0,t])} \right\}$$

are uniformly bounded in u.

Proof We note that for any $u \in U$ with norm 1 and any $v \in L^q([0, \infty))$ we have $uv \in L^q([0, \infty), U)$. To show "(a) $\Rightarrow$ (b)," we assume that $B: U \to X_{-1}$ is q-admissible. Then, by admissibility of B, we infer

$$\left\| \int_0^t T(t - s)B(uv)(s)\,\mathrm{d}s \right\|_X \leq C\|uv\|_{L^q([0,t],U)} = C\|v\|_{L^q([0,t])},$$

which implies that $Bu \in X_{-1}$ is q-admissible with admissibility constant that does not depend on u. Hence the admissibility constants are also uniformly bounded in u.

For the other implication, we assume that for all $u \in U$ with unit norm the input element Bu is q-admissible, i.e.,

$$\left\| \int_0^t T(t - s)B(uv)(s)\,\mathrm{d}s \right\|_X \leq C(u)\|v\|_{L^q([0,t])},$$

and that the associated constants are uniformly bounded in u, which is to say we have

$$\sup_{u \in U, \|u\|_U = 1} C(u) < \infty.$$

Hence the mapping $\Phi_{t,Bu} : L^q([0, \infty)) \to X$, $v \mapsto \int_0^t T(s) Bu v(s) \, ds$ allows for the bound $M_t := \sup_{u \in U, \|u\|=1} \|\Phi_{t,Bu}\| < \infty$. Using infinite-time q-admissibility of Bu, we also get a bound of the form $M := \limsup_{t \to \infty} M_t < \infty$. Furthermore, setting $\tilde{v}(s) := (q \operatorname{Re} \lambda)^{1/q} e^{-\lambda s}$, we get $\|\tilde{v}\|_{L^q([0,\infty))} = 1$. Thus $\|\Phi_{t,Bu} \tilde{v}\|_X \le \|\Phi_{t,Bu}\| \|\tilde{v}\|_{L^q([0,\infty))} = \|\Phi_{t,Bu}\| \le M$ and

$$\left\| \int_0^\infty T(s) Bu \tilde{v}(s) \, ds \right\|_X = \lim_{t \to \infty} \|\Phi_{t,Bu} \tilde{v}\|_X \le M.$$

As the resolvent is the Laplace transform of the semigroup, it follows that

$$(\operatorname{Re} \lambda)^{1/q} \|R(\lambda, A) Bu\|_X = (\operatorname{Re} \lambda)^{1/q} \left\| \int_0^\infty e^{-\lambda s} T(s) Bu \, ds \right\|_X$$

$$= \|\Phi_{t,Bu} \tilde{v}\|_X \le M.$$

Due to our assumption, $\|\Phi_{t,Bu}\|$ is uniformly bounded, which implies

$$(\operatorname{Re} \lambda)^{1/q} \|R(\lambda, A) B\| = \sup_{u \in U, \|u\|_U = 1} (\operatorname{Re} \lambda)^{1/q} \|R(\lambda, A) Bu\|_X \le M;$$

taking the supremum among $\lambda \in \mathbb{C}$ with $\operatorname{Re} \lambda > 0$ then gives

$$\sup_{\operatorname{Re} \lambda > 0} (\operatorname{Re} \lambda)^{1/q} \|R(\lambda, A) B\|_X \le M < \infty.$$

Hence B is infinite-time q-admissible by the assumed validity of the q-Weiss property for A. $\qquad \square$

Remark 5.3 Theorem 5.2 allows extending the Laplace–Carleson test for admissibility to infinite-dimensional input spaces. The practicality of this condition is yet to be seen, but this approach allows for assessing admissibility sharply in p. For example, a possible application to the heat equation could involve estimates for the norm of the trace of eigenfunctions on the boundary by means of eigenfunction expansions, as the eigenfunctions of the Laplacian with Dirichlet or Neumann boundary conditions are explicitly known for well-behaved geometries. Moreover, there is a wide array of work on norms of the appropriate traces of eigenfunctions on the boundary, see, e.g., [2, 13, 22, 43, 53].

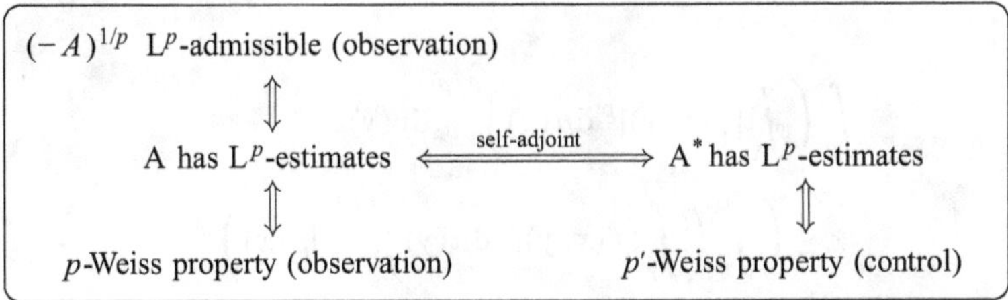

Fig. 2 The p-Weiss property for self-adjoint A, see [14]

In this context we note that the p-Weiss property is strongly related to admissibility of $(-A)^{1/p}$ for A, which was shown by Haak in [14]. If A generates a semigroup of self-adjoint operators and $\sigma(A) \subset (-\infty, 0]$, this even leads to validity of the p-Weiss property for $p \leq 2$ (Fig. 2).

Proposition 5.4 *Let A be a self-adjoint operator on a Hilbert space X with spectrum contained in $(-\infty, a]$ for some $a \in \mathbb{R}$. Then A has the p-Weiss property (for control operators) if $p \leq 2$.*

Proof The case $p = 2$ is contained in [35, Corollary 4.3] as the negative semidefinite self-adjoint operator A is the generator of a C_0-semigroup of contractions on X.

Thus we fix $p < 2$. The claim is that $B \in \mathcal{L}(U, X_{-1})$ is infinite-time L^p-admissible if and only if the quantity $\sup_{\lambda>0} \lambda^{1/p} \|R(\lambda, A)B\|_{\mathcal{L}(U,X)}$ is finite. To prove the claim, we use results from Hosfeld, Jacob, and Schwenninger [25] in conjunction with results from Haak [14].

First we note that due to the assumption of being bounded from above, the self-adjoint A generates a bounded analytic semigroup $(T(t))_{t\geq0}$, see [9, Corollary II.4.7]. We observe that the spectral theorem for unbounded self-adjoint operators allows us to prove p'-admissibility of $(-A)^{1/p'}$ as an observation operator for $p' > 2$ more directly. Indeed, as $-A$ is self-adjoint, there exists a spectral measure E such that

$$-A = \int_{\sigma(-A)} z \, dE(z), \qquad T(t) = \int_{\sigma(-A)} e^{-tz} \, dE(z), \quad t \geq 0,$$

holds. In particular, there is a representation of the norm, i.e.,

$$\int_0^\infty \|(-A)^{1/p'} T(t)x\|_X^{p'} \, dt = \int_0^\infty \left(\int_{\sigma(-A)} |z^{1/p'} e^{-tz}|^2 \, dE_{x,x}(z) \right)^{p'/2} dt;$$

cf. [6, Theorem 4.7]. For σ-finite measure spaces (X, μ) and (Y, ν), a measurable function f defined on $X \times Y$, and $1 \leq p < q < \infty$, Minkowski's integral inequality reads

$$\int_Y \left(\int_X |f(x,y)|^p \, d\mu(x) \right)^{q/p} d\nu(y)$$

$$\leq \left(\int_X \left(\int_Y |f(x,y)|^q \, d\mu(y) \right)^{p/q} d\mu(x) \right)^{q/p},$$

see, e.g., [3, Exercise 3.10.46] or [26, Proposition 1.2.22]. Applying Minkowski's inequality in (5) above yields

$$\int_0^\infty \left(\int_{\sigma(-A)} |z^{1/p'} e^{-tz}|^2 \, dE_{x,x}(z) \right)^{p'/2} dt$$

$$\leq \left(\int_{\sigma(-A)} \left(\int_0^\infty z e^{-tzp'} \, dt \right)^{2/p'} dE_{x,x}(z) \right)^{p'/2}$$

$$= \left(\frac{1}{p'^{2/p'}} \int_{\sigma(-A)} dE_{x,x}(z) \right)^{p'/2}$$

$$= \frac{1}{p'} \|x\|_X^{p'}.$$

Therefore $(-A)^{1/p'}$ is p'-admissible as an observation operator, which is equivalent to requiring that the operator $-A$ satisfies $\mathrm{L}^{p'}$ estimates in the sense of Haak [14, page 29].

Because A was required to be self-adjoint, the same estimates hold for $(-A)^*$. Hence we can apply the generalization of [14, Satz 2.1.6] given in [14, page 29], which states that the p-Weiss property for control operators is valid for A. This ends the proof.   □

Remark 5.5 The last proposition leads us to the observation that there is a noticeable difference between the cases $p < 2$ and $p > 2$ in checking p-admissibility. In the control operator setting with a system on a Hilbert space X, the case $p < 2$ is simpler than $p > 2$. By the previous argument, once we transfer the reasoning from observation operators to control operators—A was self-adjoint—we would need to require $p'/2 \geq 1$ for the Minkowski inequality to hold, which means that we would need $p < 2$.

The complementary case does not allow for such arguments. Below we give an example of a self-adjoint and negative operator that does not have the 4-Weiss property. We note in passing that in the case $p = \infty$, the property of A^* being an L^1-admissible observation operator is equivalent to A being bounded on X, see also [33].

Example 5.6 By means of the example of the one-dimensional heat equation from Sect. 2.2, we can show that a negative definite self-adjoint generator need not possess the p-Weiss property for $p > 2$.

As a continuation of our earlier computations, we prove that the resolvent condition is valid for $p = 4$. Since the control operator of system in question was not 4-admissible, this implies that the generator does not have the 4-Weiss property.

Recall that the Dirichlet Laplacian A on $L^2([0, 1])$ has the diagonal representation

$$Ae_n = -n^2\pi^2 e_n, \qquad n \in \mathbb{N},$$

where e_n are the unit vectors in $\ell^2(\mathbb{N})$. Moreover, the control operator b modeling the heat injection at the right endpoint of the interval was described by the sequence $b = \{b_n\}_{n\in\mathbb{N}}$, where $b_n = (-1)^n\sqrt{2}n\pi$.

The composition $R(\lambda, A)b$ of the resolvent and control operator is then given by the sequence

$$(R(\lambda, A)b)_n = \frac{(-1)^n\sqrt{2}n\pi}{\lambda + n^2\pi^2}, \qquad n \in \mathbb{N}.$$

Validity of the resolvent condition in the 4-Weiss property is (by definition) equivalent to uniform boundedness of the quantity $(\operatorname{Re}\lambda)^{1/4}\|R(\lambda, A)b\|_{\mathcal{L}(U,X)}$ among complex numbers λ with positive real part.

In view of $U = \mathbb{C}$ and $X = \ell^2(\mathbb{N})$, we insert the expressions for the coefficients of $R(\lambda, A)b$; it then remains to show that

$$\sup_{\operatorname{Re}\lambda>0} (\operatorname{Re}\lambda)^{1/4}\left(\sum_{n=1}^{\infty} \frac{2n^2\pi^2}{|\lambda + n^2\pi^2|^2}\right)^{1/2} < \infty. \tag{14}$$

Note that we can restrict to real λ. Moreover, it is convenient to consider $\mu = \sqrt{\lambda}$ in place of λ. Employing this substitution, after squaring the expression (14) becomes

$$\sup_{\mu>0} \mu \sum_{n=1}^{\infty} \frac{2n^2\pi^2}{(\mu^2 + n^2\pi^2)^2} < \infty.$$

A closed form of the sum $\sum_{n=1}^{\infty} \frac{\pi^2 n^2}{(b+\pi^2 n^2)^2}$ exists in terms of hyperbolic functions. Indeed, using

$$\coth x := \frac{\cosh x}{\sinh x} = \frac{e^x + e^{-x}}{e^x - e^{-x}}, \quad \csc x := \frac{1}{\sin x}, \quad \operatorname{csch} x := \frac{1}{\sinh x} = \frac{2}{e^x - e^{-x}}$$

together with the series expansions

$$\coth(\pi x) = \frac{1}{\pi x} + \frac{2x}{\pi} \sum_{k=1}^{\infty} \frac{1}{k^2 + x^2}, \qquad \csc^2(\pi x) = \frac{1}{\pi^2} \sum_{k=-\infty}^{\infty} \frac{1}{(x - k)^2},$$

see [12, page 36], and the identity $\operatorname{csch}^2(x) = -\csc^2(ix)$ for $x \in \mathbb{R}$, one can show that

$$\mu \sum_{n=1}^{\infty} \frac{2n^2\pi^2}{(\mu^2 + n^2\pi^2)^2} = \frac{1}{2}(\coth \mu - \mu \operatorname{csch}^2 \mu) = \frac{1}{2} + \frac{(2 - 4\mu)e^{2\mu} - 2}{2(e^{2\mu} - 1)^2}.$$

As $\frac{(2-4\mu)e^{2\mu}-2}{2(e^{2\mu}-1)^2} < 0$, we have

$$\sup_{\mu>0} \mu \sum_{n=1}^{\infty} \frac{2n^2\pi^2}{(\mu^2 + n^2\pi^2)^2} = \sup_{\mu>0} \left(\frac{1}{2} + \frac{(2 - 4\mu)e^{2\mu} - 2}{2(e^{2\mu} - 1)^2} \right) \leq \frac{1}{2},$$

validating the resolvent condition. Therefore the 4-Weiss property is violated for A.

The above example shows that Theorem 5.2 cannot be applied to show p-admissibility for systems arising from boundary control with infinite-dimensional input space if $p > 2$, even if A is self-adjoint. Consequently, determining admissibility with respect to sharp values of p remains difficult in this regime. Practically, the straightforward tool to derive admissibility seems to be trace estimates in conjunction with the interpolation spaces as lined out in Sect. 2.1.

References

1. R.A. Adams, J.J.F. Fournier, *Sobolev Spaces*, 2nd ed. Pure and Applied Mathematics 140 (Academic Press, Amsterdam, Boston, 2003). ISBN: 978-0-12-044143-3
2. A. Bäcker, S. Fürstberger, R. Schubert, F. Steiner, Behaviour of boundary functions for quantum billiards. J. Phys. A: Math. Gener. **35**(48), 10293 (2002)
3. V.I. Bogachev, *Measure Theory* (Springer, Berlin, New York, 2007). ISBN: 978-3-540-34513-8
4. H. Bounit, A. Driouich, O. El-Mennaoui, A direct approach to the weiss conjecture for bounded analytic semigroups. Czech. Math. J. **60**(2), 527–539 (2010)
5. C.I. Byrnes, D.S. Gilliam, V.I. Shubov, G. Weiss, Regular linear systems governed by a boundary controlled heat equation. J. Dyn. Control Syst. **8**(3), 341–370 (2002)
6. J.B. Conway, *A Course in Functional Analysis*. Graduate Texts in Mathematics, vol. 96 (Springer, New York, 2007). ISBN: 978-1-4757-4383-8
7. R.F. Curtain, A.J. Pritchard. *Infinite Dimensional Linear Systems Theory*, ed. by A.V. Balakrishnan, M. Thoma. Lecture Notes in Control and Information Sciences, vol. 8 (Springer, Berlin, Heidelberg, 1978). ISBN: 978-3-540-08961-2
8. K.-J. Engel, On the characterization of admissible control-and observation operators. Syst. Control Lett. **34**(4), 225–227 (1998)

9. K.-J. Engel, R. Nagel, *One-Parameter Semigroups for Linear Evolution Equations*. Graduate Texts in Mathematics 194 (Springer, New York, Heidelberg, 2000). ISBN: 978-0-387-22642-2

10. R.R. Goldberg, Certain operators and fourier transforms on L^2. Proc. Am. Math. Soc. **10**(3), 385–390 (1959)

11. P. Grabowski, F.M. Callier, Admissible observation operators. semigroup criteria of admissibility. Integral Equ. Oper. Theory **25**(2), 182–198 (1996)

12. I.S. Gradshteyn, I.M. Ryzhik, A. Jeffrey, *Table of Integrals, Series, and Products*. Corrected and Enlarged Edition (Academic Press, New York, 1980). ISBN: 0-12-294760-6

13. D.S. Grebenkov, B.-T. Nguyen, Geometrical structure of laplacian eigenfunctions. SIAM Rev. **55**(4), 601–667 (2013)

14. B.H. Haak, Kontrolltheorie in Banachräumen und quadratische Abschätzungen. Dissertation: University of Karlsruhe, 2005, 148 pp.

15. B.H. Haak, On the Carleson measure criterion in linear systems theory. Complex Anal. Oper. Theory **4**(2), 281–299 (2010)

16. B.H. Haak, The Weiss conjecture and weak norms. J. Evol. Equ. **12**(4), 855–861 (2012)

17. B.H. Haak, P.C. Kunstmann, On Kato's method for Navier–Stokes equations. J. Math. Fluid Mech. **11**(4), 492–535 (2009)

18. B.H. Haak, P.C. Kunstmann, Admissibility of unbounded operators and well-posedness of linear systems in banach spaces. Integral Equ. Oper. Theory **55**(4), 497–533 (2006)

19. B.H. Haak, P.C. Kunstmann, Weighted admissibility and wellposedness of linear systems in banach spaces. SIAM J. Control Optim. **45**(6), 2094–2118 (2007)

20. B.H. Haak, C. Le Merdy, α-admissibility of observation and control operators. Houst. J. Math. **31**(4), 1153–1167 (2005)

21. S. Hansen, G. Weiss, The operator carleson measure criterion for admissibility of control operators for diagonal semigroups on l^2. Syst. Control Lett. **16**(3), 219–227 (1991)

22. A. Hassell, T. Tao, Upper and lower bounds for normal derivatives of dirichlet eigenfunctions. Math. Res. Lett. **9**(3), 289–305 (2002)

23. R. Hempel, L.A. Seco, B. Simon, The essential spectrum of neumann Laplacians on some bounded singular domains. J. Funct. Anal. **102**(2), 448–483 (1991)

24. L.F. Ho, D.L. Russell, Admissible input elements for systems in Hilbert space and a carleson measure criterion. SIAM J. Control Optim. **21**(4), 614–640 (1983)

25. R. Hosfeld, B. Jacob, F.L. Schwenninger, Characterization of orlicz admissibility. Semigroup Forum **106**(3), 633–661 (2023)

26. T. Hytönen, J. Van Neerven, M. Veraar, L. Weis, *Analysis in Banach Spaces*, vol. I. Martingales and Littlewood-Paley Theory (Springer, Cham, 2016). ISBN: 978-3-319-48519-5

27. B. Jacob, R. Nabiullin, J.R. Partington, F.L. Schwenninger, Infinite-dimensional input-to-state stability and Orlicz spaces. SIAM J. Control Optim. **56**(2), 868–889 (2018)

28. B. Jacob, J.R. Partington, Admissibility of control and observation operators for semigroups: a survey, in *Current Trends in Operator Theory and Its Applications*, ed. by J.A. Ball, M. Klaus, J.W. Helton, L. Rodman. vol. 149. Operator Theory: Advances and Applications (Birkhäuser, Basel, 2004), pp. 199–221. ISBN: 978-3-0348-9608-5

29. B. Jacob, J.R. Partington, The Weiss conjecture on admissibility of observation operators for contraction semigroups. Integral Equ. Oper. Theory **40**(2), 231–243 (2001)

30. B. Jacob, J.R. Partington, S. Pott, Admissible and weakly admissible observation operators for the right shift semigroup. Proc. Edinb. Math. Soc. **45**(2), 353–362 (2002)

31. B. Jacob, J.R. Partington, S. Pott, Applications of Laplace–Carleson embeddings to admissibility and controllability. SIAM J. Control Optim. **52**(2), 1299–1313 (2014)

32. B. Jacob, J.R. Partington, S. Pott, E. Rydhe, F.L. Schwenninger, Laplace-Carleson embeddings and infinity-norm admissibility (2023). arXiv: 2109.11465

33. B. Jacob, F.L. Schwenninger, J. Wintermayr, A refinement of Baillon's theorem on maximal regularity. Stud. Math. **263**, 141–158 (2022)

34. I. Lasiecka, R. Triggiani, *Control Theory for Partial Differential Equations: Continuous and Approximation Theories*, vol. 1. Encyclopedia of Mathematics and Its Applications 74–75 (Cambridge University Press, Cambridge, New York, 2000). ISBN: 978-0-521-43408-9

35. C. Le Merdy, The Weiss conjecture for bounded analytic semigroups. J. Lond. Math. Soc. **67**(3), 715–738 (2003)
36. F. Maragh, H. Bounit, A. Fadili, H. Hammouri, On the admissible control operators for linear and bilinear systems and the Favard spaces. Bull. Belgian Math. Soc. Simon Stevin **21**(4), 711–732 (2014)
37. A. Mironchenko, C. Prieur, Input-to-state stability of infinite-dimensional systems: recent results and open questions'. SIAM Rev. **62**(3), 529–614 (2020)
38. A. Pazy, *Semigroups of Linear Operators and Applications to Partial Differential Equations*, vol. 44. Applied Mathematical Sciences (Springer, New York, 1983). ISBN: 0-387-90845-5
39. D. Salamon, Infinite dimensional linear systems with unbounded control and observation: a functional analytic approach. Trans. Am. Math. Soc. **300**(2), 383–431 (1987)
40. C. Schneider, *Beyond Sobolev and Besov: Regularity of Solutions of PDEs and Their Traces in Function Spaces*, vol. 2291. Lecture Notes in Mathematics (Springer, Cham, 2021). ISBN: 978-3-030-75139-5
41. F.L. Schwenninger, Input-to-state stability for parabolic boundary control: linear and semilinear systems, in *Control Theory of Infinite-Dimensional Systems*. Operator Theory: Advances and Applications (Springer, Cham, 2020), pp. 83–116. ISBN: 978-3-030-35898-3
42. O.J. Staffans, *Well-Posed Linear Systems* (Cambridge University Press, Cambridge, 2005). ISBN: 978-0-511-08208-5
43. D. Tataru, On the regularity of boundary traces for the wave equation. Annali della Scuola Normale Superiore di Pisa - Classe di Scienze **26**(1), 185–206 (1998)
44. G. Teschl, *Mathematical Methods in Quantum Mechanics: With Applications to Schrödinger Operators, Second Edition*, vol. 157. Graduate Studies in Mathematics (American Mathematical Society, 2014). ISBN: 978-1-4704-1888-5
45. H. Triebel, *Theory of Function Spaces II* (Springer, Basel, 1992). ISBN: 978-3-0346-0418-5
46. M. Tucsnak, G. Weiss, *Observation and Control for Operator Semigroups*. Birkhäuser Advanced Texts (Birkhäuser, Basel, Boston, 2009). ISBN: 978-3-764-38993-2
47. M. Unteregge, p-admissible control elements for diagonal semigroups on l^r-spaces. Syst. Control Lett. **56**(6), 447–451 (2007)
48. G. Weiss, A powerful generalization of the Carleson measure theorem? in *Open Problems in Mathematical Systems and Control Theory*, ed. by V. Blondel, E.D. Sontag, M. Vidyasagar, J.C. Willems. Communications and Control Engineering (Springer, London, 1999), pp. 267–272. ISBN: 978-1-4471-0807-8
49. G. Weiss, Admissibility of input elements for diagonal semigroups on l^2. Syst. Control Lett. **10**(1), 79–82 (1988)
50. G. Weiss, Admissibility of unbounded control operators. SIAM J. Control Optim. **27**(3), 527–545 (1989)
51. G. Weiss, Admissible observation operators for linear semigroups. Israel J. Math. **65**(1), 17–43 (1989)
52. G. Weiss, Two conjectures on the admissibility of control operators, In: *Estimation and Control of Distributed Parameter Systems: Proceedings of an International Conference on Control and Estimation of Distributed Parameter Systems, Vorau, July 8–14, 1990*, ed. by W. Desch, F. Kappel, K. Kunisch. International Series of Numerical Mathematics (Birkhäuser, Basel, 1991), pp. 367–378. ISBN: 978-3-0348-6418-3
53. S. Zelditch, Billiards and boundary traces of eigenfunctions. Journées équations aux dérivées partielles 1–22 (2003)
54. H.J. Zwart, B. Jacob, *Disproof of an Admissibility Conjecture of Weiss*. Faculty of Mathematical Sciences, Memorandum 1539 (University of Twente, Enschede, 2000)
55. H.J. Zwart, B. Jacob, O.J. Staffans, Weak admissibility does not imply admissibility for analytic semigroups. Syst. Control Lett.. **48**(3), 341–350 (2003)